INNOVATION IN CONSTRUCTION

INNOVATION IN CONSTRUCTION

An International Review of Public Policies

Edited by

André Manseau and George Seaden

London and New York

338.4769
I 58

First published 2001
by Spon Press
11 New Fetter Lane, London EC4P 4EE

Simultaneously published in the USA and Canada
by Spon Press

29 West 35th Street, New York, NY 10001

Spon Press is an imprint of the Taylor & Francis Group

© 2001 Taylor & Francis Books Ltd

*The opinions expressed in this work are those of the authors and do not
represent the opinions or policy of the National Research council of Canada
or any other institutions employing the authors. While every effort has been
made to ensure that the statements made and the opinions expressed in this
publication provide a safe and accurate guide, no liability or responsibility
can be accepted in this respect by the authors or publishers.*

Publisher's note:

This book has been prepared from camera-ready copy provided by editors

Printed and bound in Great Britain by Biddles Ltd, Guildford and King's Lynn.

British Library Cataloguing in Publication Data
A catalogue record for this book is available from the British Library

Library of Congress Cataloging-in-Publication Data
A catalog record for this book has been requested

ISBN 0-415-25478-7

Contents

List of figures

List of tables

Acknowledgements

Content of this volume is the result of the team effort of the members of CIB Task Group 35 on Innovation Systems in Construction. Members of this group, through a number of intense meetings, clarified the general concepts and then made specific national contributions. The editors wish to acknowledge the co-operation and the effort of all the members of TG 35, who are the co-authors of this volume.

We also wish to thank The Bartlett School, University College London; The Society of Danish Engineers, Copenhagen; Centre Scientifique et Technique du Batiment, Paris and Erasmus University, Rotterdam for graciously hosting various international meetings that helped to advance the ideas.

Many contributed to the arrangements for the book and particularly Lise Saumure for preparing this volume for the press.

Finally, the Institute for Research in Construction of the National Research Council of Canada initiated this project and maintained its support throughout the many years of the entire process. We should especially like to thank Sherif Barakat, Director General of this Institute and Chris Norris, Senior Advisor, to whom we are extremely grateful for their leading role.

Foreword

This volume is the result of an international study conducted by a special task group (TG35) created by the International Council for Research and Innovation in Building and Construction (CIB) in June 1998.

The TG 35 objective were to:

- Develop a framework for the analysis of innovation systems in construction;
- Carry out an international comparison of innovation in construction for the application in practical policy development

Significant discussions are taking place worldwide regarding the productivity of the construction industry and the role governments should play in encouraging its innovativeness. Given the national and international importance of the topic and the need to better understand the problem, CIB has set a three-year time frame to examine and report on the issue.

To achieve its objectives, the group has focused on describing the situation of innovation in the construction industry and on assessing the effectiveness of public interventions. The following questions were among the major drivers of discussions and analysis within the group:

- How to strengthen innovation in the construction industry?
- What instruments and approaches are being used by governments to promote innovation in this sector?
- What works and under what circumstances?

This volume provides, in the introduction, the analytical framework used by TG 35 to review the prevailing international innovation policies. It is followed by detailed, up-to-date descriptions of national systems of innovation and related public instruments for construction form fifteen TG 35 participating countries: Argentina, Australia, Brazil, Canada, Chile, Denmark, Finland, France, Germany, Japan, The Netherlands, Portugal, South Africa, United States of America, and United Kingdom. Common trends and differences in national approaches are presented in the conclusion together with comments as to the overall state-of-the-art in public policies related to construction.

TG 35 brought together a unique group of individuals with expertise in construction innovation theory and practice.   In their analysis they attempted to take into the account the current understanding of relationships that may exist between various economic, market, technology, human resources and other factors that influence industrial development.   George Seaden and Andre Manseau, respectively the Chair and the Co-ordinator of TG 35 prepared the introduction and the conclusions and assured the overall editorial supervision.   Various country reports identify the national contributors.

In the course of the study, extensive documentation search was carried out, using various national and international data sources.   Bibliography lists information referenced in the text as well as other documents relevant to the subject matter.

The overall "system of innovation" approach was used in the course of this study and members of TG 35 hope that it will provide a more comprehensive view of innovation issues in the construction industry world wide.

Preface

At the time of creation of CIB, in the 50's, the innovation model in most Western countries was rather simple. It was essentially a demand-pull model that was driven by the significant needs of the principal client of the construction industry: the state (at the national or regional, municipal level). In those countries, at that time, investment in built facilities expanded very rapidly and there was not much knowledge how to do it right. Hence, creation of new knowledge in general and of new technologies in particular was dominant issue, and CIB came to existence to coordinate and accelerate new knowledge dissemination worldwide. As a network of major knowledge creators (at that time, mostly national labs) the format of CIB was much in advance of traditional learned societies.

The focus of attention of most economies has now shifted from knowledge creation to more complex models of knowledge dissemination and implementation through processes and products, i.e. innovation. In some sectors of economy, such as communications or air transportation, this has resulted in significant gains in effectiveness and in productivity/quality enhancement. Construction industry world wide however has been relatively slow in the adoption of new knowledge, which has been a source of concern and frustration, expressed through several recent national reports. Yet, most of these reports, while signalling ambitious goals to be achieved, tend to be short on how to get there.

There are very many participants in the construction innovation game: knowledge generators, technology brokers, managers, regulators, construction workers, end users, etc. And all have a stake in the outcome. Rewards differ for various actors and their relationship can be sometimes supportive and sometimes adversarial. Yet , in order to increase the efficiency and the productivity of the industry we need to understand better these interactions; we need to understand the "system of innovation".

Investment in new knowledge creation, its diffusion and its implementation are all long-term, high risk ventures; hence the role of governments in funding and encouraging through public policy instruments is considered critical to success.

Since the 50's the current array of Members of CIB has expanded to include industrial enterprises, building owners, universities, consultants and others with a stake in efficiency, productivity, quality, sustainability and functionality of construction processes; and all dependent on the successful implementation of new ideas. Accordingly, CIB decided to create an international Task Group that would attempt to apply the worldwide knowledge on innovation systems to the construction industry.

CIB Task Group TG 35 "Innovation Systems in Construction" concentrated on obtaining a picture of public policies in larger and smaller, highly industrially developed countries as well as in some presently in the process of development. The result of this Task Group's excellent work is a complex mosaic of practices from 15 countries. It offers a comprehensive international perspective on government policies regarding construction innovation. Together with a review of current approaches to innovation and an extensive bibliography, this should be a useful document for policy makers in developing new solutions, for academics in developing a better understanding of the innovation problem in construction and for the industry leaders in demanding targeted action to radically improve the performance of the largest economic sector, the construction.

It is with confidence – and with some level of pride – that I would like to recommend this book to all those who think that our industry deserves all possible support in becoming as productive, efficient, sustainable, customer focussed and innovative as it should be.

Wim Bakens
Secretary General CIB

CHAPTER 1

INTRODUCTION

André Manseau and George Seaden

1.1 PUBLIC POLICY CONCERNS WITH THE CONSTRUCTION SECTOR

The very size and pervasiveness of the construction industry cause most governments to show interest in its well-being.

Together with the related sectors (manufacturers of building products and systems, designers and facility operators) it is one of the major economic activities in every nation, accounting for about 15% of national GDP. It also has a significant impact on living standards and on the capability of the society to produce other goods and services or to trade effectively. As a well-established sector, construction has been strongly determined by local tradition and culture, and geographical factors such as availability of material and climate. Construction provides houses to live in, buildings to work in and infrastructure that supports communication and transportation of the modern society.

Official statistics generally limit the construction sector to the value-added activity of firms that construct buildings and infrastructure as well as to those who install construction sub-systems (electrical works, plumbing, air systems, structure, finishing, etc.). Figure 1.1 shows that construction's (as previously defined) share of national GDP has slowly decreased during the last two decades to reach about 4-6% for almost all countries, except Japan with a 10%.

Major changes are occurring in the construction industry. Demand is shifting towards more functional buildings (with greater concern for user satisfaction and productivity), more sophisticated equipment such as intelligent devices for better control of energy efficiency or indoor environment, improved working/living conditions, and more respect for environmental constraints. National quality standards and building costs are increasingly facing international comparisons. Some of the major players in engineering services, manufacturing building products or equipment, and operation of large facilities are becoming international in scope.

Figure 1.1: Construction industry as a percentage of GDP

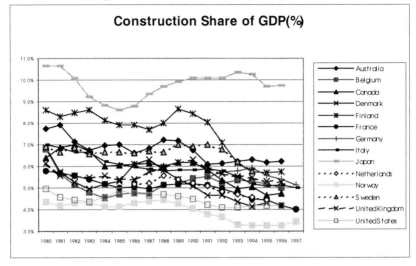

Source: OECD Statistics, International Sectoral Database (ISDB), 1999

Following a period of post-war expansion, the demand for construction has decreased in the last decades. However, the demand for machinery and equipment linked to production processes, not incorporated in the structure, continues to increase. Residential construction also slowed down as demographic pressure diminished. This may be an indication of a fundamental change that is taking place in the structure of the construction industry of many industrially developed countries (Bon, 1994). Cyclical nature of the demand tends to make such long-term predictions difficult. However, in general, spending on operating, maintaining, renovating and reusing existing facilities is growing as the percentage of the total output. These latter activities are not well captured by official statistics limited to constructors and assemblers. Some (Carassus, 1999) suggest that the industry is now more in the business of optimisation of the use of the existing stock rather than in the provision of new facilities.

Governments own and operate an important share of the built environment in every nation. Governments of the OECD group own between 10% to 25% of their total country fixed assets in a form of public works, which include buildings and general purpose infrastructure (OECD, 1997). These facilities require on-going maintenance and renovation as well as new additions, all at a significant cost, to satisfy demands for high-quality public services.

By various measures (Keys and Caskie, 1975; Seaden, 1997; Industry Canada, 1998) such as: total factor or labour productivity, customer satisfaction (Barrett, 1998), R&D intensity (Revay and Associates Ltd., 1993 and 1999) or

(Barrett, 1998), R&D intensity (Revay and Associates Ltd., 1993 and 1999) or worker's skill levels, these industries in the OECD and other countries, has lagged behind most sectors of the economy.

Over the years this has been a subject of various reports, which have advocated a more rationalised regulatory system, reduction in the adversarial regime between various participants, better risk-sharing, greater investment in R&D or improved training of labour and management. There has been gradual progress in some of these areas but no country has yet undertaken systematic changes to the overall delivery system.

1.2   DEFINING THE CONSTRUCTION SPACE

The economic space of construction is much larger than that defined by traditional "construction" statistics, limited to value-added site activities of general contractors and speciality trades (roofers, plumbers, structural assemblers, excavators, landscapers, etc.). There is a general agreement that it includes the design of buildings and infrastructure (engineering and architectural services), the manufacture of buildings products and of machinery and equipment for construction, and operation and maintenance of facilities. However, statistical data for these construction-related activities are seldom directly available, as they are usually included in other manufacturing or service industry surveys. Rough estimates of the overall importance of the industry, which have recently become available, show that it is much larger than previously perceived.

The study of innovation in the construction sector raises many challenges. The sector is a very complex arena, involving numerous agents and interactions in developing and adapting innovations. Figure 1.2 presents a systematic approach to the key actors involved in construction who can undertake innovation activities.

These actors include:

- building materials suppliers who provide the basic materials for construction such as lumber, cement and bricks
- machinery manufacturers who provide the heavy equipment used in construction such as cranes, graders and bulldozers
- building product component manufacturers who provide the subsystems (complex products) such as air quality systems, elevators, heating systems, windows and cladding
- sub-assemblers (trade speciality and installers) who bring together components and material to create such sub-systems
- developers and facility assemblers (or general contractors) who initiate new projects and co-ordinate the overall assembly

- facility/building operators and management who manage property services and maintenance
- facilitators and providers of knowledge/information such as scientists, architects, designers, engineers, evaluators, information services, professional associations, education and training providers.
- providers of complementary goods and services such as transportation, distribution, cleaning, demolition and disposal
- institutional environment actors who provide the general framework conditions of the business environment such as the physical and communication infrastructure, financial institutions and business/trade general labour regulations and standards.

Figure 1.2: Key agents, major types of interactions and framework conditions in the construction sector (source: Manseau, 1998)

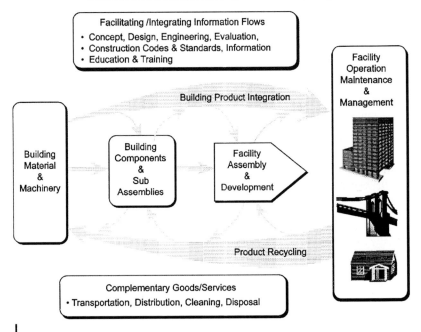

The above list provides a basic typology of construction related activities; some of those actors may be suppliers or clients of others in the production process and specific firms can be involved simultaneously in several of above activities. Some larger firms offer vertically integrated range of services from design, through manufacturing of some products or components, to building and operation of facilities.

The ultimate client, the owner of the constructed facility, can influence the innovative behaviour of various actors by identifying specific novel requirements to be supplied by the developers, building product suppliers, contractors or operators or through the general expectation of high-quality and viability, over the useful life of the project.

Construction activity has several distinct characteristics, which makes it unlike any other industrial sectors (ARA Consulting Group, 1997; Carassus, 1998 and 1999; Toole, 1998).

The site-based nature of construction activities causes the uniqueness of each construction project. It is always located on a distinct site, subject to local environmental and climatic conditions, most likely built by a different work-crew. Even two standard subdivision homes of identical design are likely to be somewhat different.

Every new or renovation/repair project is also truly a prototype. While some degree of uniformity has been tolerated in the past, with the increase in the wealth of Western industrial countries there has been an ever-growing trend in demanding custom solutions to satisfy real or perceived individual requirements. Some sources suggest (Flanagan et al., 1998) that this drive to customization as well as demands for higher quality can be achieved through the introduction of IT-supported production methods, currently used in manufacturing.

Local demands for constructed product are of extreme diversity. Industry responds to the occasional local/regional need for large hospitals, major airports, tunnels, and water treatment plants as well as to the more constant demand for single-family homes, office buildings or street improvements.

Constructed facilities tend to be very durable, lasting 25-50 years and longer. When obsolete, they are most often repaired, modernized and sometimes radically transformed to suit new requirements rather than disposed of and replaced with new, more typical for manufactured products. This also impacts on construction demand, which stresses conservative, traditional solutions and emphasises reliability. Generally however, issues of life-cycle cost or sustainability are seldom addressed, focus remaining on the initial cost of acquisition.

Aesthetic, safety, site and environmental design considerations are set not only by the builder or the owner but also by the community at large. Regulation and standards are more rigorous in construction than in most other sectors of economy, with the involvement of several levels of governments (local, provincial, national).

Construction is highly fragmented and firms, mostly small in size, tend to respond to local market needs and to control only one of the elements of the

Construction is highly fragmented and firms, mostly small in size, tend to respond to local market needs and to control only one of the elements of the overall building process. Some firms have been trying to achieve greater control of their production together with greater innovation capability through increased integration and alternate delivery systems such as design-build or build-own-transfer (BOT).

Over a number of years there has been an international trend to "industrialise" construction through greater pre-fabrication, modularization, standardisation and other manufacturing-type production techniques. There is an almost implicit wish that if construction was like manufacturing, many of its quality and productivity problems would disappear and innovation would flourish.

While many of these initiatives have shown positive results, construction, because of the variable site requirements, the durability of the product and the impact on the surrounding community, remains unique and significantly different in its characteristics from other industrial sectors.

CHAPTER 2

ANALYTICAL FRAMEWORK

André Manseau and George Seaden

There has been an increase in the general understanding of industrial innovation systems and processes (OECD, 1996 and 1997b; Hobday, 1998), with some new, interesting evidence specific to the construction industry (Bernstein and Lemer, 1996; CRISP, 1997; Winch and Campagnac, 1995). In particular, the role in innovation of the government through its various public policies and programmes has been identified. It is this general knowledge of innovation, adapted to the characteristics of the construction industry, which constitutes the analytical framework of this paper.

2.1   DEFINING INNOVATION

With the increasing openness of the world trade and globalisation there has been ever-growing interest in what makes firms truly competitive. Opinions on that matter have greatly evolved in the past twenty years and continue to be open to debate. Porter (1998) and others suggest that during the last twenty years Western companies have been responding to the international competition through continuous improvement in their operational effectiveness. At present time, Western companies can seldom use low cost of labour or of raw materials as theirs competitive advantage. Thus, re-engineering, lean production, investments in the information technology, TQM and other techniques of optimising productivity and asset utilisation have now all become parts of companies' efforts to remain/become competitive in the global market-place. Porter also suggests that continuous improvement in best practice utilisation must now be considered a pre-condition to achieve profitability and that companies have to create unique competitive positions through integration of all their competencies. To have truly lasting competitive advantage they need to offer differentiated, value creating new products to their customers.

These competitive needs as well as spectacular achievements of the high-technology sectors of the economy have driven our interest in the generation of new ideas and its implementation i.e. what is now being considered innovation. There is no generally accepted definition of innovation at the present time, however there has been noticeable convergence as to its principal characteristics.

This point can be illustrated by sample of general definitions:

- "The process of bringing new goods and services to market, or the result of that process" (Advisory Council on Science and Technology, 1999).
- "A technological product innovation is the implementation/commercialisation of a product with improved performance characteristics such as to deliver objectively new or improved services to the customer. A technological process innovation is the implementation/adoption of new or significantly improved production or delivery methods. It may involve changes in equipment, human resources, working methods or a combination of these" (OECD, 1997a).

Construction industry sources also show a variety of definitions:

- "Application of technology that is new to an organisation and that significantly improves the design and construction of a living space by decreasing installed cost, increasing installed performance, and/or improving the business process" (Toole, 1998)
- "the successful exploitation of new ideas, where ideas are new to a particular enterprise, and are more that just technology related – new ideas can relate to process, market or management" (Construction Research and Innovation Strategy Panel (CRISP), 1997)
- "apply innovative design, methods or materials to improve productivity" (Civil Engineering Research Foundation (CERF), 1993)
- "anything new that is actually used" (Slaughter, 1993)
- "first use of a technology within a construction firm" (Tatum, 1987).

These definitions may display specific biases of different sources, studies and organisations. Nevertheless certain trends and convergence can be observed. Increasingly, innovation appears to be viewed as a process that enhances the competitive position of a firm through the implementation of a large spectrum of new ideas. A recent business-level survey by A. D. Little of factors that allow a firm to be innovative, involving significant sample of companies in 8 OECD countries, disclosed a much more comprehensive and complex image of the innovation process. What distinguishes the "leaders" from the "pack" is their ability to combine marketing, internal organisation and technology. "Innovative products don't necessarily build business, often non-innovative products build business, what builds business is innovative companies" (Brown, 1998). Yet domestic and international competitive pressures are increasing in all sectors of the economy to deliver ever-greater value to the customer, mostly through innovation.

Because every construction project, new or repair, can be considered a prototype, at a new and different site and most often with a different owner, there is significant opportunity and tendency to do something new and/or distinct every time. Building practitioners and their clients have often interpreted this as innovative behaviour. But is the industry truly innovative, i.e. good at adopting new processes and products?

Unfortunately, official statistics are limited in measuring innovation. Most common indicators for assessing innovation are related to research and development (R&D) activities: R&D expenditures, number of R&D personnel, number of patents, number of publications and their citations. As many authors and OECD reports have stressed (OECD, 1996; OECD, 1997a), innovation can emerge from various sources of activities, and not only from R&D, although it constitutes an important part of innovation activities.

As discussed further in this volume, models linking R&D and innovation are complex and generally display non-linear relationship. Nevertheless, level of R&D activity has been positively correlated with the relative innovativeness of various industrial sectors, particularly high tech manufacturing sectors, and is considered a valid indicator.

Expenditures on R&D in construction (statistically limited to contractors and sub-trades) range between 0.01% and 0.4% of construction value-added for OECD countries (Figure 2.1), compared to 3-4% in manufacturing or 2-3% for all industries. Many reasons have been offered to explain the low level of investment in construction R&D; improper reporting of R&D expenses, small size of firms, lack of risk capital, conservative behaviour of clients, unsuitable government policies and many other factors. A general observation shows that very few firms in construction take advantage of current R&D or innovation programmes offered by governments. Work undertaken by TG 35 has been an attempt to study systematically construction innovation and associated R&D from the perspective of public policies and programmes.

Despite recent important efforts to measure innovation in many OECD countries, major challenges remain to be addressed. Innovation related activities include R&D as well as many other that are difficult to assess, such as developing new organisational structures, upgrading labour skills, introducing new processes or products and new marketing approaches. These innovative activities are generally pervasive throughout all on-going activities of a firm and may also be embedded in its interactions with key partners. They are difficult to isolate and evaluate. Firms do not capture and maintain records of all their efforts to innovate. Several studies have focused on measuring the number of patents and/or new products, some on assessing new manufacturing processes, but very few on new organisational processes. Yet, organisational processes are considered very important in construction, as assembly methods as well as contracting arrangements are the core activities in this industry.

André Manseau and George Seaden

Figure 2.1: Business expenditures in R&D (BERD) in construction as a percentage
of construction added value

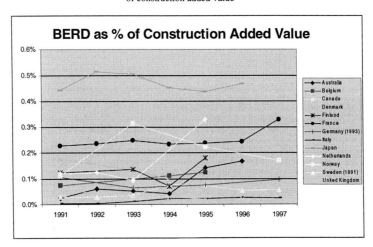

Source: OECD (2000) Basic Science and Technology Statistics and (1999) International Sectoral
Database

From the innovation perspective the important issue is how the construction firm handles the newly acquired knowledge. Two strategically distinct scenarios are available:

- Some firms consider such knowledge as an important business asset that can lead to future opportunities, possibly at improved profit margins. They are likely to invest resources outside the direct project expenditures to enhance the corporate capability through the capture of technology (and other lessons learned) from every project.
- Other firms consider lessons learned from a project as site specific because of the unique characteristics of every situation and thus of limited value in future work. It is the project team, likely to be disbanded on completion, which primarily masters this new knowledge. Some of it may be shared with others in the industry through technical press or migration of employees but there is no deliberate effort to capture and reuse such knowledge within the firm.

2.2   INNOVATION PROCESSES AND SYSTEMS

It is now generally accepted that innovation activity takes place within a "system of innovation" (Amable et al, 1997; Lundvall, 1999; OECD, 1996 and 1997b). This concept can be approached from the perspective of an industry (construction), technology (opto-electronics) or region (Silicon Valley cluster). Some suggest that high-technology industries are now truly globalised and that their innovation takes place within international alliances. While full analysis of systems of innovation is clearly outside the scope of this volume, for certain industries, particularly of the more traditional type such as construction, a "national system" is a useful unit of analysis because of common culture, legal framework, education, customer preference, institutions and many other variables that impact innovation.

Construction has been often seen as a local industry, but increasingly it is becoming national, sharing similar professional standards, building codes and partnering at a national level. International activities are also increasing, but are still limited to very large construction projects.

What is now referred to as the concept of the "innovation system" has being gradually introduced and previously known as linear models, such as the Technology-Push or the Market-Pull are still commonly used to analyse innovation. To provide a comprehensive view, these two linear models are briefly described, than three non-linear models of the innovation system approach are presented. Comments as to the relevance of various models to the innovation in the construction industry are made based on the general review of current knowledge on the topic and on the Task Group findings, presented later in the volume.

2.2.1   Technology push or science-based model

The Second World War provided a unique implementation opportunity for science and technology. For the first time ever, with the total commitment to the victory by all warring parties and with the mobilisation of national resources towards the war effort, new technology-based weaponry was rapidly developed from scientific principles/known technologies to operating products. Atomic bomb, radar, jet-propelled aircraft or mass production of Liberty ships were all developed through close collaboration of scientists, engineers and production personnel.

In the post-war years, this linear model of innovation became the reference standard that is now deeply entrenched in various policy instruments of governments as well as in the public perception of innovation. It assumed that basic (pure) research followed by applied research led to experimental development and then to new products/processes. Thus, the propensity to innovate of an industrial sector or a firm could be measured by its research intensity using several proxy indicators such as expenditure on R&D, citation analysis or education of research-qualified professionals.

Recent findings (OECD, 1997a) and OECD surveys have shown that research constitutes only one of the sources of industrial innovation among many other alternatives. Intensity of knowledge flows, levels of co-operation and effectiveness of technology diffusion may be more significant, particularly for the traditional industries such as construction (OECD, 1998). In support of this concept, as indicated further in this volume, some countries have established successful technology and best-practice diffusion networks dedicated to construction problems.

Internationally (with the exception of Japan) the level of R&D effort of the construction industry has been very low (See Figure 2.1), which has led the believers in the Technology Push model to conclude that it is a non-innovative industry.

2.2.2   Market pull model

Studies of highly successful and profitable firms (Cooper, 1998) indicate very close ties to the customer base and innovation activity linked to the actual market opportunities. The challenge of arriving at innovative products/processes that are truly competitive (i.e. meet clients' needs, have superior quality, reduce costs and present visible benefits) is met by tapping into the vast pools of existing knowledge. Close contacts are established with various knowledge sources and feedback loops are extensively used at different stages of development. Research tends to be viewed as one of the contributors in this problem-oriented innovation approach that is primarily targeted to respond to client's needs.

Kline (1985) introduced the concept as follows: "*While research in the physical and biological sciences has had an enormous impact on human societies and human lifestyles, ...research is not the direct source of innovations, and much innovation proceeds with little or not input from current research. (...) The first line of reference for innovation processes...is not research but the totality of cumulated human knowledge*".

This model suggests that employees as well as clients and customers are the dominant sources of information for innovation. Recent data (See Table 2.1) indicates that European architectural and engineering firms rely primarily on their internal sources and "trusted" peers (professional associations) and less on their clients (25%). On the other hand, innovative European manufacturing firms rely equally on in-house sources (47%) and clients (42%) and less on information from professional associations.

Table 2.1: Percentage of innovative European firms considering the
listed sources of information as very important

Sources of Information	% of innovative firms	
	Architectural & Engineering	Manufacturing
Intra-firm	55	47
Intra-group	45	25
Clients and customers	25	42
Competitors	13	16
Suppliers of equipment, materials	15	19
Consultants	16	5
Universities	5	4
Public laboratories	4	3
Computer based information	14	4
Fairs, exhibitions	20	22
Professional associations	28	8
Patent disclosure	1	3

Source: Eurostat Database, 1999 (from the Innovation survey of European Union, 1996)

2.2.3   Firm-centred knowledge networks

This is the macro/micro-economic model used as the theoretical basis of the Oslo Manual (OECD, 1997a) for measurement of innovation activity. It places the firm as the "innovation dynamo" (where economic benefits of innovation can be appropriated) at the centre of an enabling network of suppliers, competitors, clients, as well as educational, communication, financial and legislative resources. These actors and how they interact between each other in creating and exploiting innovation constitutes an innovation system.

This model has focused on the role of technology as a key source of innovation while recognising the importance of organisational change as a potential source of innovation. However, the latter is still difficult to measure. This model highlights the significance of strategic intent in a firm and of its market performance due to technologically new or improved products or production process. Innovation is expected to be "significant" and "new to the firm" (but not necessarily to the particular industrial sector) and bring enhanced performance benefits to the firm and/or the customer. The enabling network or system is taken into consideration by measuring technology dissemination, access to sources of information, internal/external barriers and potential impact of public policies.

Innovation is generally initiated for competitive reasons, to lower the unit cost of production and/or to obtain greater market share. However, development of new products or services that primarily appeal to customers' aesthetic or quality perception or personal taste and thus can provide firms with significant competitive advantage, are not considered innovative by the Oslo methodology.

The OECD model was devised following extensive studies of advanced manufacturing and high-technology sectors of the economy and it may not be fully applicable of other industrial groupings. So far, there has been very little in-depth analysis of various innovation framework factors related to the construction industry.

2.2.4   Production systems

Implementation of new ideas happens through interaction between workers within organisational constraints and structures. Recent work (Amable et al, 1997) suggests that certain features of production systems may be particularly conducive to innovation while others tend to suppress it. The following factors and their consequences are considered as contributing to a positive climate for innovation:

- organisational flexibility (leading to) $\rightarrow$ rapid response to changes and innovation
- employee reward structure connected to corporate profitability $\rightarrow$ greater acceptance of technological changes
- good and safe work environment $\rightarrow$ streamlined production systems
- general policy of full-employment $\rightarrow$ enhanced investment in productivity
- markets open to domestic/international competition $\rightarrow$ changes in sourcing of supplies, optimisation of work-processes, technological changes.

On the other hand, the following factors are considered to have detrimental effect on the innovation climate:

- work organised around strict functional definitions $\rightarrow$ slow and difficult response to technological changes
- frequent lay-offs and technology related unemployment $\rightarrow$ resistance to productivity enhancing initiatives
- salaries based on market rates or collective agreements $\rightarrow$ little employee interest in quality or productivity

- acceptance of unsatisfactory work practices → obsolete equipment not replaced
- relatively high level of general unemployment → investment in mass production
- national/regional barriers to trade → reduced pressure to innovate.

This model, based to a great degree on the intensity of employee's involvement and participation suggests that the Taylor-type mass-production organisation is not conducive to innovation and that new, more flexible work structures need to evolve to encourage creation of new products or processes.

This methodological approach appears of a particular relevance to the construction industry. No formal research has been done but it would appear that most of the listed detrimental factors are currently present in the industry. There are also a few of the positive elements, since shortage of skilled labour has encouraged investment in new equipment and open market has maintained high level of competition. Some of these factors can be observed in the analysis of Swedish construction industry (McKinsey Global Institute, 1995) that found high costs and low productivity, mainly due to fragmented and inflexible work practices, low level of domestic competition and very strict, performance-driven building regulations.

2.2.5   Complex product systems (CoPS)

Another interesting model that has recently been applied to the construction sector is the Complex Product Systems (CoPS). This model finds its origin in the project management literature and in the defence industry, in the 1950's, but has only been systematically applied to innovation processes in the early 1990s. The CoPS model Involves close interactions and negotiations between a relatively small number of key players and a system integrator who can rapidly change or modify major components of the project. The model is also characterised by its project orientation, involving a number of different firms that are required for producing a complex and usually unique or small batches product (Hobday, 1998).

One of the first applications to innovation processes has been in the flight simulation industry (Miller et al, 1995). Another early application, this one directly related to the construction sector, was to manufacturers and installers of home automation products (Tidd, 1995). Among key results, Tidd showed that the CoPS approach required managing across traditional product division boundaries and strong interfirm linkages.

Winch (1998) has discussed potential application of this model to the overall construction sector and Miller (2000) has recently applied the model to large engineering projects. This model could become increasingly applicable to a large portion of the construction industry with the current trend towards integration and partnerships (design-build, BOOT, manufacturers-installers-services, public-private partnerships).

2.2.6   Summary

In reviewing various models presented in this section, we observed the difficulty of describing the complexity and the multi-dimensional aspects of all innovation activities. Applying such models to the construction industry makes the task even more difficult since none of them appear to have the overall "best-fit" to the specific characteristics of this economic sector.

Nevertheless, we also observed that these models are mutually complementary, representing various perspectives and different drivers of innovation. Some models focus on specific drivers, such as science-technology capability of a firm, its organizational context or its client's needs. Other models have a broader macro-view, where the interaction between a number of different players and the nature of connecting networks is taken into consideration.

In certain industrial sectors, particularly with high technology content, these different models of innovation appear to be aligned, reinforcing each other. Innovation, economic growth, and increases in profit margins appear to happen because very interactive, complex business arrangements were created and there is agreement as to the overall objectives among all the key players (clients, suppliers, competitors, regulators and researchers). In most instances, such positive environment can be traced back (sometimes 10-20 years) to an array of public policy initiatives supportive of innovative climate (Porter, 1990).

It could be argued that during the post-war "golden period" of construction, between 1950's and 1970's, such congruence of public and private objectives prevailed in many countries. This period was also noticeable by significant intensity of construction innovation. Since then, the economic environment, the characteristics of the industry and the public policy context have significantly evolved. As reported elsewhere in this volume, many countries are now seeking an appropriate model to stimulate greater innovative activity in construction.

2.3   INFLUENCE OF POLITICAL AND SOCIAL STRUCTURES

It is generally agreed that construction is an industry that plays a key role in the creation of the asset base of a country. It can be reasonably assumed that the public policy regime for this sector will reflect the political, social, economic and cultural values of a given country.

Groupings of different political/constitutional and socio-economic structures have been developed (Amable et al, 1997), presented here with their potential impact on the national construction issues:

- Countries with a more centralised government structure (Japan, France, United Kingdom, Netherlands, Denmark, Finland) always have a national "construction" ministry which has been able champion the particular needs of its constituency and promote customised innovation-enhancement policies in construction. There is evidence of public concern with specific characteristics of the construction industry and of recent policy changes. Research funding is being redirected from products to processes, government acquisition practices are being modified to stress value over price, industrial collaboration is being promoted and technology/knowledge brokers are being introduced.

- Countries with a federal type of constitution (USA, Germany, Canada, Australia) and a more decentralised government structure tend to place the responsibility for construction at the state (province, Lander) level. Generally, there is no central focus or a champion for the construction industry; instead there is a multitude of various agencies dealing with particular concerns such as acquisition of public works, technology development, safety, consumer protection or losses due to the natural disasters. As well and possibly as a consequence of the lack of a national point of convergence there is no unified industrial representation of the construction point of view to the senior levels of government. There does not appear to be significant public policy interest in construction innovation separately from the general issue of industrial development.

For construction, there are significant differences between national attitudes towards the role of the government as the regulator or the principal customer, in the liberalisation of domestic markets, in labour relations, education and training, in the legal regime and in the methods of financing (Winch and Campagnac, 1995).

Therefore another approach is to look through the perspective of the "social systems of innovation" proposed by Amable et al. (1997). Authors analysed a large number of various parameters that encourage or hinder innovation including technology factors as well as production issues of management systems, labour motivation, regulatory processes, training and funding. They identified amongst the highly industrialised OECD countries four different innovation systems:

- The Market-driven system, present in the USA, UK, Canada and Australia, assumes that the market allocates resources in the most efficient manner through the bidding process and that government, even though a significant buyer of construction goods and services, is just another participant in the market place. Construction practitioners in these countries display significant distrust in government's ability to influence economic development and they believe that public intervention should only take place in the case of an obvious market failure. Regulation is to be kept at the minimum, dealing primarily with safety and consumer protection. Labour arrangements are flexible with high mobility. There is an assumption that innovation occurs due to opportunities created by the competitive forces in the market place and by systematic deregulation. Public sector actors are considered inherently conservative in their acquisition practices, resisting innovation. Commercial negotiations are adversarial in nature creating concern of liability when new practices are introduced.

- The Government-led system, which can be observed in France, Germany, Italy or Netherlands, sees government, with its large purchasing force and social responsibility, playing a central role in the market place. Publicly sponsored projects are often used to initiate/demonstrate new technologies that are then disseminated to other practitioners. There is significant level of regulation of all aspects of construction. Labour arrangements tend to be inflexible with little mobility. Public policy instruments are seen as essential elements of innovation, with government being perceived as a valuable partner. Commercial negotiations tend to emphasise existing linkages and government may intervene to achieve the desirable socio-economic goals.

- The Social-democratic system, more prevalent in the Scandinavian countries, shares many of the characteristics of the government-led system with a particular emphasis on the tripartite (industry/labour/government) approach of solving industrial issues. Labour arrangements show greater level of flexibility and consideration of competitive forces. Innovation is state supported, expected to balance social, economic and environmental values.

- Finally, the Meso-corporatist system, typical of Japan, is based on the presence of very large corporations. Domestic competition in construction is intensive but a certain amount of market sharing exists, which has allowed individual companies to achieve relatively high margins. Government expects reinvestment of excess profits in innovation, because innovativeness is considered as important corporate and national value. Public policy is focussed on supporting large companies, expected to lead in continuous technological and quality improvements. Labour flexibility within organisations is relatively high in a context of lifetime employment. Public policies on technology directions, regulation, training and innovation related issues are arrived at through consensus negotiations between public sector and major industrial participants.

CHAPTER 3

INTERNATIONAL PRACTICE IN CONSTRUCTION INNOVATION SYSTEMS

André Manseau and George Seaden

3.1 INTERNATIONAL TASK GROUP

Fifteen countries participated in the study, based on their willingness and interest in the project. While, a priori, no specific selection criteria were established to ensure significant international representation, in fact a good sample of developed, developing and less industrially developed countries was obtained, as well as broad geographical distribution. Table 3.1 indicates the distribution of participating countries with regard to their regional representation and GDP per capita.

Table 3.1: Participating countries by regions and GDP per capita (in US$)

	Less than $5,000 per Capita	Between $5,000 and $20,000	More than $20,000 per Capita
Europe		Portugal (10,269)	Denmark (30,748) Finland (23,309) France (23,843) Germany (25,468) Netherlands (23,270) UK (21,921)
Asia			Japan (33,265)
America	Brazil (4,930)	Argentina (9,070) Chile (5,271)	Canada (20,082) USA (28,789)
Australia			Australia (21,971)
Africa	South Africa (3,331)		

Source: United Nations, Statistics Division, 1997 data.

The first section of each country's chapter, presented in this volume, aims to briefly describe the general national context influencing its construction industry. Key characteristics of the construction industry and related businesses are presented, as well as the current situation of innovation in this industry. The second section describes various approaches and major public programs that address innovation in construction. Each country chapter's presents a scheme on major public programs available to the construction sector using the pattern shown in Figure 3.1.

Figure 3.1: Framework for presenting public policy instruments

Innovation stage	*Name*	*Annual Resources*	*Objectives*	*Means*	*Contribution to innovation*
Programs to support R&D					
Programs to support advanced practices and experiment					
Programs to support performance and quality improvement					
Programs to support taking up of systems and procedures					

Programs that support R&D usually refer to the government funding of extramural activity or to the performance of R&D in construction-related public sector agencies. Programs that support demonstration projects or promote advanced practices are covered under the "support advanced practices and experimentation" category. Next are programs that support performance and quality improvement such as evaluation methods, indicators of performance, benchmarking, quality awareness or diffusion. Finally, are programs to support taking up of systems and procedures such as building codes or to promote adoption of standard practices.

Following a brief description of each major program, authors of country chapters were asked to provide their expert opinion on program contribution to innovation in construction in their countries. This theme is more specifically developed in the third section of each paper. Major changes are discussed, as well as issues and factors that have recently influenced innovation and industry performance.

The last section of each chapter analyses major trends in types or approaches of public instruments; what are major the challenges for the future and what should be prioritized? Promising or interesting new approaches to public interventions are indicated.

3.2   MEMBERS OF TG 35 – COUNTRIES AND CONTRIBUTING AUTHORS

- Argentina - Dora Mabel Zeballos, Universidad de Buenos Aires
- Australia - Keith Hampson and Karen Manley, Queensland University of Technology
- Brazil – Franciso Cardoso, Universidade de Sao Paolo, Marco A. Rezende, Universidade Federal de Minas Gerais, Mercia Barros, Universidade de Sao Paolo and Roberto de Oliveira, Universidade Federal de Santa Catarina
- Canada - André Manseau and Aaron Bellamy, National Research Council of Canada
- Chile - Alfredo Serpell, Pontificia Universidad Catolica de Chile
- Denmark - Henrik L. Bang, Danish Building Research Institute, Sten Bonke and Lennie Clausen, Technical University of Denmark
- Finland - Tapio Koivu and Kej Mantyla, VTT Building Technology
- France - Elisabeth Campagnac, Ecole nationale des Ponts et Chaussées and Jean-Luc Salagnac, Centre Scientifique et Technique du Bâtiment (CSTB)
- Germany - Thomas Cleff and Annette Rudolph-Cleff, Centre for European Economic Research (ZEW)
- Japan - Ryoju Tanaka, Shin Okamoto and Tomoya Kikuoka, Japan Association of Representative General Contractors (CRITC)

- Netherlands - Joris Meijaard, Erasmus University Rotterdam
- Portugal - Fernando Branco and Adriana Garcia, Technical University of Lisbon
- South Africa - Rodney Milford, Chris Rust and Mabela Qhobela, Council for Scientfic and Industrial Research (CSIR)
- United Kingdom – UK TG 35 Team coordinated by Graham Winch, University College London
- USA - Harvey Bernstein and Richard Belle, Civil Engineering Research Foundation (CERF), and André Manseau, National Research Council of Canada

CHAPTER 4

INNOVATION POLICY AND THE CONSTRUCTION SECTOR IN ARGENTINA

Dora Mabel Zeballos

4.1 DESCRIPTION OF THE SITUATION

4.1.1 The national context of science and technology

The characteristics of the economic development in Argentina together with the extended domestic market's protection processes and long stages of political and economic instability did not promote the implementation of technological conducts which would innovate industrial companies. The lack of technological demand and the indifference of companies to improve quality and productivity caused disconnection among production, technological development and basic investigation with the resulting isolation of scientific groups.

Science and Technology public programs have been created over a century ago, and more as a derivation of politic, cultural and ideological decisions than as an answer to an effective demand of knowledge by the productive sector.

The country has allocated more than $10,000 million in the last 25 years but the expected results were not achieved due to the erratic economic policies and the non-effective control of the actions carried out.

Investment made in scientific and technological activities has increased in the nineties; even so it has not been able to surpass the 0.50% of the Gross Domestic Product, corresponding 0.35% of the investment to the public sector and 0.13% to the private sector. The aim meant by the Legislative Power for the sector was to progressively increase the Gross Domestic Product's percentage till reaching 1% in the year 2000 but the economic crisis of the last years has not allowed to achieve this.

The National Government's central organ regarding scientific and technological policy is the State Department of Science and Technology (Secretaria de Estado de Ciencia y Tecnologia -SECYT-). In 1996, the Science and Technology National System was restructured and the SECYT, previously under the National Presidency, was transferred to the Culture and Education Ministry. Two new agencies were created. The first one is the Scientific and Technological Committee (Gabinete Cientifico Tecnologico - GACTEC -) with a mission to define national priorities for the sector as well as to distribute resources assigned to its different agencies and institutions. The second agency is the

National Agency of Scientific and Technological Promotion (Agencia Nacional de promocion Cientifica y Tecnologica – AGENCIA -) which is an institution exclusively devoted to the promotion of sector's specific activities.

Science and technology development activities are carried out by institutions that can be distinguished in two groups. First, we have the university system and the COCICET (Consejo Nacional de Ciencia y Tecnologia – National Science and Technology Board) essentially oriented toward research and science promotion, with a wide coverage in all areas of knowledge. The second group is composed of a series of sector technology agencies that includes: INTA (Instituto Nacional de Tecnologia Agropecuaria – National Agricultural Technology Institute), CNEA (Centro Nacional de Energia Atomica – National Atomic Energy Centre), INDEP (Instituto Nacional de Desarrollo Pesquero – National Fishing Development Institute), and ANLIS (Administracion Nacional de Laboratorios e Institutos de Salud – National Laboratories and Health Institutes Administration). This distinction does not mean either that no sector based technology activities are carried out in the first group or that no basic research are made in the second group. These institutions also carry out, to a further or lesser extent, promotion activities.

Furthermore, in the year 1997 the Executive Power created the Small and Medium-Sized Business Department (Secretaria de la Pequena y Mediana Empresa) as organ of application of Small and Medium-Sized Business (SMEs) promotion Act, enacted in 1995. This organ of application has the fundamental mission of impelling general extent policies through the creation of new support instruments and the consolidation of those already in existence. This Department promotes re-organisation, modernisation, and reengineering of SMEs and facilitates SMEs access to loans, to technological modernisation as well as to human resources training.

4.1.2 The construction industry sector

Many authors have shown that the Argentina's construction industry is mostly composed of firms with very limited technological capabilities. This situation can be explained by a combination of several factors including the followings: its nature of a quasi artisan traditional industry, a rather captive local market, and a strong influence of public investments which accentuates the intensive use of manual labour and conditions technology changes in firms. However, signs of technological actualisation have started in some dwelling constructions firms in the recent years. The incorporation of new equipment and tools as well as new prefabricated components (panels, fixtures, roofs, etc.) enables and simplifies the traditional construction by promoting the development of production rationalisation processes. This changing context urges for rearrangements of management structures that require higher skilled technicians and labour. Moreover, quality certification program is being incorporated in the production of some material (concrete, steel and others) and INTA has elaborated rules to promote construction participation in the ISO 9000 quality certification. We can also observe that a great part of technological advances seen in construction firms comes from technological innovations developed by product suppliers (materials and components or machinery and tools).

4.2   PUBLIC POLICIES THAT PROMOTE INNOVATION IN ARGENTINA

In the context of considerable political-economic changes that have been developing internationally at the end of the century, decision-makers of the country have been making successive and different efforts to attempt to accompany international movements and trends.

Act 21.617 of Technology Transference is enacted in 1977 which is substituted by Act 22426/81, so as to make flexible the negotiation process and technology transference by decompressing some clauses and restrictions expressed in the previous Act.

In accordance with the development structure set up, numerous programs oriented to promote innovation and competitiveness of the industry have been implemented in Argentina.

4.2.1   General programs available to all sectors

In 1990, Act 23.877 of Promotion and Fostering of Technological Innovation is enacted. The act sets forth the creation of a series of fostering instruments that will be implemented in time.

In the framework of the above-mentioned Act, the Technological Modernisation Program I (TMP I) is created. Its goal is to develop and strengthen Argentine companies' competitiveness through processes that connect the national scientific-technological system with productive sectors. In order to achieve this specific funds from the Inter-American Development Bank are employed. This program co-ordinately operates two funds: 1) Technological Development Fund Argentina (Fondo de Desarrollo Tecnologico Argentina – FONTAR). It finances technological innovation and modernisation projects. Its goal is to improve goods and services companies' competitiveness and promotes PYMES technological training, 2) National Science and Technology Fund (Fondo Nacional de Ciencia y Tecnologia – FONCYT) which subsidies investigation projects and activities which results are of public domain and/or give rise to pre-competitive technologies.

PMT-I has been an important support for institutional changes recently introduced within the Science and Technology sector. Having the possibility of incorporating additional funds from the Inter-American Development Bank, the SECYT has been able to modify the operating structure of the different institutions producing a new encouragement system for the promotion of innovation and interaction of public institutions with the private sector.

Nowadays, PMT-II, which will be a sequel of PMT-I, is in the process of being approved and implemented. It will extend the scope of promotion alternatives by supporting the successful instruments of PMT-I and carrying out spending priorities of the multi-annual Plan of Science and Technology. With regard to innovation promotion, it will introduce incentives and co-financing loans for companies and institutions as well as a new scheme of venture capital in order to support new undertakings of technological basis. As regards scientific-strategic promotion, the formation of human resources and Investigation and Development (I + D) groups in prior areas as well as large strategic projects in critical areas will be supported.

Other public instruments of innovation promotion and fostering, of a general-direct character, include investment loans to support quality certifications of national and international standards (IRAM, ISO 9000 AND 14000, ISO-IES, ISO-CASCO, etc.). These programs are promoted by BICE and other financial entities authorised by the Banco Central de la Republica Argentina. Other programs that provide technical assistance and labour training are also available, as well as programs that support creation of companies associations and co-operation networks between companies and institutions. These latter programs are promoted by both the Mining, Commerce and Industry Department and the Social Security and Labour Department as well as by official technological institutions like INTI, INTA, INET.

Public universities are actively participating in the process of scientific-technological development, which are related to the social-economical sector.

Within the scope of the University of Buenos Aires (UBA), one of the most important academic institutions in the country, the Promotion Board of Scientific and Technological Investigation (Consejo de Promocion de la Investigacion Cientifica y Tecnologica – COPICYT) was created in 1994 to be specially in charge of the permanent production and evaluation of the policies and activities carried out in UBA. Actions oriented to promote technological innovation include different instruments such as: the Network of Technology Transference; Developments and Services of UBA as a structure of official management for the transference operations through an interrelated system; the Units of Transference established in each School of UBA; the National Laboratories of Investigation and Services (LANAIS), in agreement with SECYT-CONICET, which offer specialised scientific and technical services to the scientific community and to productive sectors; UBATEC, which is a joint stock company of linkage and technology transference from the scientific-academic sector towards national, regional (Mercosur) and international public and private companies; a series of instruments of technical co-operation and transference of technological knowledge concerned with university and third parties (companies, organisations or public or private institutions, other universities or individuals) such as: general agreements, specific undertaking agreements, services to third parties, apprenticeships, post-graduate and extension courses, etc.

4.2.2 Programs that have indirect effects on innovation

In 1995, Act 24467 of Small and Medium-sized Business was enacted which purpose is to promote the growth and development of SMEs, through the creation of new support instruments and the consolidation of those in existence.

Within the scope of this law, the Small and Medium-sized Business Department was created in 1997. Its main duty is to consolidate and extend the productive poles in the interior of the country, develop regional economies and facilitate the introduction of SMEs within the international markets. For this purpose, this Department is in charge of promoting modernisation, re-organisation and reengineering in SMEs, facilitating the access to credit and the formation and training of managers, technicians and professionals, in order to encourage productivity and competence among companies.

Many new instruments are being implemented in that direction, including: the Reciprocal Guarantee Companies (Sociedades de Garantia Reciproca –SGR), programs that support SMEs (PRE), programs that support development of exports (PYMEXPORTA, EXPORTAR Y PREX), the regime of fiscal credits for companies that invest in its workers training, and a new program on "incubator of firms".

4.2.3 Programs that support innovation in the construction sector

There is no specific program that directly promotes innovation in construction. However, some programs indirectly affect innovation in the construction industry.

Among these "indirect" programs, we can identify the National Fund for dwelling (Fondo Nacional para la Vivienda – FONAVI-) which was created by law number 21.581 in 1972 – and modified in 1977 – is still in force. This program provides funds for the construction of dwellings with a social interest. Among its rules, it encourages technological development and innovation by promoting special housing complexes that use industrialised systems and which that have approved technical aptitude certificates (certificados de aptitud tecnica –CAT-). FONAVI was managed and implemented by the National Dwelling Department (Secretaria de Vivienda de la Nacion) until 1996. That year, a State reform was initiated which introduces an important transfer of responsibilities from central towards the provincial and municipal communities. FONAVI has become a distributed fund directly managed and implemented by Provinces. From this moment, the social control of this fund has been lost and the intention of promoting technological development on the basis of the execution of dwelling plans, became a duty of each Province.

4.3 MAJOR TRENDS IN INNOVATION POLICIES

Generally, public policies implemented in Argentina are oriented to develop and improve the enterprising technological capacities, supporting the productive sector – with plans specially directed to SMEs – in their human resources training, technical assistance and the execution of investigation and development projects. In every implemented action, the technological scientific and academic public institution participation is highlighted. Within this process, companies contribute resources to the scientific-technological sector, whereas the State facilitates its access to credits on favourable conditions and the financing of technological investment projects with long-terms amortisation.

4.4   THE EFFECTIVENESS OF PUBLIC POLICIES ON INNOVATION

Table 4.1 summarises major programs available to the construction sector to support innovation.

Table 4.1 Major programs available to support innovation

	Programs to support R&D	**Programs to support advanced practices and experiment**
Name	Technological Modernisation program I/II: FONCYT (Fondo Nacional de Ciencia y Tecnología – National Science and Technology Fund). CONICET-SECYT projects. Universities Projects.	Technological Modernisation Program I/II: FONTAR (Fondo de Desarrollo Tecnológico – Technological Development Fund) Technology, Development and Services Transference Network UBATEC Special agreements: University – Company
Resources	Inter-American Development Bank (IDB) State Department of Science and Technology (SECYT)	Inter-American Development Bank (IDB) University of Buenos Aires National Universities
Objectives	Support research projects, development of pre-competitive technologies and related of the public domain Support for education of human resources and R&D groups in critical and priority areas.	Finance technological innovation and modernisation projects to improve companies' competitiveness and promotes Small and Medium-Sized Business' (SMEs) technological training. Support new undertakings of technological basis. Promote University-Company relationships and technology transfer from the academic sector to public and private companies.

Table 4.1 (Continued) Major programs available to support innovation

	Programs to support R&D	**Programs to support advanced practices and experiment**
Means	Development of R&D projects.	Joint projects with university researchers and staff from companies/institutions.
Contribution to innovation	Through R&D activities of pre-competitive and advanced Technology.	Through companies technological training.
Name	Loans for investments and quality certifications. National Investigation and Services Laboratories (LANAIS-UBA). PYMEXPORTA/EXPORTAR/PREX programs.	Reciprocal Security Partnership SMEs re-organisation program Fiscal Loan Regime Various programs from National Universities.
	Programs to support performance and quality improvement	**Programs to support taking up of systems and procedures**
Objectives	Promotion and fostering of technological innovation and companies competitiveness. To provide specialised services to fulfil requests from scientific, technological and productive sectors. To promote exports development	To promote SMEs modernisation, re-organisation, and reengineering To facilitate access to loans, education and training of technical and professional staff and managers to encourage productivity and competitiveness.
Means	Specific loans for quality certifications and specific development projects with the support of official technological institutions. Co-financing of consulting services on management, productivity and development activities. Creation of supporting specialised services centres.	Creation of companies between SMEs (participant partners) and large companies (protective partners). Specific advising in technical and economical-financial aspects. Courses on labour training. Agreements: general, specific. Services to third parties, apprenticeships, post-graduate and extension courses, etc.
Contribution to innovation	Through training activities and specialised services	Through training activities and services.

Since the beginning of the nineties, a large proportion of companies in Argentina have faced a strong increase of competitive pressures as a result of the new conditions of the Argentinean economy within the global world. Limited innovative capacity of companies, and particularly SMEs, can be explained by limited training efforts that have been made since changes begin. Public policies have been also slow to change and programs to promote technological innovation in our country are quite recent. Moreover, one the greatest problems is the scarce diffusion of information on available instruments. Therefore, the construction sector does not participate, generally speaking, in those new policies because they ignore them or because there is a lack of interest.

CHAPTER 5

CONSTRUCTION INNOVATION AND PUBLIC POLICY IN AUSTRALIA[1]

Keith Hampson, and Dr. Karen Manley

5.1   INTRODUCTION

Over the past few years there has been rapidly growing interest in developing a more robust, internationally competitive construction sector in Australia[2].   All major stakeholders are involved in associated initiatives – industry, government, research sector and interest groups.   There has been a significant improvement in the level and quality of communication between major stakeholders, which is yielding initiatives that promise to lift future sector performance.   However, it is early days yet and concrete improvements are modest at this stage.   The performance of Australia's construction sector remains far short of its potential.

Australia's construction sector operates against a background of industry fragmentation, intense competition, falling profits and new challenges including IT advancements; increasing public interest in environmental protection; increasing demand for packaged construction services; and moves toward private-sector funding of public infrastructure.

Innovation and innovative behaviour are seen as key opportunities to raise the sector's performance and meet new challenges.   However, to date, innovation in the Australian construction sector has been marked by poor innovation levels, government programs of limited effectiveness and poor uptake of those innovation programs that are available.

---

[1]   The authors are indebted to the Australian Expert Group in Industry Studies for access to OECD databases.

[2]   Throughout this chapter, the term 'construction sector' refers to a broadly defined construction industry which includes all the segments comprising the building and construction industries, including contracting firms, up-stream materials and equipment suppliers and the property segment. The term 'construction industry' is used to denote a narrower definition of the construction sector which comprises only contracting firms. Most construction data is collected on this basis. Further, the term 'construction' is used as shorthand for 'building and construction'.

5.2   THE AUSTRALIAN CONTEXT

5.2.1   Background, institutions and culture

5.2.1.1 Background[3]

Australia is a Federation with six states (capital cities in brackets) – New South Wales (Sydney), Victoria (Melbourne), Queensland (Brisbane), South Australia (Adelaide), Western Australia (Perth) and Tasmania (Hobart) and two territories – Northern Territory (Alice Springs) and the Australian Capital Territory, where Canberra, the nation's Capital is situated. Australia's construction sector is concentrated in Australia's three largest states - New South Wales, Victoria and Queensland – in terms of number of businesses, employment and income (ABS 1996-97).

Australia is a developed nation occupying approximately seven and a half million square kilometres situated between the Pacific Ocean to the east and the Indian Ocean to the west. The area of Australia is almost as great as that of the United States (excluding Alaska), about 50% greater than Europe (excluding the former USSR) and 32 times greater than the United Kingdom.

Australia lies between latitudes 10 degrees south and 43 degrees south, and between longitudes 113 degrees east and 153 degrees east. Figure 5.1 shows a map of Australia.

The Australian continent is almost bisected by the Tropic of Capricorn and subsequently experiences great variations in climate. Climatic conditions range from hot dry deserts inland, to tropical rainforests in the north, to snow-capped mountains in the south. The climatic extremes experienced in Australia and the huge distances between many major regional centres pose challenges for Australia's construction sector, yet also create the potential to develop competitive advantage internationally in certain niche markets – for instance, in relation to cyclone-resistant facilities or remote-area construction using advanced information technologies.

Australia's population density of 2.3 people per square kilometre (in 1998) is one of the lowest in the world, although most Australians live in densely populated cities on the coastal strip – especially on the eastern seaboard.

Australia's 19 million inhabitants enjoy a GDP per head of US \$20,505 (1998-99 figures)[4]. A 1997 world ranking of the top 20 national economies by 'Real 1997 GDP per Capita (1995 U.S \$)' shows Australia ranked 15th at US \$20,380 and the United States ranked 8th at US \$29,142 (*Wall Street Journal* website).

There are three levels of government in Australia: Federal, State and local. Both the State and the Federal systems of government derive from the British

---

[3]   Much of the data in this section is based on the *Australia Now* website and *The Commonwealth Online* website.

[4]   All data is shown in Australian dollars, unless specified otherwise. In several places, Australian dollars have been converted to United States dollars using the exchange rate at 31 December 1999, which was 0.6538 \$US per A\$.

Westminster system, although many features of the Federal Constitution (including the federal structure) are based on the United States Constitution. In the twentieth century, Australia has been characterised by a strong party system and adversarial style of politics between the government and opposition. The activities of all levels of government impact on the construction sector.

Figure 5.1 Map of Australia

The Federal Government has responsibility for matters relating to 'workplace relations, environmental and other standards, education and training, trade-related matters and financial and corporations law' (ISR 1999: 11). The state and territory governments have responsibility for 'building regulation and the planning approvals system, although elements of the system are delegated to local government. This level of government is also responsible for the registration of builders and some trades, and accrediting and registering professionals' (ISR 1999: 11).

Historically, much of Australia's infrastructure was constructed following World War II, with major road, energy and water supply schemes being undertaken through direct government funding. Housing developments also boomed in most major Australian towns following substantial post-war immigration. Upgrading and extension of infrastructure is now providing new funding challenges, with increased involvement from the private sector gaining momentum.

5.2.1.2 Institutions[5]

The Australian construction sector operates in an institutional context that is highly cluttered. Adopting a broad view of the sector's boundaries, various types of institutions have an impact on firms in the sector, these include research institutions; agencies that set standards and regulate performance; providers of vocational training; industry associations; clients; unions; and interest groups. Key actors in the Australian construction sector include:

- Approximately 20 university departments active in research related to construction
- hundreds of standard-setting agencies and regulators including key actors such as: Australian Building Codes Board, Standards Australia, National Association of Testing Authorities Australia, Local Councils and Builders Licensing Authorities
- various Federal, State and Territory government portfolios covering: education and training, corporate and consumer affairs, industry development, trade, competition policy and public works
- the Commonwealth Scientific and Industrial Research Organisation (CSIRO), particularly the Division of Building, Construction and Engineering
- hundreds of industry associations including major players such as the Housing Industry Association, Master Builders Australia, Civil Contractors Federation and Australian Industry Group
- hundreds of property trusts, superannuation funds, finance and banking institutions and real estate agencies, and
- hundreds of training providers, including the: Australian National Training Authority, Construction Training Authority, State Industry Training Advisory Bodies and private training providers.

This multitude of actors arises from both the fragmented nature of the sector and Australia's federal system of government which invites duplication (through overlapping and/or unclear responsibilities), and variations in standards that create numerous inefficiencies. Federalism contributes to a sector culture marked by poor communication and rivalry. The fragmentation of the sector and proliferation of policy actors makes coherent policy formulation difficult, and

---

[5]   This section is based in part on ISR 1999.

hence restricts the development of coherent approaches to improving the sector's innovation performance.

5.2.1.3 Culture

The performance of the Australian construction sector is further constrained by 'a focus on short-term business cycles and a project-to-project culture' (ISR 1999: 8). Construction contracting in Australia is regarded as a competitive and high-risk business (Uher, 1994). This competitiveness is largely due to the fragmented nature of the sector and cost traditionally being the prime factor in the tender selection process (Hampson and Kwok, 1997). In 1997 a major joint government/industry workshop was held to develop a strategic plan for the construction sector in Queensland. The workshop resulted in a number of key elements of the sector's culture being identified, these were:

- The use of a subcontracting system to create flexibility and to cope with the cyclical nature of the sector
- adversarial nature of contracts and high levels of litigation
- unethical dealing between contracting parties
- perpetuation of demarcation lines between professionals, trades and their representative bodies
- a lack of coalition through strategic alliances, joint ventures, consortia or partnerships
- poor utilisation of new technologies, with less than 15% of small to medium-sized firms in contracting using computers
- as absence of an export culture and collaborative approach to lifting the sector's export performance
- a reliance on a casual labour employment process with little long term training
- price-based competition that fails to sufficiently account for quality or whole-of-life costs
- little investment in research and development and an unsatisfactory innovation record, and
- lack of investment in human resource development or any other medium to long-term investment that has little short-term pay-off .

5.2.2 Australian construction sector overview

Construction activity is undertaken by both the private and public sectors in Australia. The private sector is engaged in all three categories of construction (residential, non-residential building and engineering), and plays the major role in residential and other building activity. The public sector plays a key role in initiating and undertaking engineering construction activity, and building activity relating to health and education (*Australia Now* website).

        The Australian construction industry is highly fragmented with 94% of all
businesses employing less than five people in 1996-97 (latest available data).
Together these businesses employ just over two-thirds of all people working in the
industry, but earn slightly less than half of total industry income (ABS 1996-97).
        The construction industry occupies a significant position in the Australian
economy and plays a vital role in sponsoring economic growth. The figure below
shows the industry's long-term contribution to Australian national output.

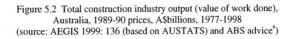

Figure 5.2 Total construction industry output (value of work done),
Australia, 1989-90 prices, A$billions, 1977-1998
(source: AEGIS 1999: 136 (based on AUSTATS) and ABS advice[6])

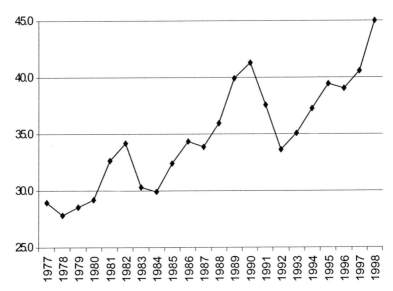

        Over the past two decades, the Australian construction sector has grown at
an average annual rate of approximately 2.6 percent in real terms. Over this
period, the industry, as conventionally defined, has mostly contributed a relatively
stable proportion to GDP of between 6.5 and 7%. Such a contribution means the
industry has a significant impact on national output. Even so, it is likely that this
impact has been understated because it is based on a narrow definition of the
industry (AEGIS 1999: 135).

        There has been considerable focus in Australia over recent years on the
performance of a *broadly defined* construction sector, which includes participants

---

[6]   Data based on the activities of project-based firms – see discussion to follow.

often overlooked in traditional analysis. Whereas traditional analysis is based on the activities of project-based firms - firms that design, engineer and construct buildings and major projects, some new analyses also take into account the activities of:

- The supply network - firms that manufacture and distribute materials, products, machinery and equipment, and
- the property sector - firms that invest in and manage property.

The sector, viewed in this way, can be summarised as follows for 1996-1997.

Table 5.1 The Australian construction sector, key statistics, 1996-1997 (source: ISR 1999: 8)

Sector	Contribution to GDP	Number of Firms	Total Employment
Supply Network	6.8%	165,300	452,900
Project-Based Firms	3.6%	46,600	182,400
Property Sector	4.0%	16,500	94,100
Total	*14.4%*	*228,400*	*729,400*

These data indicate that the sector's total contribution to GDP was actually around 14% in 1996-97, rather than around 7% as indicated by conventional analysis. Further, the supply network, including materials, product, machinery and equipment suppliers, is collectively the largest sub-sector, by output, employment and number of businesses.

Recent analysis undertaken by the Australian Expert Group in Industry Studies (AEGIS) for the Federal Government presents data based on a greater disaggregation of the sector. These data are older, however they provide a useful view of the relative contribution of the sector's segments. Income and employment perspectives are shown below.

The role of service providers in the sector is highlighted, with trade services providing approximately one-quarter of sector income.

Table 5.2 Australian construction sector income, by segment,
A$millions, 1995-96 (source: AEGIS 199: 57 (based on ABS data))[7]

Sector Segment	Income ($Million)
On-site Services (Trade Services)	21,898
Client Services (Engineering, Technical, etc.)	8,607
Building & Construction Project Firms	34,250
Materials and Products Supplies	18,608
Machinery and Equipment Supplies	2,803
TOTAL	*86,166*

The employment table shown below highlights the size of both the trade services segment and the material and product supplies segment. Together, they account for nearly two-thirds of sector employment.

Table 5.3 Australian construction sector employment, by segment,
1995-96 (source: AEGIS 1999: 59 (based on ABS data))[8]

Sector Segment	Employment
On-site Services (Trade Services)	220,000
Client Services (Engineering, Technical, etc.)	102,000
Building & Construction Project Firms	108,000
Materials and Products Supplies	222,000
Machinery and Equipment Supplies	30,000
TOTAL	*682,000*

The Australian Bureau of Statistics collects detailed construction industry data on the activities of prime contractors and sub-contractors. Looking at the industry defined in this narrower way, data is available based on three broad industry segments – residential building, non-residential building and engineering construction, and by client type – the public and private sectors.

[7] Figures drawn from the relevant ABS publications have been adjusted to: a) reflect only that income earned by the construction sector and b) avoid double counting. See AEGIS 1999: 57 for sources and methods employed. 'Income' is based on a number of measures. See original source notes.

[8] Figures drawn from the relevant ABS publications have been adjusted to: a) reflect only that income earned by the construction sector and b) avoid double counting. See AEGIS 1999: 57 for sources and methods employed.

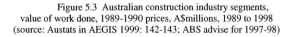

Figure 5.3 Australian construction industry segments,
value of work done, 1989-1990 prices, A$millions, 1989 to 1998
(source: Austats in AEGIS 1999: 142-143; ABS advise for 1997-98)

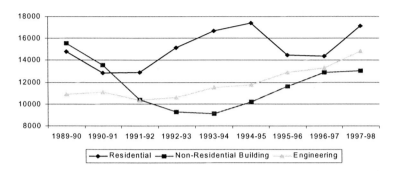

Engineering construction activity shows a largely consistent upward trend over the period. Residential construction also shows an upward trend overall, with a period of strong growth between 1992 and 1994 and strong growth returning in 1997. Declining activity was recorded in 1990 and 1995. Non-residential building construction activity reached a low point in 1993 and has been increasing consistently since then.

The figure below examines the same data by client type.

These charts show very clear trends. The private-sector is by far the most important client for residential construction. Private-sector clients also dominate the non-residential building sector, although to a lessor degree. In contrast, public-sector clients dominate the engineering construction sector, although this dominance has been declining since 1993 as a result of privatisation trends in Australia.

Keith Hampson, and Dr. Karen Manley

Figure 5.4 Australian construction industry segments,
value of work done for the public and private sectors, 1989-90 prices, A$millions, 1989-1998
(source: Austats in AEGIS 1999: 142-143; ABS advise for 1997-98)

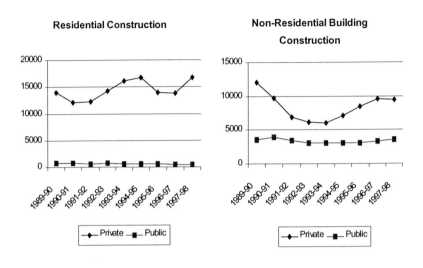

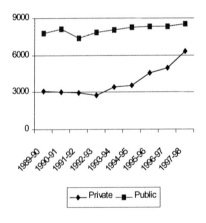

Figure 5.5 Construction expenditure per capita, US$, top 20 countries, 1998 (source: ENR 1998)

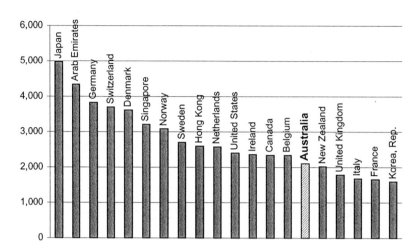

The chart shows the world's 20 largest construction spenders per capita; Australia is ranked 15th.

Figure 5.6 Total construction expenditure, US$ million, top 20 countries, 1998 (source: ENR 1998)

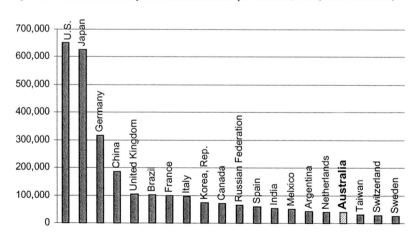

The government – at local, state and federal levels - is a large construction sector client, as well as industry regulator. The government can affect the volume of construction work by directly influencing consumer demand and more indirectly by manipulating its fiscal and monetary policies.

Finally, the charts below review Australian construction expenditure in the international context.

Of the world's 20 largest construction spenders in absolute terms, Japan and the U.S. spend far in excess of the remaining 18 countries. Australia is ranked equal 17th, making it a significant market globally.

5.3   INNOVATION IN THE AUSTRALIAN CONSTRUCTION SECTOR[9]

Overall, the Australian construction sector has a relatively poor innovation record, as revealed in R&D statistics.[1] The AEGIS study mentioned earlier noted the following about resources devoted to construction R&D in Australia:

Over the period from 1992-93 to 1996-97, R&D expenditure on construction … averaged only 1.4% of total R&D expenditure in Australia. This is significantly less than the share of construction in total output, which averages around 6.5 to 7% of GDP [possibly up to 14% adopting a broad definition of the sector – see earlier discussion][10] (AEGIS 1999: 60).

Further, in a ranking of Australian businesses by R&D expenditures, the Federal Government's *R&D Scoreboard '98* indicates that only one construction firm is among the top 20 private sector research performers (although another diversified company in the top 20 has some operations in the sector). Of the 325 companies listed by the *R&D Scoreboard* only 21 or 6.4% are in the construction sector (ISR 1999: 19).

Another key measure of the sector's R&D performance involves R&D intensity measures. The Australian sector's R&D intensity over the past 13 years is reflected in the figure below.

Despite significant growth in R&D intensity during the first half of the 1990s, the comparative performance of the construction sector remains poor, nationally and internationally. The figure below compares construction performance with that of the manufacturing sector.

---

[9]   This section focuses on R&D statistics. Other innovation indicators are not presented due to the relative absence of such data and time constraints.

[10]   R&D share for construction is based on construction defined as a socio-economic objective of the R&D.

Figure 5.7 expenditures by construction contracting firms as a proportion of construction value added, Australia, 1984-1996 (source: OECD 1999b, main industrial indicators)[11]

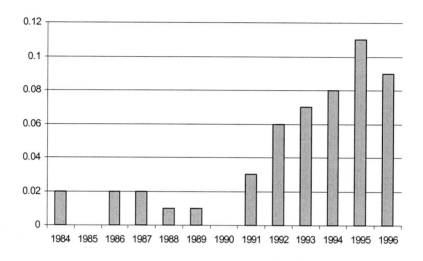

Figure 5.8 Business expenditure on R&D as a proportion of value added, construction and manufacturing, Australia, 1984-1996, % (source: OECD 1999b, main industrial indicators)

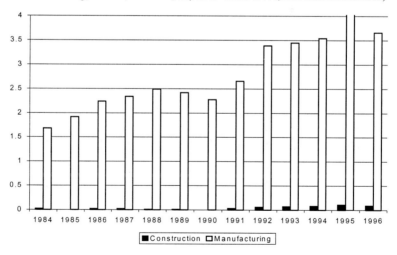

[11]   No data available for 1985 and 1990.

The construction industry's performance lags so far behind that of the manufacturing sector, that the two sets of data can barely be plotted on the same chart. In 1996, (latest available data) the manufacturing sector's R&D intensity was 40 times that of the construction industry. The comparative performance of the Australian construction industry is also very poor in the international context.

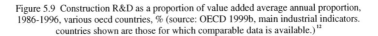

Figure 5.9 Construction R&D as a proportion of value added average annual proportion, 1986-1996, various oecd countries, % (source: OECD 1999b, main industrial indicators. countries shown are those for which comparable data is available.)[12]

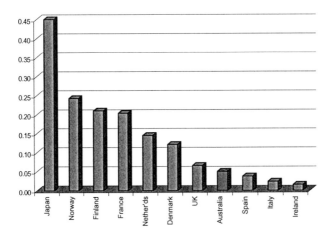

Of the countries shown, Australia's construction industry had the fourth lowest R&D intensity over the period. It may be that Australia's relative performance would be even worse if the sector were more broadly defined.

A recent study reviewed R&D per head of population for the 'built environment' sector. The study defined the 'built environment' sector to include the construction sector (broadly defined), together with energy utilities, water supply, sewerage, drainage, transport, and public order and safety services. For this very broad sector, in 1996, R&D per head was equivalent to US $50 in the US and US $14 in Australia (Cebon et al 1999: 3).

Australia's poor R&D performance is partly due to relative inactivity by particular segments within the construction sector. In 1996-97, R&D expenditure by the sector (broadly defined) was $43.97 million (AEGIS 1999: 62). Figure 5.10 provides a breakdown by segment.

---

[12]   Four values have been estimated for this series: Australia 1990; Denmark, Ireland and the Netherlands 1996. Each estimate is the average value across the remaining years in each country's series.

Figure 5.10 R&D Expenditure by construction sector segments
(source: AEGIS 1999: 62 (based on ABS unpublished data))

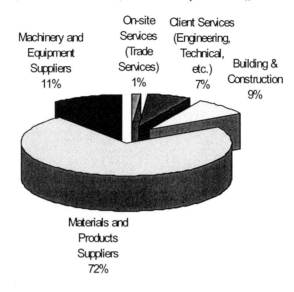

Clearly the materials and products segment (involving up-stream supply) is the key R&D driver within the sector. The very poor performance of the trade-services segment can be attributed to the very fragmented structure of the segment and very low profit margins (ISR 1999: 51).

Hampson (1997) examines a number of other factors underlying the Australian construction sector's poor innovation performance. The following key innovation constraints were noted:

- Financial commitment – construction innovation involves high levels of risk, including whether an innovation is transferable across projects
- Time – key employees are needed for successful innovation, however with construction firms 'running so lean' such people often have little free time, and
- Intellectual property – the benefits of construction innovation tend to be highly dispersed, reducing firm-level innovation incentives (Hampson 1997: 9).

This work was extended by research conducted by AEGIS (1999: 196-197), which uncovered the following impediments to innovation:

- Site-based production – the temporary nature of production processes disrupts innovation processes, in large part because of short-lived relationships

- Project size and complexity – large and complex projects involve significant communication challenges and invite 'disparate and discordant effort
- Risk of failure – the sector exhibits a preference for established practices, based in part on bad experiences historically in relation to public safety issues
- Competitive bidding contracts – result in an overdeveloped sense of cost and an underdeveloped sense of value and the importance of innovation, and
- Changing finance systems – the trend for industry contractors to provide finance further squeezes resources, which might otherwise be devoted to innovation.

It is difficult to canvass the costs of Australia's lack of innovation in the construction sector given the limits of the present study. Nevertheless, Hampson (1998: 2) reports that a 10% reduction of costs in the sector would have a significant impact on long term GDP growth in Australia.

Another study notes that turnover per employee in the construction sector is amongst the lowest of all industry sectors in Australia and that this 'indicates clear prospects for increasing efficiency and productivity through innovations' (Cebon et al 1999: 3).

Analysts and public policy makers in Australia are well aware of the poor innovation performance of the construction sector. A number of programs have been designed to improve this situation (see next section), with there being evidence of an increased focus on finding effective solutions over the past year or so.

The discussion thus far has focused on *business* innovation. The importance of public-sector R&D expenditures is reflected in the chart below.

Figure 5.11 R&D Expenditure of 'construction' as a socio-economic objective, 1996-97, % of total expenditure (source: AEGIS 1999:61)

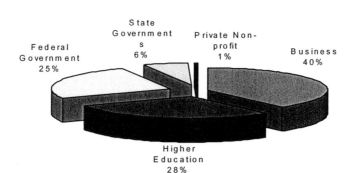

Although the business sector provides the largest single source of funds, the combined government sectors provide the majority of funding. The state governments play a relatively marginal role compared to federal government spending, which is both direct and funnelled through the higher education system. The table below tracks total public sector spending over time.

Table 5.4 Public sector R&D expenditures as percentage of total R&D, various years
(source: derived from AEGIS 1999: 61)

Year	Percentage
1992-93	74.7
1994-95	62.5
1996-97	59.8

Despite public expenditures accounting for a very high proportion of total R&D expenditure, the relative level of public-sector commitment has been falling in recent years. The Federal Government's recent *Building for Growth* report on the construction sector suggests that:

> ...public investment in R&D is critical. The present level is probably too low for the size and importance of the building and construction industry (ISR 1999: 20).

A recent paper provides qualitative insights into public-sector construction-related R&D expenditure. The paper by Marceau, Manley and Hampson (1999) formed part of an AEGIS report for the Federal Department of Industry, Science and Resources and it's Building and Construction Action Agenda program. The paper was based on a survey of 23 public-sector R&D organisations (mainly university departments) in Australia active in construction research (representing 85% of the total population of such organisations). Data were principally gathered in relation to the 2 largest projects undertaken by each organisation between 1996 and 1998. In relation to these 'nominated projects', some of the study's key findings were that:

- 75% of nominated projects were completed, or expected to be completed, in three years or less
- 50% of nominated projects attracted funding of over $100,000
- 75% of nominated projects were partially or fully funded by external funders – 39% by companies, 23% by Australian Research Council

(ARC) grants, 20% by government departments, 11% by the Construction
Industry Institute (CII), 7% by industry associations
- the level of funding provided by government departments and the ARC
  appeared to be substantially higher than that provided by companies and
  industry associations, while CII funding levels appeared to lie somewhere
  in between
- in-kind contributions from project funders were significant and pervasive
- 86% of respondents with pre-existing relationships with primary funders
  rated that relationship as either important or very important in obtaining
  funding
- 66% of nominated projects (both completed and still in progress) had
  resulted in at least one journal publication at the time of the survey
- 64% of nominated projects had resulted in at least one workshop or
  conference paper; and
- constraints noted by respondents in adoption processes indicated the
  presence of industry resistance to change.

Based on these data and historical observations, the report concluded that
in recent times there may have been some improvement in both the relevance to
industry of R&D projects undertaken by public sector research organisations and in
the extent of linkages maintained with the construction sector. However, the report
also noted that these was substantial scope for further improvement (see Section 3).

5.4   DESCRIPTION OF INITIATIVES IMPACTING ON
CONSTRUCTION INNOVATION

Very little research has been conducted in this field in Australia. There appears to
be no comprehensive prior examination of the impact of innovation-related policy
mechanisms on the construction sector, nor does there appear to have been any
systematic investigation of the sorts of policies that might be most useful in the
special circumstances evident in the construction sector. Hence, the research
conducted for this paper has identified the need for in-depth research to be
conducted into innovation policy programs and their effectiveness in the
construction sector in Australia.

Given the absence of any prior investigations and the time limitations of
this paper, this section presents only an overview of the issues. The discussion is
framed around the institutions administering public programs which do, or could
potentially, impact on construction innovation.

5.4.1   Federal Department of Industry, Science and Resources

There is no separate Department of Construction in the Australian Federal Government (the Department of Housing and Construction was abolished in 1987 as part of a portfolio restructuring program). The Federal Department of Industry, Science and Resources (ISR) administers a series of generic innovation programs aimed at lifting the rate of investment by Australian business and a specific program targeted at further development of the construction sector. The separate state governments also provide some broad industry support, though this is generally not specific to construction and is largely uncoordinated with federal government initiatives. The key innovation-related programs administered by ISR are summarised in Table 5.5.

Table 5.5 Key innovation programs available to firms in the construction sector
(source: ISR advice and ISR 1999)

Innova-tion Stage	Program Name	Funding A$	Objectives and Means	Contribution to Innovation in Construction
	R&D START Program (ISR)	$739 million over 4 years	To increase R&D by small to medium companies, usually not in profit. Provides assistance of up to 50% of project costs through grants, and loans for early commercialisation. Also provides assistance for small business to develop their R&D capability.	Use by the construction sector is minimal.
	Renewable Energy Equity Fund (ISR) in conjunction with the Australian Greenhouse Office (AGO).	$20 million over 10 years	Provides a one-off sum of Federal funds to be matched on a 2:1 basis with private-sector capital. The total capital sum flowing to renewable energy research is expected to be $30 million. Program delivered by private sector fund managers.	This program aims to assist companies at the seed, start up or early expansion stage of investments involving the commercialisation of renewable energy technologies.

Table 5.5 (Continued) Key innovation programs available to firms in the construction sector
(source: ISR advice and ISR 1999)

Innova-tion Stage	Program Name	Funding A$	Objectives and Means	Contribution to Innovation in Construction
Programs to Support Commer-cialisation of R&D	Innovation Investment Fund (ISR)	$130 million allocated to Round 1 fund managers. $100 million for Round II fund managers. Over 10 years.	Facilitate venture capital investment in small to medium size innovative companies. The Federal Government has invested in venture capital funds at a ratio of 2:1 (government /private) over the 10 year life of the VC funds. These funds invest in companies commercialising technology with annual revenue of $4 million or less.	To date, investment in construction related activities is minimal.
Programs to Support R&D	R&D Tax Concession (ISR and Australian Tax Office)	About $400 million per year in revenue forgone	To increase business R&D through a 125% concession on eligible expenditure. Principally used by profitable firms.	Use by the construction sector is minimal.
	Co-operative Research Centres (ISR)	Provides funding over 7 years. Funding per centre averages $2.5 million/year. and ranges between $1.4 – 4.6 million. Total government funding has averaged $140 million/year since 1990 and is ongoing.	Brings together key national and international public research agencies and industries to: enhance linkages; develop new technologies; promote technology transfer; and promote industry-relevant postgraduate training. Aim of CRC Program is to enhance capture of research benefits through strengthening the links between researchers and the users of research.	There are 65 CRCs, but none in the Construction sector. One focussing on construction has been targeted in the Action Agenda (see below). Seventh selection round due 2000.

Table 5.5 (Continued) Key innovation programs available to firms in the construction sector
(source: ISR advice and ISR 1999)

Innova-tion Stage	Program Name	Funding A$	Objectives and Means	Contribution to Innovation in Construction
Programs to Support Techno-logy Diffusion	Technology Diffusion Program (ISR)	$101.8 million over 4 years 1998-2002	Accelerate the spread of new technology through key sectors. Program assists with the non-research costs associated with collaborative research and identifying opportunities for technology diffusion.	Use by the construction sector is minimal. Greater use of the fund is targeted under the Action Agenda (see below).
Multi-stage Programs	Building and Construc-tion Industries Action Agenda	$3.6 million over 3 years, 1999-2001	Involves a suite of initiatives specifically aimed at the construction sector (see Section 4)	Tailored to construction sector needs.

Although the Renewable Energy Equity Fund could potentially have a significant impact on the construction sector (broadly defined), the only program specifically targeted to the construction sector in the above table is ISR's recently launched Action Agenda process. This move towards tailored assistance for the construction sector is likely to have a major impact on the sector (see Section 4).[13]

5.4.2   Australian Research Council

The Australian Research Council (ARC) provides research funding to Australian universities. Australia has 38 universities, of which probably two-thirds carry out some sort of research related to the construction sector. Most research is carried out in schools or departments of construction management, building, or civil engineering. Australian universities tend to operate autonomously and independently, though some attempts have been made to share research experiences in the built environment domain (see Alliances section below).

The ARC provides research funding for pure and applied research projects of up to 3 years duration to individual university researchers or teams of university

---

[13] For a more comprehensive listing of the full-range of innovation-related programs made available to industry by Australia's Commonwealth and State Governments, see ISR 1999.

researchers, in some cases in conjunction with industry partners. The Council's total program budget for 1999 was $359million. Several grant programs are operated, these include: the Large Grant Scheme (largest scheme); the Small Grants Scheme (provides block grants for dispersal by universities); the ARC Research Fellowships program (in part to provide career opportunities for early career researchers); and Strategic Partnerships with Industry: Research and Training (SPIRT) Scheme (supports collaboration with industry).

Only $750,000, or 0.2% of total ARC expenditure was allocated to the building and construction field of research in 1999 (ISR advice and ARC advice). Other research focusing on management practices in the sector, for example, may be classified under other fields.

5.4.3   Australian Building Codes Board

The Australian Building Codes Board is the result of an agreement between the Federal Government and the governments of the six Australian states and two territories. In part, the Board aims to promote performance-based regulations and undertake research likely to result in product and process innovations for the construction sector. One of the main research programs being undertaken is the Fire Code Reform Research Program. This program has total government funding of $2.5 million over the period 1995-1999. It is funded by both government and industry and is designed to introduce a cost-effective, fully engineered approach to fire safety regulations for incorporation into the Building Code of Australia. Several projects related to fire research and experimental testing have been undertaken. Any further government funding will be contingent on the results of these projects.

5.4.4   Commonwealth Scientific and Industrial Research Organisation

The Commonwealth Scientific and Industrial Research Organisation (CSIRO) is Australia's principal public-sector research agency. In turn, CSIRO's Division of Building, Construction and Engineering is Australia's principal research agency for the building, construction and related industries. The Division provides strategic and tactical research, development, consulting and testing services for the sector. It has produced about 1500 reports on the sector and has a strong reputation for research and innovation across a wide range of scientific and engineering disciplines. The annual expenditure of the Division is approximately $26 million, of which some $9 million is provided by industry. The Division's staff level is currently almost 300. Major laboratories and research groups are located in Melbourne and Sydney, with a new research team in Brisbane working with Queensland University of Technology (QUT) (see Alliances section below).

The Division has built a base of expert staff who are among international leaders with extensive experience in their fields. This combined expertise covers a diverse range of experts including sociologists, mathematicians, geographers, architects, engineers, building economists, computer scientists and urban planners. The Division works across a wide range of areas including planning, performance-based standards, structural engineering and fire safety, and development of new products and systems.

Research and development work is being undertaken in co-operation with major companies, professional and industry associations and government agencies that service the building, construction and engineering sector. The Division has an extensive and well established contact network with other leading research institutions, universities, professional and industrial associations, government agencies and private companies both nationally and internationally.

5.4.5   Relevant centres and alliances in the construction sector

Initiatives of several university and public sector research institutions are breaking new ground in collaborative construction research structures, with the potential for major benefits to construction innovation.

5.4.5.1   *The Construction Research Alliance*

The Construction Research Alliance (CRA) was formed in 1998 to better serve the research needs of the Australian construction sector. The genesis of this Alliance was an informal relationship commenced in 1994 between Queensland University of Technology's School of Construction Management and Property and CSIRO's Division of Building, Construction and Engineering. A second agreement extending the alliance until 2002 was recently signed.

The Alliance also brings together RMIT's Department of Building and Construction Economics, and the Construction Industry Institute (Australia). It draws on member institutions' international networks – comprising industry, academic and research organisations.

CRA is currently partnering on construction research projects valued at more than $1.3 million.

5.4.5.2   *Construction Queensland*

Construction Queensland (CQ) is a public sector body which was formed in 1996 as a grouping to link participants involved in the Queensland construction sector – involving government, industry associations, contractors and sub-contractors, designers and research providers. CQ promotes information sharing and networking. It has received substantial industry and government support over the past two years. Its first three years will be supported by public sector funding, with industry contributions expected to support it in subsequent years. CQ has recently identified six core focus areas for its work.

1.   Integrated Industry - To increase the effectiveness of Construction Queensland activities through co-operative integration.
2.   Employment & Training - To identify and develop future skill requirements (for a world competitive industry).
3.   Industry Sustainability - To develop a long-term sustainable industry.
4.   Equitable Delivery - To develop win/win procurement systems where all participants benefit.
5.   Industry Innovation - To develop a world competitive industry.
6.   Export Development - To increase capability of the construction sector to export its products and services to a world competitive standard.

CQ is now well positioned to take advantage of a broader involvement of industry representation – now including the civil engineering sector in addition to the earlier principal focus on building construction. ISR views this Queensland initiative as a positive approach bringing together industry and government. In current discussions taking place throughout Australia, the CQ and CRA models are viewed as significant achievements in unifying the thrust to more focused R&D efforts linking industry and government with R&D providers.

5.4.5.3   AUBEARes

The Australasian Universities Building Education Association's research section (AUBEARes) was formed in 1995 and aims in part to:

•   Facilitate research in the construction sector, and
•   Promote co-operation between universities, industry and government in the design and construction industries.

The linkages developed through AUBEARes have largely been amongst academic researchers.

5.4.5.4   *The Construction Industry Institute Australia*

The Construction Industry Institute (CII) Australia was formed in 1995 modelled on the very successful CII in the US. The CII co-ordinates research on behalf of sponsoring industry partners – comprising facility owners, contractors and specialist sub-contractors and material and equipment suppliers. Queensland University of Technology provides a small in-kind contribution to support the CII. The actual research is carried out by appropriate university and public sector research institutions – primarily CSIRO.

5.4.5.5 *The Australian Centre for Construction Innovation*

The Australian Centre for Construction Innovation (ACCI) was formed in 1998 as an outgrowth of the Building Research Centre at the University of New South Wales. It focuses on physical material testing and research into site safety and use of IT in the construction process. ACCI has a loose linkage of researchers in the construction area across the UNSW and other universities.

> The public-sector's role in most of these initiatives is via support for universities and CSIRO. In both cases, the government provides the majority of funding.

5.4.6   Standards Australia

Standards Australia is a recently corporatised public sector organisation that still receives marginal support from the Federal Government. Standards Australia promotes construction innovation through the structure and focus of its day to day operations. Additionally, Standards Australia operates specialised programs to support innovation in the construction sector, such as the Australian Design Awards. The awards are intended to promote excellence in product design. Products are assessed according to a range of criteria including functionality, aesthetics, ergonomics, creativity, safety, usability and environmental considerations.

> The initiatives discussed above, together with a number of more general policies, are classified below according to the degree to which they are aimed directly at the construction sector, and the extent to which they are innovation-specific.

Table 5.6 Categorisation of initiatives that potentially impact on construction innovation

Dimension	Initiative Directly Impacts on the Construction sector	Initiative May Indirectly Impact on the Construction sector
Innovation Specific Initiative	• Renewable Energy Equity Fund • Building and Construction Industries Action Agenda • CSIRO, Division of Building, Construction and Engineering • Construction Research Alliance • Australasian Universities Building Education Association's research section • Construction Industry Institute (CII) Australia • Australian Centre for Construction Innovation	• R&D Tax Concession • R&D Start Program • Innovation Investment Fund • Technology Diffusion Program • Co-operative Research Centres • ARC Grants
More General Initiative	• Australian Building Codes Board • Construction Queensland • Standards Australia	• Fiscal Policies • Monetary Policies • Employment policies • Export development incentives

The table may look well balanced, however problems emerge principally because:

- Broadly targeted innovation initiatives are not sufficiently exploited by firms in the construction sector, and
- Most of the innovation-specific initiatives which directly impact on the construction sector are alliances that rely, in part, on the availability of other government programs to support their activities.

Indeed, despite this range of initiatives, historically Australian governments:

> ...have not clearly articulated an industry development vision for the building and construction industry, and this is especially true for the non-residential segment' (ISR 1999: 35).

As a result, until the Building and Construction Industries Action Agenda was formulated in 1999, there was no coherent policy vision for the sector.

5.5   EFFECTIVENESS OF GOVERNMENT PROGRAMS

In Australia, the key government department involved in developing programs to facilitate growth in the construction sector has admitted that...

> ... it is valid to argue that in the past the Commonwealth has not been particularly adept in co-ordinating its various policy arms in delivering policy to the industry (ISR 1999: 39).

However, the current climate in Australia is one in which renewed and innovative efforts are being made to improve the construction sector's performance. A major part is being played by programs set up to encourage communication between relevant parties – such as the Federal Government Action Agenda process and the Queensland Government's Construction Queensland venture. Such initiatives are showing early signs of success in developing an industry vision and encouraging co-operation and understanding between stakeholders in the pursuit of innovation and improved performance.

Similarly, CSIRO has become involved in new alliances, such as that with QUT, which are likely to improve the outcomes of its research efforts, again probably primarily through closer relationships with key stakeholders.

More general programs, such as the CRC and ARC programs and the R&D Tax Concession and the START program have had little impact on the construction sector. In part, this is likely to be because they are not targeted to the sector. However, there are additional problems. Both the CRC and ARC programs have been criticised for the lack of construction-related research proposals receiving funding (Hampson 1998). More generally, there has been extensive criticism of the reduction in the R&D Tax concession from 150% to 125% in 1996 (e.g. Business Council of Australia, 1999; AI Group, 1999; Taxation Institute of Australia, 1998). Recent Australian Bureau of Statistics (ABS) and OECD data indicate substantial falls in R&D activity since the rate of concession was reduced (ABS 8104.0 and OECD 1999 Main Industrial Indicators). Also, the ABS data indicate that the reduction in R&D activity has been greatest amongst small firms, suggesting that the R&D START program, which was designed to improve innovation amongst smaller enterprises, may not be effective.

Further, the ISR-commissioned paper, 'The Building and Construction Product System: Public Sector R&D' (1999) highlighted that although alliancing arrangements appear to be on the increase, industry/public-sector research collaboration still needs to be improved to support higher innovation rates in construction. For instance:

- Although extensive linkages with industry were reported in the study, the level of funding provided was quite low compared to that received from other sources;
- Over half the ideas underpinning public-sector research projects were generated within the public-sector institutions, as opposed to being generated by industry, and
- While diffusion of research was extensive, as measured by publication activity, the adoption of results by industry was more problematic with many respondents indicating the sector was resistant to change.

5.6   FUTURE DIRECTIONS

The Federal Government released an Action Agenda to enhance the construction sector's performance in May 1999. The Agenda was the culmination of extensive discussions between major stakeholders over an 18 month period. Key initiatives in the strategy include: an international benchmarking study; a heightened focus on education and training; enhancement and acceleration of innovation; and efforts to increase diffusion of information technologies. The Agenda process resulted in the articulation of future policy directions to support sector growth. In respect of innovation, the major recommendations of the Action Agenda were to:

- Develop innovation indices more relevant to the construction sector;
- Seek assistance from AusIndustry to better market the current array of programs to the construction sector;
- Seek the IR&D Board's assistance in reviewing the sector's use of the R&D Tax Concession; and
- support initiatives that recognise innovation excellence and raise awareness of innovation as a driver of productivity and growth within the sector.

The Action Agenda concluded that for innovation based on technology, the sector must concentrate on material technologies that are increasingly recognised as having major economic importance. It will be essential, for example, to know the performance and behaviour of new and recycled materials particularly for special applications. Knowledge of the performance of these materials will then have to be embodied in standards to allow those who specify building and sub-systems to predict them accurately under all conditions.

Further, current industry challenges suggest that process innovation should focus on decision support systems, process re-engineering and adopting concurrent construction and "lean" construction" methodologies.

Indeed, the Australian innovation and R&D landscape is in the process of being fundamentally restructured through:

- Client demands on manufacturers to supply 'bundled' products and services (Marceau and Manley *forthcoming*)
- pressure by the Australian Government on the country's largest public R&D institution, the CSIRO, to develop stronger relationships with industry and increase levels of external funding
- the ARC directing increased proportions of its funding towards industry-linked research
- the development of national and international construction research alliances, and
- the extension of existing export development incentives which support the construction sector.

These initiatives both reflect and promote greater recognition by the Australian construction sector of the value of enhanced collaboration between users and providers of R&D. However, they also indicate that private-sector funding will need to increase, both to take advantage of new opportunities and to compensate for reduced public-sector funding. To date the private-sector has been playing a more active role in funding the provision of infrastructure (under bipartisan pressure from pro-development governments), however the level of private-sector R&D funding has been slow to respond to old pressures and new opportunities.

Finally, the Federal Government has responded to calls from the construction sector and major R&D institutions for increased government support to establish a national CRC in the construction area. Such a collaborative research centre would be the first CRC specifically centred in the construction sector (65 CRCs are currently operating). Australian Government support for a CRC in construction is viewed by all participants in the Australian construction sector as a most positive public policy initiative which could substantially enhance the level of medium and long term R&D in the sector.

The construction sector is on the cusp of a new era in realising the benefits of a close industry/government/research institution relationship for a stronger national and international industry through improved innovation performance.

CHAPTER 6

PUBLIC POLICY INSTRUMENTS TO ENCOURAGE CONSTRUCTION INNOVATION: OVERVIEW OF THE BRAZILIAN CASE

**Francisco Cardoso, Marco A. Rezende,
Mercia Barros, and Roberto de Oliveira**

6.1   INTRODUCTION

This chapter mainly aims to present and to analyse the policies adopted by the Brazilian government to promote technological innovation in the construction sector.   Knowing there is a great demand for building construction and especially for housing, the main investments oriented to innovation have preferably been accomplished in the latter and, for that reason, it will be the analysis and consideration focus in the present work.

The work is divided in four sections. Section 6.2 briefly describes the context of the Brazilian's construction industry. Section 6.3 characterises the process of economic and technological development the Country has been going through in the last years, focusing on the innovation process in construction.   In section 6.4, there is a focus on the current and more significant Government policies that are oriented to technological innovation in construction, particularly in the construction of buildings, trying to show its impact on the whole productive sector. In this section, we attempt to answer the following two questions posed to all the countries participating in the TG35 Group:

* What instruments and approaches are being used by Governments to promote innovation in this sector?
* What works and under what circumstances?

In order to answer these questions, data were collected from academic works focusing on Government policies oriented to innovation and also to valuable information gathered directly from persons in charge of several projects in progress - to whom the authors would like to thank for the collaboration - and also the information supplied by several research and development supporting agencies in the Country.

Finally, the last section presents the conclusions of the work along with answers to another question posed by the TG35 Group:

* How to strengthen innovation in the construction industry?

6.2   CONTEXT

6.2.1   Technological and economic Brazilian development

This first section of the work shows the context in which technological innovation is happening in Brazilian Construction.

Brazil, after a period of great economy growth which happened from the beginning of the sixties to the end of the seventies, went through a period of great economic instability, high inflation, and literally no economic growth that lasted the whole 80s, extending into the early 90s.

That period was characterised by a series of financial difficulties, when the Government could not honour its commitments and not even refinance its debt; this also led to an increasing inflation, and to an expressive recession; as result, there was a series of economic plans that tried to stabilise the economy again.

Among the several plans proposed, in 1994, one was implemented that managed to keep the economy stability until now, at least from the inflationary point of view. However, the great foreign debt and the need of structural reforms in several productive sectors of the Country, and the Government itself, are still hindering factors that make it difficult to plan and forecast for medium and long terms. Completing this picture, the Government policies that have been establishing very high interest rates, is one more obstacle to be overcome for fully resuming growth.

In spite of all those difficulties, the 90s were also marked by internationalisation that caused the market to open to foreign companies and products, thus contributing to an expressive change in the national economy because, threatened by foreign competition, all the productive sectors were compelled to modify and to modernise their production relationships, aiming to increase their product competitiveness.

With that, a new phase is being consolidated of industrial growth oriented to production relationship modernisation. The construction industry, characterised by several authors for wastes of different natures, for high production costs, and for sustaining and keeping disqualified labour employed, due to the economic and social importance that it represents in the productive group, could not help participating in the process of the Country modernisation because, besides being responsible for the Country modernisation process, it is also responsible for high percentage of the gross domestic product (*GDP*), and intimately related with countless other activities, not just the ones concerning construction. It is within this modernisation process which the Construction Industry is going through that the next topic will be approached.

6.3   CONSTRUCTION AND TECHNOLOGICAL INNOVATION: RECENT EVOLUTION AND CURRENT PRACTICES

6.3.1   The years of growth and the crisis

Inserted in the economic context described above, construction has gone through different development phases and Government policies.

In the beginning of the 60s, according to the works of Farah (1988), from Fundação João Pinheiro (1992) and of Vargas (1994), structural changes happened in the whole Brazilian society, with significant repercussions in the Construction industry.

Infrastructure was implemented to make industrialisation possible, strengthening the heavy construction subsector, with great projects in the area of transports, energy, mining, and metallurgy; in addition, there was a strong urbanisation process that led to the development of the material and component construction subsectors particularly due to Government intervention, through the Welfare Institutes, of the People's House Foundation and, in 1964, through the creation of the National Bank of Housing (*BNH*), that aimed at mass production of housing units.

That period of economy investments and growth went on into the 70s, beginning to give signs of gradual fall starting in the late 70s, with an intensifying recession in the middle 80s. During that period great housing blocks were built, marking an important stage of the construction of buildings history in Brazil, mainly because it allowed the introduction of technological innovations leading to industrialisation. Between 1976 and 1982, for instance, the housing financial system financed annually, on average, the construction of nearly four hundred thousand new dwelling units. In the beginning of the seventies, motivated by the high demand provided by Government resources, the Building Construction sector was motivated mainly to increase productivity, so that it was possible to produce a great number of housing units in a short period of time.

The introduction of "innovative" constructive systems or "industrialised" systems, based mainly on the pre-fab components, mostly brought from other countries, was the answer given by the building companies to the demand.

According to Castro (1986), more than 50% of the technology used in those sites was imported and, for its adaptation to the national conditions, investments were required for technological research, carried out by the Government as well as by the manufacturers, and by construction firms. By the end of that period and the beginning of the eighties, the picture of resources supply begins to change, then marking the beginning of the crisis.

At that time, two factors began to substantially modify the the construction scenario. On the one hand, there was a political opening in the Country, allowing the Government to enforce an action to allow for popular dwelling construction; on the other hand, there was an increasing economic crisis that hindered the liberation of financial resources, mainly because the financial system used until then was completely compromised.

The need to build more dwellings, given the existing deficit, and the shortage of resources led the search of a more intense construction cost reduction. This last factor has consequently led to the need of adopting new constructive technologies; that is to say, of technological innovations.

So as to solve the housing problem, the State triggers an action oriented to new construction technologies and thus promoted several seminars and discussions that involved the whole productive chain. Farah (1988) reminds that at that time the experimental sites located in *Naramdiba* (Bahia state), in 1978, and in *Jardim São Paulo* (Sao Paulo state), in 1981, in which technological innovations were tested. The action aimed at achieving the modernisation of the sector so as to meet the objectives of the governmental action, of low-cost and of large-scale production, to cater for the low-income population.

The experimental sites were created to accomplish an evaluation of the proposed new technologies and, starting from there, to promote those that could be used.

For that occasion the outstanding role of *Financiadora de Estudos e Projetos - FINEP* (Financial Support of Studies and Projects), was stressed. *FINEP* is a public company belonging to the Ministry of Science and Technology - *MCT*, of the Federal Government, that aims to promote technological development and innovation in the Country, in compliance with the goals and priorities established by the State.

FINEP, that was acting in the housing area since 1976, in the research and technological development area, had a complementary action in connection with the National Bank of Housing - *BNH*, and with the National Council of Scientific and Technological Development - *CNPq*, an organism also belonging to the Ministry of the Science and Technology.

One of the actions of *FINEP*, in the late seventies, resulted in the definition of some research lines, with a priority to support the projects oriented to the low income population. A co-operation protocol was signed between *FINEP* and *BNH* for information exchange about researches in the area. Based on the activities developed in this period, in 1978 the Integrated Program of Habitation and Sanitation was approved, through which *FINEP* supported (together with *BNH*) several researches of the two areas chosen.

Through that action it was possible to set an important partnership gathering *FINEP*, *BNH* and *IPT* (Institute of Technological Researches of the State of Sao Paulo) for the development of an evaluation methodology for new constructive processes, that aimed at the establishment of evaluative parameters for the units produced in the experimental sites.

That partnership, in spite of undergoing ups and downs, has persisted until now, as it will be seen further on, because the evaluation of new constructive technologies is still an unsolved subject in the ambit of the Construction Industry in Brazil and new actions with that same objective are still under way.

Unfortunately that Governmental action did not last long because, although financial resources still existed for carrying out works with innovative processes and constructive components and of there being Government agents really engaged in the incentive process to the financing of those types of construction and of there still being actions as the partnership established with *IPT*, a more effective technological innovation policy never existed in the sense of involving all the actors in the same direction.

The Government itself was fragmented in relation to that subject, having groups that defended the maintenance of conventional construction and they only gave room to innovation when the lack of financing money forced the search for lower costs. Thus, the experience of technological innovations introduction ended up by characterising it as a punctual event, because as soon as the favourable conditions provided by the Government to the use of innovations ceased, there was a retraction in their use on the companies' side and the whole process formerly described was abruptly interrupted with the worsening of the economic crisis which, added to the problems of construction financing fund management, caused the wreckage of the whole housing financial system operative until then.

According to Castro (1986) for the construction industry of material, component, and equipment, the investment in production amplification oriented to the housing demand was idle in the following years, mainly due to the countless pathological problems that appeared in the dwelling units, as a consequence of the absence of an adequate technological development.

With the lack of financing, the companies that had opted for development and the use of new technologies were forced to re-set their strategies. The flimsy innovation experiences ended up, and since 1986 the sector has been complaining about the lack of a financing structure for construction in the Country.

6.3.2 The current scenario

The current construction configuration begins with the crisis experienced since the late seventies, which has caused expressive changes in the sector.

A great competition resulted from the decrease in the number of works (in 1987 the number of licensed works was just 47% as compared to 1980). IBGE (1989) pointed out that companies had to work towards cost-reduction as the only possibility of maintaining their profitability and staying in the market.

That cost-reduction was sought mainly through production rationalisation. According to IPT (1987) and (1988), the companies try to obtain productivity gains and to minimise costs as well as time through production rationalisation, without disrupting the productive base that characterises the sector. To reduce the waste of time and of material, some of the main conventional construction bottlenecks are fought, such as: lack of articulation among the several design types and between office and site; absence of quality control; bad work conditions as a factor for low productivity; site disorganisation, etc.

Although the building companies were concerned with their costs and with identification of new actions so that they could keep their competitiveness in the market, on the Government side, only *FINEP*, besides other fostering organisms for essentially academic research, through scholarships, such as *CNPq, FAPESP* - Foundation for Supporting Research of the State of Sao Paulo (*Fundação de Amparo à Pesquisa do Estado de São Paulo*) and *CAPES* – Foundation Co-ordination for Personnel of University Level Improvement, maintained some action aiming at the development of the construction industry.

In mid 80s, *FINEP* programs were elaborated in the field of social development. For the housing area four researches lines were established with emphasis on the formation of human resources, as described below:

- production process of the urban space (land use legislation; studies on urban land profit, and on the real estate market)
- habitation production process (the construction industry and the construction material industry; technological researches seeking low-cost construction)
- the Government, policies for investments in housing; equipment of collective use
- dissemination (exchange promotion in researches; creation of informative bulletins; surveying the "state of the art" in the sector).

The Program of Housing, Sanitation, and Urban Development was also elaborated, with the priority of choosing, in each of the following subsectors:

- researches aiming at subsidising the responsible agents performance for policies planning and implementation
- case studies and pilot-experience aiming at the solution of local problems, with special concern to the low income population
- projects aiming at simplified technologies diffusion for users.

In 1987, a new Program, *PROURB* - Urban Development Program - was approved by *FINEP*; it concerned the areas of Housing and Urban Development. Its research lines were defined from a comprehensive debate with the scientific and technological communities, and with other financing agencies such as: *CNPq*; Ministry of Housing, Urbanisation and Environment; and Federal Savings and Loans (*CEF - Caixa Econômica Federal*). It should be remarked that after *BNH* was closed in 1986, *CEF* was appointed to co-ordinate all activities concerning the research related to that bank.

The implementation of those programs were made both in terms of public policies oriented to the housing area, and in the new material and constructive systems development.

In the area of new materials, studies concerning the use of vegetable fiber (*sisal* and coconut) were financed and the use of lightweight aggregate in construction, developed by the Centre for Researches and Development of Bahia - *CEPED* - and for the use of fine grained concrete in construction, accomplished by the School of Engineering of Sao Carlos - *EESC/USP*.

Concerning constructive systems, it is worth mentioning the outstanding research accomplished by several organisms, among them the following stand out *PCC-USP* (*Departamento de Engenharia de Construção Civil da Universidade de São Paulo*), with the development of new constructive and dissemination methods and processes for the technical environment; *FAU-USP* (*Faculdade de Arquitetura e Urbanismo da Universidade de São Paulo*) with the elaboration of a catalogue of constructive typologies potentially usable in low-cost housing in several areas of the State of Sao Paulo; *IPT* (Institute of Technological Researches of the State of Sao

Paulo), establishing minimum performance criteria for the production of housing to low-income strata; *CETEC* (Foundation Technological Centre of Minas Gerais), with the elaboration of a catalogue of constructive typologies potentially usable in low-cost housing in several areas of the State of Minas Gerais.

In spite of the efforts accomplished by several researchers in the area (notably between the end of 1989 and the beginning of 1993), sponsored by *FINEP*, countless difficulties existed to establish a true policy oriented to the Technological Innovation of the sector.

Even so, two researches were supported: the first, developed by the Foundation for Technology of the State of Acre - *FUNTAC*, addressing the search for alternatives of low-cost housing for the Amazon Region; and, the second, accomplished by the Latin American Institute - *ILAM*, of Sao Paulo, the latter aiming at designing and building prototypes adapted to the conditions of certain urban areas of the Brazil.

A more systemic approach from the Government happens through the Ministry of Social Action within which the National Program of Housing Technology - *PRONATH* (BRASIL, 1991) was created, linked to the Brazilian Quality and Productivity Habitat Program *PBQP-Habitat* (*Programa Brasileiro da Qualidade e Produtividade do Habitat*), having as a guideline the technological and managerial modernisation of housing production. In this program, it is clear that actions aiming at the sector "technological innovation" are indispensable, and the following should be stressed:

- to strengthen the productive structure of the sector concerning its technological and managerial capability
- to implement policies of labour training that enhances the introduction of technological innovations
- to motivate the use of new technologies for housing production
- to strengthen laboratory and research infrastructure for technological development and services rendering
- to foster the creation of new laboratory units and research groups, advisory and technological consultants
- to develop courses on technological innovation
- to develop mechanisms of technology transfer.

Regarding the Technological Innovation Program - *PRONATH* (Brasil, 1991) it shows the following central objective: "*to improve the knowledge range and available technologies in the Country, in the design, material manufacturing, and component areas, as well as in construction, operation, and maintenance of housing construction*".

Aiming at that, a proposition was made to minimise the current "technological bottlenecks" in the subsector to reach the following goals:

- developing technological innovations in the design, material, component, and constructive systems areas
- training all the national productive sector for the use of design, production, construction, operation and maintenance new technologies

- promotion and large-scale application of new technologies in housing construction.

In 1992, now in the competence of the Ministry of Social Welfare, the Program was reformulated concerning the actions planned for the biennium 93/94, becoming less comprehensive and more adapted to the financial constraints of the Country. The objectives of "final cost reduction of low-income housing" and of "housing quality improvement" started to prevail over others and the actions to be developed aim at technological innovation, the use of appropriate technologies of local conditions, and the fight against construction waste.

From this new definition, several initiatives were adopted in the ambit of the Program, for example, the accomplishment of a show room in Brazil, gathering alternative constructive systems; and also, a Co-operation Agreement with the Italian Government, in the year of 1989. Within this agreement, a "National System of Technological Development Applied to Housing" project was elaborated and approved in 1991, whose objective was the financing of equipment for construction materials and component tests, aiming at creating conditions, in national laboratories, to improve the quality and productivity of housing production, especially the one of social interest. The project intended to establish ten evaluation centres counting on equipment and technical personnel, besides the assemblage of a reference system of technologies at national level.

Besides, there was also the popularisation of the Ministry of Welfare Normative Instruction no. 4, that established general guidelines for systems and constructive components approval in programs administered by the National Secretary of Housing and by the Ministry itself. This document, of great relevance for the sector, was approved by actors such as the Private Insurance and Capitalisation National Federation Companies - *FENASEG*; Brazilian Association of the Companies of Real Estate Loans and Savings - *ABECIP*, and Federal Savings and Loans - *CEF* and, although it was not actually implemented, it generated a lot of discussion and debates in the construction sector, strengthening the need of such guidelines as well.

For a presidential decree of July 1993 a new program of Government action was also created, aiming at technological innovation, called Program of Technology Diffusion for Low Cost Housing - *PROTECH*, linked to the General Secretary of the Republic Presidency and supported by eight Ministries, among them the Social Welfare and the Science and Technology Ministries (*MCT*), getting resources from of the Federal functional properties alienation (Decree no. 1,036 of 04.01.94).

PROTECH aimed at the diffusion of new construction technologies for low-income housing by means of financing the construction of some pilot units, and the establishment of a diffusion centre from those units. That group was named Technological Village. Eight of those villages took part in this program, but only five were concluded at an average of a hundred dwelling units each; ten technologies were selected for each village. A valued aspect of this program was the community participation and of representative society sectors for choosing the technologies to be adopted. Many Technological Villages were built in several Brazilian States, such as Parana, Minas Gerais, Bahia, Sao Paulo, among others.

Parallel to these initiatives, a wide debate was promoted by *MCT*, in April 1993, through the Brazilian Academy of Sciences - *ABC*, about housing research in the Country. From this debate, some issues emerged: normalisation and certification areas as priority subjects, assembling and memory recovering of the experiences accomplished, quality and productivity management in construction, new materials and technologies development, and sector trade formation and training. As a conclusion, the need to promote sector re-articulation due to the action of multiple institutions also emerged (FINEP, s/d).

Facing this scenario, it was necessary to implement a Housing Technology Program - *HABITARE*, by *FINEP*, to plan and to carry out the strategy in this area, especially concerning the contribution to the public policies formulation, implementation, administration, and evaluation in the field of science and technology. More information will be provided in the second part of this work; however, one action to be remarked in this program (for its relevance in the innovation technological process) is a research initially conducted by *IPT* and now by *COBRACON* (Brazilian Committee for Construction, from *ABNT* - Brazilian Association of Technical Norms) that will allow for minimal norms for innovative construction system performance evaluation. For that, the Ministry of Science and Technology has allotted resources from the National Program for Privatisation to FINEP within the "Low Cost Housing Technology."

Finally, we attention should be drawn to the last Government policies concerning the State of Sao Paulo's Low-Income Housing Quality Program, the so-called *QUALIHAB* (started in 1996); the *Programa Brasileiro da Qualidade e Produtividade do Habitat – PBQP-H* (Brazilian Quality and Productivity Habitat Program), of the *Secretaria Especial de Desenvolvimento Urbano - SEDU* (the State Secretariat of Urban Development, directly related to the President of Brazil) (1998); and the Competitiveness Forum of the Civil Construction Industry, co-ordinated by the Development, Industry, and Foreign Trade Ministry (*Ministério do Desenvolvimento, Indústria e Comércio Exterior – MDIC*) (2000).

In the second part of this work, more information about these and other Government actions in progress that aim directly or indirectly at construction modernisation will be discussed and answers will be provided to the following questions:

- What instruments and approaches are being used by Governments to promote innovation in this sector?
- What works and under what circumstances?

6.4   PUBLIC INTERVENTIONS

This section discusses the major public interventions that are now contributing to technological innovation in the Brazilian Construction sector, as shown in Table 6.1:

- programs to support R&D: *FAPESP, CAPES* and *CNPq* actions
- programs to support advanced practices and experimentation: the HABITARE Program
- programs to support performance and quality improvement: the *QUALIHAB* Program, the *PBQP-Habitat* Program and the Competitiveness Forum of the Civil Construction Industry
- programs to support the adoption of systems and procedures: *COBRACON / ABNT* Actions.

6.4.1   FAPESP

6.4.1.1   Objectives

The *Fundação de Amparo à Pesquisa do Estado de São Paulo – FAPESP* (Foundation for Support to Research of the State of Sao Paulo) is one of the greatest Brazilian technological and scientific financing organisms. Its major objective is to provide grants and other activities related to technological and scientific research, local and international interchange, and information in the State of Sao Paulo. It is an autonomous organism[1].

6.4.1.2   Annual resources

FAPESP annual budget is one hundred and sixty thousand million dollars (*FAPESP,* 1998), and it finances not only actions concerning the construction industry, but also academic research programs. The Health sector is the first one, with 21% of the funds, followed by Biology, with 16%, Engineering – 16% (all fields together, among which Construction), Social and Human Sciences – 12%, and Physics – 9%.

Its resources come from the VAT - value-added tax, once 1% of this tax of the State of Sao Paulo is destined to technological and scientific research.

The construction industry projects are granted about 1% of the total amount of funds, or something like one and a half million dollars per year (*FAPESP,* 1998).

---

[1]   In reality FAPESP is largest state financial backer in Brazil where similar institutions exist such as in states of Alagoas, Bahia, Ceara, Goias, Maranhao, Mato Grosso do Sul, Minas Gerais, Para, Paraiba, Parana, Pernambuco, Piaui, Rio de Janeiro, Rio Grande do Norte, Rio Grande do Sul, Santa Catarina e Distrito Federal. From its importance is given the focus.

6.4.1.3   Means

The *FAPESP* actions concerning construction industry projects include the sponsoring of about one hundred and eighty projects, from which two thirds are grants (for young scientists and for Master of Science, Doctoral and Post-doctoral studies). The other third is related to research funds for specific projects, laboratory equipment, library acquisitions, congress organisation, participation in congresses, invitation to foreign researchers, book and journal publications, for example.

The construction projects are related to almost all types of studies, focused on building and the habitat problems, which goes from building economics and architecture, to very technical issues; from studies carried out in laboratories to those conducted by means of field or library investigations; from individual to integrated projects, involving different groups from different universities or research centres.

Even if *FAPESP* has normally financed universities or research centres, from the mid nineties onwards, it has also attempted to finance industry projects where innovation appears as a central subject, mainly for small and medium firms, by means of *PITE - Inovação Tecnológica em Parceria* (Technological Innovation by Partnerships), and *PIPE - Inovação Tecnológica em Pequenas Empresas* (Technological Innovation in Small Firms) Projects; the former is based on the partnership between the innovative firm and a research centre, and the latter is the same, but specifically oriented to small firms.

6.4.1.4   Contribution to innovation

It is very difficult to objectively estimate *FAPESP* actions for contributions to innovation in the sector.

The grants allotted to young scientists (fifty per year) are investments for the future; the results of those allotted to master of science (thirty per year), doctoral (thirty-five), and post-doctoral (five) students are more obvious, even if we do not have indicators to measure their impact. Nevertheless, one can say that the *FAPESP* system is one of the most effective and efficient of the Country, and that it is very selective, with a very strict control.

Some reflections may arise about the investments directly related to technological projects and also to the promotion of innovation.

Thus, on the one hand, if we regard the amount of specific projects, normally those with more resources, related to construction, it can be easily seen that their relative participation in all the money allotted is less important in other Engineering fields. In fact, this kind of project responds to 37% of the financial investments of the Foundation, percentage that decreases to 10% if we regard Engineering, and to only less than 7% in the construction case.

On the other hand, the number of industry financed project based on innovation (*PITE* and *PIPE*) is almost zero (according to FAPESP, 1999, only one of *PITE* Projects are related to construction subjects). Two may be the reasons for that: either the idea is not very attractive to "construction" firms, or the technical board of the Foundation prefers to finance more "advanced" areas, like electronic, chemistry or new material. We suppose that the first is the real reason, even if "Construction" is not one of the major research strategic themes defined by the technological and scientific policy of the Country.

The conclusion: construction has to look for other financing sources, and not only for grants. If it reaches the same average figures of Engineering, or 10% of the total funds, this can represent the increase of three hundred and eighty thousand dollars per year, or more 26%. In the same way, if construction firms, in partnership with the academic sector, get 1% of the total amount of funds allotted to innovation, this will represent some thing like one hundred and thirty six thousand dollars a year, or additional 9%.

It means that the construction industry could get from *FAPESP* something like two million dollars per year, or 36% more than it is granted nowadays. It can also be supposed that the contribution to innovation of the total investment would grow in the same proportion.

6.4.2   CAPES's actions

6.4.2.1   Objectives

Foundation Co-ordination for University-Level Personnel Improvement - *CAPES* is a public organism linked to the Ministry of the Education - *MEC*. Its main objective is in helping *MEC* in master degree policies, co-ordinating, and stimulating - by means of scholarships, aids and other mechanisms - the formation of highly qualified human resources for teaching at universities, research, and the attainment of public and private sectors professional demand. *CAPES* also manages a specific program that is of interest to this work called *PADCT* (Scientific and Technological Development Support Program) whose goal is to widen, improve, and consolidate the Nation's scientific and technological competence in the university environment, research centres, and private enterprise; this is done through integrated projects that impact scientific and technological development. This program results from World Bank funding to the Brazilian Government through the Ministry of Science and Technology, *MCT*. FINEP and *CNPq* also act as *PADCT* administrators, and their aims are wider than the ones mentioned above.

6.4.2.2   Annual resources

As the data for the scholarships specifically granted to the construction area could not be accessed, they can be estimated by adopting the same percentile of incidence of that area in relation to the total amount allotted by *FAPESP*, which is 1%.

In fact, it seems that no significant mistake will be committed, once the relative percentile of Construction Engineering in relation to the total are similar in the two cases: 16.0% on average in FAPESP (1998) and 16.6%, 15.9% and 18.9% in *CAPES*, respectively for master, doctorate and post-doctorate grants (1999 and 2000 - until June - data).

If the value of 1% is adopted, it will provide the following numbers (1999 and 2000 average): annual number of master scholarships: 88; annual number of doctorate: 69; annual number of post-doctorate abroad: 3; total annual scholarships: 160.

Those values correspond to a monthly average expenditure of US$76,000 or a little more than nine hundred thousand dollars a year (1999 and 2000).

Besides scholarship awarding, *CAPES* also finances part of the Graduate Programs infrastructure (*PROAP* and *PROF*) and it also has fostering programs. According to the same previous reasoning, the average annual values granted to construction would be of six hundred and seventy thousand dollars.

On the whole, *CAPES* direct annual support to construction is estimated to be almost a million, six hundred thousand dollars.

The amount of *PADCT* resources oriented to construction could not be calculated, either. In all, the Loan Contract between the Brazilian Government and the World Bank foresees, for the first stage of *PADCT*, the amount of US$ 310 million, which must get US$ 50 million as a private sector counterpart, leading to a grand total of US$ 360 million.

6.4.2.3   Means

The essential means for innovation contribution for *CAPES* is through scholarships besides financial support for specific research projects, laboratory equipment, library acquisitions, congress organisation, participation in congresses, invitations to foreign researchers, book and review publications, etc.

Besides that, as manager of programs such as *PADCT*, *CAPES* could also act indirectly in the financing of companies and specific projects in science and technology.

However, it must be remarked that among the *PADCT* components, just one of them, the so called *CDT* - Component of Technological Development, which aims to promote technological development of companies and to increase investments from the private sector in R&D, through fostering the partnerships between the academic and the productive sector aiming at the improvement of the global performance of the Brazilian innovation system and the diffusion of technology, favouring the construction sector.[2]

The *PADCT* funds managed by *CAPES* are limited to the following: allotments for human resource training and development project formation; scholarships in the Country and abroad; aids for events and trips both in the Country and abroad; resources for equipment and permanent material to

---

[2]   The other two components are: 1) *CCT* or Science and Technology Component that aims at promoting and financing Research and Development/R&D and human resources in areas of national development relevance (Chemistry, Chemical Engineering as *QEQ*; Geo-Sciences and Mineral Technologies as *GTM*; Science and Material Engineering as *CEMAT*; Environmental Sciences as *CIAMB*; Biotechnology as *SBIO*; Applied Physics as *SFA*; 2) *CSC* or Sectorial Component that consists in promoting and financing activities oriented of services rendering such as supporting to Brazilian sector of R&D reform and improvement process.

laboratories; acquisition for science and research schools and centres; air ticket provision.

6.4.2.4   *Contribution to innovation*

As in the case of *FAPESP* actions, it is very difficult to objectively estimate the contributions to innovation of the *CAPES* actions, including the ones originated by *PADCT* resources.

6.4.3   The CNPq actions

6.4.3.1   *Objectives*

CNPq, National Council for Scientific and Technological Development, is a Foundation linked to *MCT* - Ministry of the Science and Technology - to foster research. Its mission is to promote scientific and technological development and to support researches, necessary to the social, economic and cultural progress of the Country.

To carry out its mission, *CNPq* accomplishes three basic activities: fostering, research support and science and technology information and diffusion. Among them, the first two are of special interest.

CNPq also manages a specific program on this line, the Training of Human resources for Strategic Activities Program - *RHAE*, whose goal is to improve competitive conditions in Brazil for the international scenario, by improving the technological capacity in themes selected according to their strategic relevance, according to *MCT* guidelines; their clients are public or private companies, producing goods and rendering service.

6.4.3.2   *Annual resources*

In all its programs, *CNPq* granted in 1998 a total of 933 scholarships for Civil Engineering, that represented 2,2% of the overall. Once again, adopting the same percentile of incidence in the construction area in relation to the total adopted by *FAPESP*, which is 1%, the total number of scholarships reached 424, and almost two thirds of them are either for master or doctorate programs.

In terms of values, the estimated total invested in construction, in 1998, was three million dollars.

6.4.3.3   *Means*

As mentioned before, two of the three basic activities of *CNPq* are of interest here: fostering and supporting research.

Thus, fostering is the main action developed by *CNPq*, for promoting scientific and technological development in the Country, aimed essentially   at   preparing   human   resources   and   for   supporting   the accomplishment of researches, through the granting of scholarships. The direct support to research seeks to promote and to stimulate the production of the necessary knowledge for the economic and social development of the Country, to the confirmation of the cultural identity, besides the rational and non-predatory use of its natural resources.

The fostering action is organised in Basic and Special Programs. Basic Programs are those oriented to the planned use of the fostering instruments, according to the traditional areas of knowledge, among which Civil Engineering in general, and Construction in particular. The Special Programs are those that correspond to strategic areas and multidisciplinary fields, as well as those of regional features, whose action concerns any area of knowledge. They are characterised by the perspective of medium range, focusing on induction mechanisms, for inter-institutional articulation and for the relevance criteria incorporation, in compliance with Government orientations contained in sectorial and regional policies that require strategic contributions from science and technology.

Besides that, in the area of the *RHAE* Program, *CNPq* manages a specific budget destined to grant scholarship to several modalities (3 to 24 months long): in-Country and abroad internship; to welcome specialist visitors; for industrial technological development; for in-country training and abroad.

The 1998 percentile shows, however, that the supporting modalities, at least in the case of Civil Engineering, took place mainly through the fostering actions through scholarships (66% of the total value) or of "productivity in research scholarships", granted to renowned researchers (25%), than those that support research, directly, or indirectly, through support coming from external researchers (total of 9%).

6.4.3.4   *Contribution to innovation*

It is once again very difficult to objectively estimate the contributions to innovation from *CNPq* actions.

6.4.4   The HABITARE Program

6.4.4.1   *Objectives*

To contribute, through support to researches in the area of science and Technology, to the solution of the Brazilian housing problem and for the modernisation of the construction sector, always having in mind integration with environmental concerns (FINEP, s/d).

To achieve the general objective the following action lines were defined:

- Setting of co-operative research networks
- dissemination and evaluation of the available knowledge
- development of new technologies
- integration with the productive chain

- stimulation and consolidation of research institutions' partnerships with companies of the sector
- management of quality and productivity
- development of normalisation
- decrease in construction environmental impact
- proposition of urbanisation criteria and infrastructure aspects
- innovative procedures of housing management
- evaluation of public policies
- post-occupancy evaluation.

6.4.4.2   *Annual resources*

There is a variation of resources from one year to the other. In average it totals something close to US $816,000.00. In the last years the total resources invested were: 1996 – US $530,000.00;   1997 – US $1,055,000.00; 1998 – US $786,000.00; 1999 – US $896,000.00. For 2000, the forecasted value is R$2 million or US $1,111,000.00 (not reimbursable). The resources of the non reimbursable financing come from *FNDCT* - National Fund for Scientific and Technological Development. *FNDCT* is managed by the Special Presidential Secretariat.

It is a Fund established so as to provide financial support to priority programs and projects for Brazilian scientific and technological development.

6.4.4.3   *Means*

HABITARE is a *FINEP* (Financial Support of Studies and Projects) program which is a public company belonging to the Ministry of Science and Technology - *MCT*, responsible for financing projects concerned with the technological development of the Country.
      In this program, based on the action lines previously presented, two basic types of projects exist: the non-reimbursable, from which financial return is not expected, and the ones of applied research, in which it is mandatory to involve the company that will use the technology to be developed or improved, and financial return is expected.
      The projects of applied research are of on-line application, that is, they may apply at any time. The required information must make possible for FINEP not only to confirm the adherence of the proposals to the operational policies and action lines, but also the financial conditions of the proposing company, the warranties offered, the internal coherence of the proposal in terms of objectives, budget, methodology, deadlines, etc.
      The interest rates, paying off periods and amortisation are defined case-to-case, within the possible institution limits (usually more favourable than market values).
      There is no previous definition of the type of technology to be financed, but it is necessary to define its importance for the sector or for the productive chain of the financed action. It is possible to finance projects of

quality management and of managerial administration, as well as basic education on general contents to workers, so as to complement a technological development project.

The non-reimbursable projects are launched by means of public submission - each them contemplating specific themes - and they are open to the participation of: Research Institutions, non-profit Associations and Technical-Scientific Societies - both public and private. Companies of the sector can participate under research institutions co-ordination, and offer financial counterpart (that can be financed).

In compliance with the proposed themes in the submission phase, the institutions elaborate projects that, besides the usual presentation of a project of R&D itself, they should inform the monitoring and evaluation processes of the projects and - which should be emphasised - the strategies through which the research results are passed to the productive sector.

The criteria for judgement and choice of the projects to be selected may vary according to the submission, but they are basically:

- performance qualification - approach related to the researcher's capacity or the group's, research methodology and institutional infrastructure
- intrinsic merit of the project - likelihood that the researches will lead to new discoveries or progress inside science and technology or that will cause impact in the area or in areas of science and technology
- usefulness or relevance of the research - it considers the likelihood that the research can contribute to the technological development and solution of social problems
- impact on the infrastructure of science and technology - potential that the researches contribute to better understanding or improvement of the quality, distribution or efficiency of the scientific and technological research, education and human resources
- budget - the consistency of the proposed budget in relation to the submission objectives, to the institutional and the researchers' team capacity
- financial participation - financial participation of the productive sector in the project
- resulting projects for researches supported by HABITARE.

6.4.4.4 Contribution to innovation

There is no numeric data about the impact of those actions on the productive sector, and much less of the specific impact over the technological innovation. However, it is possible to verify that the Program has indeed contributed to the technological improvement in the sector constituting not only one of the small financial sources for the study and development of innovations, but also a great opportunity of effort concentration of the Government' organisms, research institutions, non-governmental organisations, class union, and private enterprise around that subject.

Unfortunately, the reimbursable projects of applied research have not received great acceptance from the sector. Nonetheless, its contributions have been more punctual in helping to develop some products and specific processes.

It is expected, in the next years, with the search of better quality and productivity for the companies, that there will be a larger demand for projects in that modality. On the other hand, the non-reimbursable projects have had a good adhesion of the research institutions, and, among their contributions to innovation, the following stand out:

- clustering of several organisms and innovation-oriented institutions in joint projects;
  the program has been working under the consultation of a co-ordinating group, composed by main agents that act in the area, and in that way it has allowed those institutions wider possibilities for the development of the sector.
  In that sense, one of the projects financed by the program resorted to a very interesting research strategy for the first time in a continental dimension country, and several scattered research centres, like Brazil: the concentration of several research centres and universities around a single project, optimising the resources and enlarging the research scope and its results in the productive sector.
  This is the research "Alternatives for reducing material at construction sites", involving researchers from more than sixteen Brazilian universities, whose main goals were to collect rates for material and components loss at the sites, to identify the causes and, based on the results obtained, to propose alternatives to reduce these wastes (see: http://www.pcc.usp.br/Pesquisa/perdas/)
- establishment of dissemination/information networks on innovation processes and products in the sector;
  the concern of the program with its knowledge transfer to the productive sector has lead to the creation of informal networks of knowledge diffusion among the research centres and the productive sector. Besides that, an information network was created for the sector, formed by several linked nucleus, and centred in an Internet site
- technological innovations financial proposals;
  several projects were financed aiming at both new products development and of the improvement of existing processes and products, some of them with the collaboration of several companies of the sector, and, consequently, with the immediate diffusion of the researches results.
- financing for new technologies evaluation;
  evaluation of some advanced experiences of introduction of new constructive systems, which is fundamental for future decision making about the use of those systems, and, consequently, for their dissemination.

Finally, it should be noted that the limit of those results is also found in the development of broader Government projects and policies, where *PBQP-Habitat* is an example of the productive sector itself (see below).

6.4.5   The QUALIHAB Program

6.4.5.1   *Objectives*

The State of Sao Paulo's Low-Income Housing Quality Program, the so called *QUALIHAB*[3], is a Program to support performance and quality improvement, that is being implemented in the State of Sao Paulo to the local supply chain of the building sector. The Program is based on the purchasing power of a common client of these actors, the low-income housing office of the State, *CDHU (Companhia de Desenvolvimento Habitacional e Urbano* – State of Sao Paulo's Housing and Urban Development Company.

Through the *QUALIHAB Program, CDHU* aims at optimising the quality of housing with regard to products and services that have been used in their conception and implementation, from partnerships with the main actors of the sector through agreements. These agreements ponder: the implementation of specific quality program, the maximisation of the rate profit x cost (direct costs and costs of exploitation) and the satisfaction of the customers.

Although this Program is not meant to encourage innovation itself, it has had a very important effect, as will be seen below.

6.4.5.2   *Annual resources*

The *CDHU* contracts, on average, forty thousand housing units per year, on three hundred financial operations. Its resources come from the State of Sao Paulo's budget and, especially, from VAT -value-added tax (up to 95%), once 1% of this tax is destined to the so-called social interest housing. *CDHU's* annual budget is about three hundred million dollars.

Despite those values, the resources of the *QUALIHAB Program* are not so important, as they represent only seven hundred thousand dollars per year, or less than 0.3% of the total budget. It must not be forgotten that it is not a program oriented to support R&D, not even to support advanced practices and experiments. In fact, the firms have done the most important investments, and it is difficult to correctly estimate the counterpart amount of money they have actually spent. Small firms, that normally adhere to the Program, usually claim that their external investments to comply with all the *QUALIHAB* requirements are of about thirty thousand dollars, without taking into consideration the internal costs (the salaries and upgrading skills of the employees that work in the reorganisation process of the firm, besides machinery, such as computer soft- and hardware, for instance). If the number of firms involved in the Program up to now is estimated, nearly three hundred can be figured out, which means that these firms have invested something like nine million dollars to support performance and quality improvement, or three million dollars per year, since 1998.

---

[3]   QUALI is a contraction of quality, and HAB, of housing.

6.4.5.3   *Means*

The *QUALIHAB Program* is based on "deals", even if it has a local scope (the State of Sao Paulo represents almost 40% of the Brazilian *GDP*). A particular characteristic of these agreements is that they have been negotiated by the *CDHU* with the different unions acting. For instance, after a negotiation period, the general contractors' union, which is the most affected agent up to now, set up an agreement with *CDHU* in order to create a quality assurance system that has been included in *CDHU's* procurement processes since July 1998. This system is based on the ISO 9000 Standards and it includes the "level of qualification" concept.

It is easy to understand the *CDHU* power as a client that is able to improve quality and innovation in all the supply chain of the local housing providers sector. It should be noted that Brazil has undergone a very particular moment, with low inflation and important economic growth; from this context, paradigms have been changed, and the barriers that regulate competitiveness in many sectors, including the building construction have been gradually removed.

Some of these new features can be remarked concerning the economic and commercial measures through: the new requirements from customers, both public and private, that express themselves further in terms of quality, delay decrease, and demand of services, among them innovation; the tendency of decreasing prices of the housing units; the opening of markets to international competition.

The lack of financial resources and the deep crisis in the housing financing system, that have affected the private market as well, are also new important constraints, now connected to the financial dimension of the environmental concerns.

It is said that intervening changes in the Brazilian real estate market caused an unprecedented growth in the competition among firms. Indeed, two thirds of the State of Sao Paulo general contractors building sector considered "competition" as their most prominent problem in 1997, preceding others as interest rates, being equivalent to problems linked to manpower or suppliers.

It is within this context that it is necessary to understand *CDHU* actions: the financial power of the State, that collects funds thanks to a specific legislation, combined with a strong competitive environment, that has driven all actors of the sector to react and to accept a new logic of efficiency, ruled by quality management and assurance principles, and by the search for innovation.

In spite of its comfortable position in this process, the *CDHU* did not want to impose rules. Instead, and this is one of the most prominent features of the *QUALIHAB Program*, it has made the decision to conceive and to put into practice a program based on "deals". What were they concerned about? In short, they consist of commitments negotiated with the employers' different actor unions, witch have established two main points:

1)   a set of "technical referential" requirements, adapted to each case, based on concepts and tools of quality management and assurance

principles which *CDHU* progressively requires in the procurement phase of its contracts, and
2)   deadlines to answer them, in a short period of time (in the particular case of the general contractor's agreement, it took from January 1998 to January 2000 to be accomplished).

Indeed, the *CDHU* well understood that the only way to get some concrete and positive results is to efficiently contract the distinct actors of the sector, represented by their unions. To recover the delay in relation to the technical capacity, organisational ability, and bound to management, the partnership public housing "buyers" x private actors has imposed itself.

What were and still are interests for industries, engineering firms and contractors in this setting? Very simple: in a context of strong competition it means to achieve advantages in the procurement phase of the *CDHU* operations. Besides, even though the rule of the "least price" remains valid, all those that have been "certified" by the Program economising. For example, general contractors no longer need to proceed to quality controls while using products certified by *QUALIHAB*, provided that they are already qualified.

The State purchasing power connected to private sector partners' expertise, weaknesses, and needs has led to a common goal: housing quality; this quality is perceived in its multiple features: architectural, constructive, environmental, performing, innovative, etc.

Going deeply into the details of the Program organisation, it could be said that the *QUALIHAB Program's* negotiations with different unions began in 1994. In all the cases it has conducted to progressive processes, with gradual deadlines, that foresee a period of two and a half or three years for their complete implementations. The Program is organised in two committees: "Material, Components and Systems" and "CAE – Construction, Architecture & Engineering." The unions that take part in the first one are those that represent producers of steel for reinforced concrete, cement, lime, prefabricated products, ceramic products, tubes and plastic components, industrialised sealers, electric components, for instance. The "CAE" committee is more heterogeneous, and it has included, among others, unions that represent general contractors, architects, engineers, construction managers, laboratories, topographic services, and foundation subcontractors firms.

So far, almost twenty different agreements have been signed in both committees, and all of them are based on the principles shown above. So, every actor's technological and financial limits have been respected.

For instance, in the "materials, components, and systems" agreements, the technical clauses were based on the objectives and principles of product certification. Before requiring that only certified products should be employed in *CDHU* projects, the Program foresaw some intermediate levels, based on the idea of progressive quality control of production lines. It also foresaw the implementation of new standards and codes, adapted to the Brazilian scenario.

In their turn, most of the agreements of the "CAE" committee are based on the objectives and principles of the ISO 9000:1994 Standards, as is the *QUALIHAB* Qualification System for General Contractors (*QUALIHAB-GC*). Going deeper into the presentation of this System, it can be said that, in an agreement signed in November 1996 between *CDHU* and the two general contractors' unions, the *QUALIHAB*-GC as a gradual process (with its series of partial deadlines) had a period of three years foreseen for its complete implementation. The System reached its first upper level "A" in January 2000.

The *QUALIHAB*-GC System has four levels of achievement and, according to them, the quality management system of the firms is evaluated and ranked in a progressive and continuous way. It has also another particular point: the qualifications are made only by third-parties audits carried out by accredited external independent auditing service organisations. It is consisted of eleven requirements that are related to some of the twenty chapters from paragraph 4 of the ISO 9001:1994 Standards. It can be considered as a preparatory model to the ISO 9002:1994 certification.

The most important principles of the *QUALIHAB*-GC are (Cardoso (1997) and Cardoso *et al.* (2000)):

- The step-by-step progression, that allows general contractors to adjust themselves to the accorded requirements, offering them the necessary time for self-development, while creating the educational conditions that induce them to progress in the improvement of their quality management system
- the proactive characteristic, that aims to create an environment that leads firms to a certain degree of qualification
- the fact that the assignment of a qualification is a privilege of the Assignment Commission of an accredited organisation which is composed of representatives of the general contractors and of the customers (*CDHU* and civil society)
- the guarantee of anonymity of the firm until the end of the evaluation process and the use of the transparency principles, the independence of those who agree to the qualifications and the collective decision process adopted in this case.

It can be said that a French successful experience in adapting the ISO 9000 Quality Management Systems requirements to the Building scenario inspired the Brazilian system: the QUALIBAT System (Sycodés 1996/97; Archambault, 1995).

In conclusion, as from January 2000, to participate in the *CDHU* submissions, general contractors should have the qualification level "A", the highest. In mid 2000, there were almost two hundred and fifty firms qualified by *QUALIHAB*-GC, among which more then one hundred level "A".

Figure 6.1 illustrates the major ideas related to the agreements celebrated between *CDHU* and the actors' unions, taking the General Contractor's as an example.

Figure 6.1 The interaction between the actors of the supply chain of the housing
construction sector and the CDHU, consolidated by means of an agreement.
Example for the General Contractors case (*QUALIHAB*-GC).

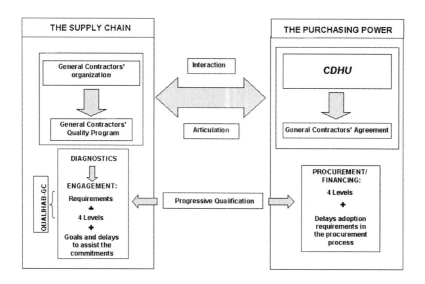

6.4.5.4   *Contribution to innovation*

Since 1998, the *QUALIHAB* Qualification System for General Contractors has
been adopted by almost two hundred and fifty general contractors, and is
deeply changing their relationship with their suppliers, with a very important
forward linkage effect in the supply chain; the backward linkage effect,
affecting the relationships between general contractors and their clients in a
positive way, is also an unanswerable fact.

More than this, other actors in the supply chain are also concerned
with this movement. For instance, architects and engineering firms have just
followed in the same way, proposing a progressive system, similar to the
general contractors'. Construction managers and foundation subcontractors
firms already have their progressive systems, which have qualified almost
thirty-five firms of both types (mainly foundation ones).

Industrialists are also implementing quality principles and tools in
their production lines; new standards and codes have been developed; the
technical capacity of the laboratories have been stimulated as well, among
other actions.

Not only the *CDHU* projects, but also many others all over the
Country, including the private sector, have been affected in a positive way by
all these actions. More than that, customer's needs are now more respected
than they used to be.

Nevertheless, its is not easy to estimate the contributions of the
QUALIHAB Program to innovation. It is undeniable that it has changed at the
same time the way companies understand "quality", such as *CDHU* and the
most important actors of the building sector. The firms' internal and external
processes, and their relationships are changing, since new procedures have

been established, new technologies can now be tested in a much more "controlled" environment due to the right of anonymity, and so on.

In conclusion: the *QUALIHAB Program* is a very important driving force to all the actors, acting as a motor of innovation, that now begins to spread all over the Country, by means of the national Program *PBQP-Habitat*, as will be seen below.

6.4.6   The PBQP-HABITAT Program

6.4.6.1   *Objectives*

Like the *QUALIHAB Program*, the *Programa Brasileiro da Qualidade e Produtividade do Habitat – PBQP-Habitat* (Brazilian Quality and Productivity Habitat Program) aims to improve quality and innovation of the social housing sector. The major differences between *QUALIHAB* and *PBQP-Habitat* are:

1)   *PBQP-Habitat* is a national program, dealing with national projects, looking for national solutions for the commons problems found in the supply chain, all over the Country
2)   it understands housing in a broad way, meaning habitat, and deals with the implementation of services other than housing units themselves, as streets and roads, utility networks (for water, electricity, etc.)
3)   it has the search for innovation as one of its central points. Most of the major actors of the sector are concerned with the optimisation of the quality of housing and its environment concerning products and services applied in their conception and execution.

It was created in 1998, as an evolution of de "building" subjects related to the *Programa Brasileiro da Qualidade e Produtividade – PBQP* (established in 1992), and is nowadays one of the Programs of the *Secretaria Especial de Desenvolvimento Urbano - SEDU* (the State Secretariat of Urban Development, directly related to the President of Brazil), that is in charge of its co-ordination.

The underlying principles of the purchasing power of the State and of the partnerships with the main actors of the sector have also been adopted here. Nevertheless, now there is not a single client like *CDHU*, but a set of them, in the three levels of the Brazilian Government field of action: municipal, regional (each of the 27 Brazilians States) and federal ones. The most important is *Caixa Econômica Federal - CEF*, a Federal Bank of Savings and Loans, with an annual budget of more than 2.2 billion dollars, which financed two hundred and eighty thousand new houses in 1999.

The *PBQP-Habitat Program* was created in December 1998, and its major objective is to "*give support to the Brazilian effort toward modernisation of the housing construction sector, trying to increase the competitiveness of its products and services, stimulating projects that could increase quality and productivity in the sector*". (www.pbqp-h.gov.br)

It has some additional objectives:

- To stimulate the relationship among the actors of the supply chain
- to promote international relationship among South American countries
- to collect and to make available information about de Program itself
- to stimulate the quality assurance of material, components, and systems
- to stimulate the quality assurance of construction, architectural and engineering services
- to promote actions to increase the level of qualification of the manpower, from the traders to the directors of the firms
- to promote the establishment and the diffusion of standards and codes
- to fight disrespect to the Codes
- to support innovation
- to promote managerial capacity of the public organisms that are in charge of the social housing sector, in all the Government levels (municipal, regional and federal).

6.4.6.2   *Annual resources*

The *SEDU* does not construct housing itself; as was seen above, *CEF* has an annual budget of 2.2 billion dollars. One can consider that part of this amount is invested in innovation, by means of some "experimental" operations, where innovation is carried out through controlled monitoring.

In their turn, the resources of the *PBQP-Habitat Program* are of about five hundred thousand dollars per year. As it belongs to the *QUALIHAB Program*, *SEDU* is not oriented to support R&D, nor to support advanced practices and experiments. In fact, the firms are induced to do the most important investments and, in spite of the similarity of both programs, as the *PBQP-Habitat Program* has just begun, we can not forecast its effects yet, as with *QUALIHAB*.

6.4.6.3   *Means*

The *PBQP-Habitat Program* is also based on "deals", negotiated by the public "buyer" in each city or region with the different local actor's unions, respecting of the local characteristics of the supply chain, level of technical and economical development, for example.

In fact, the Program is based on the action of each "regional co-ordination", all of them co-ordinated by the central structure, in Brasilia, the capital of the Country. Sixteen of the twenty-seven Brazilian's states are already members of the *PBQP-Habitat* (on May 2000). Even having their "local" focus, the local agreements should respect some national rules defined by the national union of the actors concerned, in a process co-ordinated by the *PBQP-Habitat Program* staff.

Here the same technical, social, and economical aspects used to explain the "success" of the *QUALIHAB Program* can be used. It is to justify the intended effect to be caused by the *PBQP-Habitat Program* on the supply chain, which is being progressively spread throughout the Country, in a negotiated and co-ordinated process. The particular moment of the economy should be noted, in which many factors could be explained, such as: the new constraints that influence the competitive stakes of the building sector; the new requirement demonstration on behalf of customers, mainly public ones; the lack of financial resources; the growth of competition among firms.

Once again, in spite of their comfortable position in this process, *SEDU, CEF* and the local public "buyers" involved in housing projects did not want to impose rules. On the contrary, and taking advantage of *CDHU* good experience, they are once again making the decision to conceive a program based on agreements and to put it into practice. There are already at least three States where the local "buyers" are very involved with the Program, and where many local "agreements" have already been signed: Rio de Janeiro, Bahia and Pará (in the south-eastern, north-eastern and north regions, respectively).

The main difference of the *PBQP-Habitat* agreements signed between the local public and private "buyers" of Low Income Housing sector, in the *PBQP-Habitat* ambit, and those signed in the *QUALIHAB* ambit, is that in the *PBQP-Habitat* case the local goals and delays are defined depending on the local conditions (a diagnostic phase will demonstrate this), even if the "requirements" are the same all over the Country.

The interests for industries, engineering firms, and contractors that are reacting in a positive way to the ideas of the Program are exactly the same as they are in the *QUALIHAB* case: to take advantage in the procurement phase, but now in another scale, as all the Country is concerned. It means that a general contractor qualified level "B" in Rio de Janeiro is eligible to participate of the procurement phases at level "B" in the State of Pará, and *vice-versa*.

The organisation of the *PBQP-Habitat Program* has some similarities with the *QUALIHAB Program*. Besides being based on the "local" structure and adopting the same dual structure of the *QUALIHAB Program*, based on both committees, "Materials, Components and Systems" and "CAE – Construction, Architecture & Engineering". Due to its federal scope, the National Forum of the Industry of Materials and Distributors is managing the first committee and the National Federation of General Contractors the second one.

The PBQP-Habitat Program comprises twelve Projects. Each of the twelve Projects is trusted to a manager, and has particular objectives and goals to achieve. Among them, seven can be highlighted: the *"PBQP-Habitat* Qualification System for General Contractors"; the Project "Regional Program"; the "Co-operation Project with France"; the Project "Support to the Use of Alternative Materials and Systems"; the Project "National System of Technical Approval"; the Project "National Goal of the Housing Sector"; and the Project "Improvement of the Technical Normalisation".

The *"PBQP-Habitat* Qualification System for General Contractors – SiQ-GC" is very similar to the *QUALIHAB* Qualification System for General Contractors – *QUALIHAB*-GC. It is closer to ISO 9002:1994 standards and keeps the central idea of four progressive levels of qualification.

The Project "Regional Program" is promoting an international relationship between Brazil and the southern countries of South-America (Argentina, Chile, Paraguay and Uruguay), working with themes related to quality and productivity (*habitat*), where innovation appears as one of the central subjects. In March 2000, a Regional Forum was created in the fifth meeting of the group, in Buenos Aires. The major objectives of the Forum are: integration of qualification process in all levels; stimulation of the development in the supply chain; development of common norms and codes; development of common mechanisms for technical approval; stimulation of forms of co-operation with other economic blocs, like ALCA or EU. Consequently, the innovation is directly related to the Forum's concerns.

In its turn, the co-operation Project with France, supported by the Inter-American Development Bank, is one of the most important concerning innovation. The French partner is *CSTB – Centre Scientifique et Technique du Bâtiment*, the most important French building research institution. Thanks to this Project, Brazil had five technical missions in 1999 dealing with subjects such as quality management, innovation in construction, and site organisation. In 2000, five exchange programs in France of members of the *PBQP-Habitat* staff took place (with twenty participants so far)[4].

The Project "Support to the Use of Alternative Materials and Systems" was developed between 1998 and 1999, to offer housing "buyers" and in particular *CEF* a technical way to evaluate the performance of innovation solutions proposed by the actors of the sector in operations financed by this bank. To understand the importance of this Project, it can be said that there are inadequacies in the Brazilian standards, and the consequences are not only the difficulties in evaluating ordinary solutions, but also mainly innovative ones. The Project created a direction that aims to evaluate these innovation solutions, reducing the risks involved. It is a "prototype" of a Technical Approval System.

The Project "National System of Technical Approval", that is not yet in progress, will allow the development of an "integral" Technical Approval System, in accordance with international rules, mainly with those of the South-American countries. It will be a very important tool for innovation.

The last two projects - "National Mobilisation Goal of the Housing Sector" and "Improvement of the Technical Normalisation" – are very closely related. In fact, the National Mobilisation Goal of the Housing Sector, the rate of conformity with the national standards of the major products that are used in housing construction, is expected to grow to 90%, from 1998 to 2002. It seems to be a simple task or even obvious for those from a developed country; however it is a great challenge that should be overcome by the less developed and poorly structured ones, like Brazil. In fact, Brazil does not even have all the necessary standards, many industry products do not follow the existing ones and the required justice system is not so effective to enforce customers' rights, for example. Attaining the fixed goal, and, at the same time, developing the standards is one of the most important aspects of the *PBQP-Habitat Program*, and many actions have been carried out in this way, since 1998. The engaged actors from the supply chain expect to meet the deadline.

---

[4]   The amount of money accorded by the Inter-American Development Bank to the *CSTB – Centre Scientifique et Technique du Bâtiment* has not been considered in this document (Annual Resources).

6.4.6.4   Contribution to innovation

Since 1998, the *PBQP-Habitat Program* has tried to organise the actions that are carried out all over the Country concerning quality, productivity and innovation in housing construction. Many things have already been done, but many others still lag.

As *PBQP-Habitat* is trying to make use of the successful experiences conducted all over the Country, where *QUALIHAB* seems to be the most important, and if the actions already implemented are considered, it can be said that its impact on innovation will be very important.

For instance, we estimate that, from January to June 2001, up to one hundred general contractors' firms will be certified by the *PBQP-Habitat's* SiQ-GC. This will deeply change their relationships with their suppliers, with a very important forward linkage nation-wide effect in the sector.

Industrialists are already progressing in the search of standardisation and compliance with the accorded codes. The Project "National Mobilisation Goal of the Housing Sector" is working very well, with a reasonable reduction of codes non-compliance.

Laboratories capabilities seem to be one of the major problems in this process, as they cannot for the moment respond to the whole need of experiments, and tests that have to be conducted. Important investments should be made by the sector to overcome this weakness in the production chain.

It can also be expected that architecture and engineering firms will join this movement in the coming months.

The international experiences were been important, concerning Brazil neighbours and France, thanks to the Regional Program and to the technical co-operation Project.

Its is not easy to estimate the future contributions of the *PBQP-Habitat Program* to innovation. Nevertheless, as in the *QUALIHAB Program*, it is undeniable that it has already begun to change the way the major actors of the building sector think "quality". The internal process of the firms and the external ones, concerning to the relationship between them and the way they take customers' necessities into consideration, are beginning to change; new procedures have already been created, new technologies can now be tested in a much more "controlled" environment, and so on.

In conclusion the *PBQP-Habitat Program* is also a very important driving force to all the actors the innovation of products and processes. Its major characteristic, which means the great *passport* for its success, is the partnership between the public and the private forces involved in housing construction (namely *habitat*).

6.4.7 The Competitiveness Forum of the Civil Construction Industry

6.4.7.1 Objectives

The Competitiveness Forum of the Civil Construction Industry is another federal project that is related to the improvement of the supply chain of the sector. As its name suggests, it is a forum, where the major actors of the sector, co-ordinated by the Development, Industry, and Foreign Trade Ministry (*Ministério do Desenvolvimento, Indústria e Comércio Exterior – MDIC*), are put together to identify problems related to their relationships and also internal ones, so as to solve them, to expand the competitive capacity of the sector (*MDIC* (2000a), (2000b) and (2000c)).

      The Forum has also other objectives, related to generation and income of employment, to regional development, to innovation promotion and to foreign trade promotion.

      The Construction Industry is one of the twelve forums organised, and it was started in May 2000.

6.4.7.2 Annual resources

In this case, nothing can be said about investments, as the Forum has just been established.

6.4.7.3 Means

After a first phase, when a diagnostic has been done and the strategic plan of actions established, it can be supposed that, from the moment all the actors of the supply chain are trying to improve their competitive capacity, promoting innovation will be one of the ways to do this.

      Nevertheless, so far, it is not clear through which actions this innovations will happen, but we can suppose that organisms like *FINEP*, the financial supporting organism will finance them.

6.4.7.4 Contribution to innovation

There has been no direct contribution so far. The only product is a diagnostic of the sector, and a strategic plan of actions, with objectives and goals, defined according to the problems identified.

6.4.8 The COBRACON / ABNT actions: technical norms for evaluation of innovative constructive systems for habitations

6.4.8.1 Objectives

The objective is the elaboration of norms for performance evaluation of new material, components and housing construction systems.

Addressing all the stages of the productive process - from the initial research - allowing a larger flexibility and a better adaptation to its development - going through process and product improvement up to easy acceptance and adoption of new technologies.

6.4.8.2   Annual resources

The total cost of the project, expected to be developed in 18 months, from January of 2000 to June 2001, involves resources around US$ 80,000.00, that would lead to an annual average of US$ 53,000.00.

6.4.8.3   Means

The project is being developed by the Brazilian Association of Technical Norms (*ABNT*) – a non-profit civil entity. *ABNT* is, in Brazil, the association oriented to Technical Norms elaboration in several areas whenever necessary. The association works by means of committees that are established in connection with each productive sector. In the case of construction, the committee in charge is *COBRACON* - Brazilian Committee of Construction - that elaborates the norms for the sector.

For that specific work, *COBRACON* is receiving financial support from Federal Savings and Loans (*CEF*) together with *FINEP*.

6.4.8.4   Contribution to innovation

One of the problematic aspects for the use of new technologies is the absence of norms for approving new ones in the Country. The financial agents, and even the Government institutions, and NGOs feel insecure about financing constructions with innovations due to doubts concerning their performance.

When ready, it is intended that those norms are a safe tool on which the Government's financial supporters, NGOs, or private entrepreneurs could base their work, bringing a positive impact to the productive chain in the sense of facilitating the adoption of technological innovations in the sector.

6.5   EFFECTIVENESS OF PUBLIC INTERVENTIONS

The current and more significant Government policies oriented to technological innovation in construction are synthesised in Table 1, along with an attempt to assess their contributions to the sector. This table also summarised our answer to the questions:

- What instruments and approaches are being used by Governments to promote innovation in this sector?
- What works and under what circumstances?

Data from that table show an annual investment from the Government of US$ 8.15 million, mainly being invested in human resources formation (75%), that is not meant as an investment in technological innovation.

On the other hand, in this last section, the conclusions of the work are presented through which we will try to answer the last question posed by the participating countries:

- How to strengthen innovation in the construction industry?

Before proceeding to the answers, it is appropriate to say that, Government programs, although timid, have been providing conditions for technological evolution of the sector; there are also the initiatives from the private sector. Many authors that have been studying the productive chain behaviour involved with the construction of buildings, among them, Picchi (1993), Cardoso (1993 and 1996) and Barros (1996) have identified a series of other factors, belonging to the market, favourable to the technological development and the innovation implementation in the construction sector such as:

- increasing demand from consumers that derive from: Country democratisation, establishment of *PROCON*'s (public institutions for consumer protection), and the new code for consumer's defence, which prioritises consumers' rights from the market point of view
- influence of the sectors of heavy construction manufacturing industry - to face the crisis, some heavy construction companies decided to operate in the building sector carrying influences and procedures from a part of the industry that always had better organisational and technological procedures
- more demanding labour - with the widespread democratisation of the Country, and the increase of labour rights from the 1988 Constitution, labour became more demanding by itself, concerning safety and construction procedures.

It is worth mentioning that the sector of construction material and components that has been presenting a growth not only qualitative, but also quantitative since the sixties, and has been modernising in a very consistent way ever since.

During the last decades it was possible to find a great increase in the number of manufactured products on-site. The basic materials such as sand, cement and lime, for example, are giving way to industrialised mortar. The mortar coatings themselves are being replaced by polymeric products, for a faster application and better performance.

Even though the Brazilian construction productivity remains low (McKinsey, 1998) and if material losses in works are still significant (Agopyan *et al.* 1998), evolutions have happened in that sense. The movement towards quality management systems implementation in companies, especially in construction firms is a reality.

From a general point of view, the numbers from the sector continue thoroughly favourable towards its modernisation. The housing deficit of the

Country remains high (5.5 million dwellings to be built and 8.5 million dwellings demanding urgent improvements).   The participation of the construction industry in the *GDP* has kept an average of 6% of the total, constituting one of the highest as a single economy sector. Besides, if all the construction macro-complex (comprising material and components) are considered, that participation jumps to almost 15% of *GDP* (FIESP, 1999).

More than 60% of the national gross investment is done in construction while machinery and equipment come to 25%; the remainder is stock.   Linkage Index sets Construction Sector ranking as fourth in the Brazilian Economy.   Backward linkage index is the main effect for this performance; it provides US$ 27 billion; for the forward linkage index this number is USD 2.8 billion.   The private sector has the first place in the backward linkage index; that is why the construction sector is called the "economy locomotive" (de Oliveira & Cardoso, 1999).

Also remarkable is the level of direct and indirect employment generated by this sector. In 1989, the industry employed 3.9 million peoples, representing 6.9% of the occupied labour in the Country and 26.3% of the total employment in industrial activities, according to data from the Brazilian Institute of Geography and Statistics (*IBGE*, 1989). Considering that each direct job in construction generates 2.68 indirect jobs, the total level of activity reaches more than ten million people.

It is important to remark that Brazilian construction is one of the sectors less dependent on imports, which means, according to the Government's effort in search of credit in the trade balance, a thoroughly favourable aspect to its development.

However there are factors that have been hindering the innovation process, among them the following:

- difficulty in forecasting the future market behaviour
- great number of taxes and other social contributions for each registered worker, that has been motivating workers' precarious and informal hiring and to the growth in the subcontractors number, not only for increasing productivity and quality of the work process, but as a form of cheating labour laws, and of cost-reduction
- great trade rotation that hinders training
- backward vision by some managers in the sector, who do not see innovation introduction as an investment, but as cost-increase, and who do not perceive technological innovation as a strategy for profit increase
- low capitalisation of the sector with the existence of a great number of companies with small capital, and therefore, with difficulties in making investments.

We do not have all data to estimate the level of innovation and the R&D investments in the sector. On the other hand, although the sector has many problems to solve and the level of these investments is lower than desirable, the environment is favourable to innovation.

Table 6.1 Main public Brazilian intervention synopsis of technological innovation support in the field of construction in progress in July 2000, described in section two of this work.

Innovation Stage	Name	Resources to Construction	Objectives	Means	Contribution to Innovation
Programs to support R&D	FAPESP	One and a half million dollars per year.	To finance grants and other activities related to technological and scientific research, local and international exchange and information in the State of Sao Paulo.	The sponsoring of about one hundred and eighty projects, which two thirds are grants. The other third concerns research funds for specific projects, laboratory equipment, book acquisitions, congress' organisation, congress' participation foreign researchers, book and review publications, etc.	It is very difficult to objectively estimate the contribution to innovation due to FAPESP actions. The grants accorded tin 1998 were: to young scientists – fifty; to master of science studies - thirty; to doctoral studies - thirty-five; to post-doctoral studies – five. The FAPESP system is one of the most efficient and effective of the Country, and it is very selective, with a very rigorous control. 60 other research projects (1998) for specifics projects, laboratory equipment, book acquisition, congress' organisation, congress' participation, foreign researchers, book and review publications, etc.
	CAPES	One million, six hundred thousand dollars a year (estimation excluding special programs such as PADCT).	Highly qualified human resources formation for the teaching at universities, research and the attainment of the professional demand of both public and private sectors.	Grant of scholarships and graduate programs. Infrastructure support.	It is very difficult to objectively estimate the contributions to innovation due to CAPES actions.

Table 6.1 (Continued) Main public Brazilian intervention synopsis of technological innovation support in the field of construction in progress in July 2000, described in section two of this work.

Innovation Stage	Name	Resources to Construction	Objectives	Means	Contribution to Innovation
	CNPq	Three million dollars a year (estimated).	To promote the scientific and technological development and to support researches necessary to the social, economic and cultural progress of the Country.	Development of three basic activities: fostering (grant of scholarships; 91 % of the budget), research support and science and technology information and diffusion.	It is very difficult to objectively estimate the contributions to innovation due to *CNPq* actions.
Programs to support advanced practices and experimentation	*HABITARE*	Eight hundred thousand dollars	To contribute, through support researchers it in the area of science and technology, to the solution of the Brazilian housing problem and to construction modernisation, always having in mind the connection with environmental concerns.	Financial support for companies' proposals for technological development.	• Clustering of several organisations innovation-oriented institutions in joint projects. • Creation of diffusion /information networks on innovation in the sector. • Financing proposals for technological innovations. • Financing for new technology evaluation.
	The *QUALIHAB Program*	Seven hundred thousand dollars per year.	To optimise the quality of housing with regard to products and services employed in their conception and execution, from partnerships with the main actors of the sector established through agreements.	Based on "agreements".	• Three hundred firms have already implemented their Quality System, mainly general contractors. • A very important forward linkage effect, effect in the building sector; the backward linkage effect, affecting the relationships between general contractors and their clients.

Table 6.1 (Continued) Main public Brazilian intervention synopsis of technological innovation support in the field of construction in progress in July 2000, described in section two of this work.

Innovation Stage	Name	Resources to Construction	Objectives	Means	Contribution to Innovation
Programs to support performance and quality improvement	The *PBQP-Habitat Program*	Five hundred thousand dollars per year.	To support the Brazilian effort of modernisation for the housing construction sector, trying to increase the competitiveness of its products and services, stimulating projects that could increase quality and productivity in the sector.	Based on "agreements".	Three States with agreements already signed. Integration with countries of Southern South-America. The SiQ-C System. The Project "National Mobilisation Goal of the Housing Sector". The technical co-operation Project with France.
	The Competitiveness Forum of the Civil Construction Industry	Nothing up to now.	To improve the sector supply chain and also the generate employment and incomes, regional development, to promote innovation and foreign trade.	Based on an initial and a strategic plan of actions, with objectives and goals, defined according to the problems identified.	Diagnostic, plan of action, objectives and goals.
Programs to support implementation of systems and procedures	*COBRACON / ABNT*	Fifty thousand dollars	To elaborate norms for evaluating housing new materials, components and construction systems.	Multidisciplinary research team co-ordinated by *COBRACON* with financial support from *CEF* and *FINEP*.	To facilitate the adoption of new technologies as soon as guidelines are offered for its safe use.

Therefore, to motivate innovation in Brazilian construction, and thinking of the actions that should be kept by the State, or its support or articulation, these actions would be oriented to:

- incentive new products and processes research and development, allotting an amount of resources in R&D to the sector in compliance with its economic and social importance
- human resources formation in all education levels
- increase Government and its institutions' induction action as buyer (purchasing power)
- incentive technological competence development, and local and regional producers, with environmental protection
- incentive better integration of the productive chain with co-ordinated modernisation actions in progress
- nation-wide incentive to technological co-operation actions between research institutions and companies by means of co-operative networks, as well as abroad, mainly with Southern South America
- companies refinement aiming at competitive and innovating capability increase, including financial support for R&D investments
- better capability and valorisation for research institutions and universities, both working together with the private sector
- improve laboratory capability in the Country, with help from the public sector, linked to the research institutes and universities, or from the private sector
- refine technical normalisation, with emphasis on norms oriented to performance, still incipient in Brazil
- incentive the certification of traditional products and processes
- develop new products and processes technical approval mechanisms
- make public the tested and approved innovations, that can reach the less privileged layers of the population, which by themselves build three-fourths of the housing production of the Country, through self-help-construction[5]
- valorise consumers' protection mechanisms and increase their understanding in relation to their rights
- the warranty of an efficient and effective law functioning system.

As shown in the work, the Brazilian Government has been promoting important actions in that sense, but there is still a lot to be done. One percent of its investments in a sector that responds for, at least, 6% of *GDP* is too little.

On the other hand, it is not the Government task to promote construction industry innovation by itself and not even to be the main agent of the process. The roles and responsibilities have to be shared:

- private enterprises (makers, manufacturers, designers, managers, dealers, etc.), by themselves have been promoting concrete and integrated action; they should improve their understanding about the

---

[5]   This production is estimated around 800,000 new dwellings a year.

meaning of innovation, of its strategic role, and of the benefits it provides, by investing more on its development; although there is no data, the percentage of the sector gross revenue invested in R&D is, certainly, very low; could there be a distinction between manufacturers and others? Do they reasonably invest in R&D?

- should the financial agents and the insurance companies more and more motivate the use of innovative solutions that accrue technical and economic benefits, with performance warranty; why constitute a performance insurance system that does incentive and does value the innovations of assured performance?

- research institutes and universities should more and more look for innovation as a clear and tangible goal, by integrating with the productive sector, without losing sight of consumer's interest and society's in general

- society and consumers should enforce the Government, its institutions and the companies, claiming their rights and valorising innovative initiatives that bring them benefits.

In spite of all these difficulties, the Brazilian scenario is quite promising, above all for the development sense that it presents, more than for the absolute numbers themselves, as they are still low, and very far from desirable.

CHAPTER 7

CANADIAN PUBLIC POLICY INSTRUMENTS THAT AFFECT INNOVATION IN CONSTRUCTION

André Manseau and Aaron Bellamy

7.1 CANADIAN CONTEXT

Construction has fewer long-established traditions in North America than in many other countries. While North American construction practices reflect their European roots, they are also influenced by other cultural backgrounds. In general, the North American processes are considered more flexible and less regulated than in other regions such as Europe and Japan. Replacement is generally chosen over rehabilitation. Because emphasis tends to be on cost efficiency and speed of construction In North America, there have been numerous innovations in machinery, techniques and equipment. This tendency toward innovation and speed of construction, coupled with the continent's abundance of raw materials, has ensured that the productivity and costs of North American construction are very competitive with other developed nations.

7.1.1 Key characteristics of the Canadian construction industry

The Canadian construction industry and its related sectors have a significant impact on the national economy, accounting for about 13-14% of total GDP. The economic activity of builders and related workers (electrical, plumbing, air system installers, structure assemblers, external envelope works, plasters, etc.) accounts for about 5% of GDP. From partial information on import/export of building products and engineering services gathered annually by Statistics Canada, we estimate that the manufacture of building products and construction equipment represents another 5% of GDP, engineering and architectural Services about 1%, and service firms involved in building/infrastructure operation and maintenance account for about 2-3% of the national GDP. Exact and complete data for related sectors are not available from official statistical sources.

    The Canadian market for construction is characterised by three key factors, which have a profound impact on the industry's structure and performance:

- The harsh climate creates special requirements for buildings and infrastructure and causes an important decline in construction activity during the winter.

- In this extraordinarily large and geographically diverse country, construction projects are subjected to a range of climates, from temperate to polar across maritime, prairie and mountainous zones.
- The product is invariably complex and highly customised. Site - and project-specific construction requires those products and processes are adapted to local conditions and regulations.

The result is an industry that is highly fragmented, specialised and composed primarily of small -- even very small -- companies. In 1997, the industry comprised more than 150,000 contractors (general and trade) with about half being firms of only one person. These contractors employed 755,000 workers (about 5% of the total Canadian work force). In that year, the total Canadian construction market stood at approximately US $72 billion[1]. Roughly two-thirds of this activity was carried out by the contracting industry. The balance was undertaken by the in-house resources of organisations not primarily engaged in construction, such as utilities and governments (Industry Canada, 1998).

In addition, about 4,000 companies in the building products industry employed more than 110,000 workers, produced over US $11.7 billion worth of products, and had a positive balance of trade in 1994 of close to US $1.46 billion.

Engineering and design firms are not captured by construction industry statistics, as they are part of the services sector. A special survey of consulting firms, conducted by Statistics Canada in 1996, showed that approximately 25% of their total output relied on construction, representing US $1.16 billion in 1992 (Industry Canada, 1996).

The construction industry has evolved to allow it to operate even in a feast-or-famine market. It is easy to enter and the risk is distributed among many players. Only a strategic nucleus of key employees is kept on staff over the long term.

Nevertheless, Canada has established a strong and efficient construction industry, with a solid reputation for reliability and innovative design. There is a significant technology parallel with the United States, which, particularly in the north, shares many of Canada's geographic and climatic conditions. Niche strengths in Canada include cold-weather engineering and construction, the design and construction of hydro-electric power projects, the repair and protection of salt-affected structures, and management of the indoor environment.

A large share of construction activities is conducted on behalf of governments, both national and provincial. Canada is among top OECD countries in terms of government share in construction assets, with nearly 25% of all fixed structures belonging to the state (OECD, 1997).

Since 1990, much of the industry has endured a prolonged period of stagnation, except for the engineering construction sector, which is the least cyclical of construction sectors. Large infrastructure projects last-up to a decade. Activity has recently increased in some sectors; particularly value of residential construction has increased of about 10% per year from 1995 to 1997 (Statistics Canada, 1999).

---

[1]   The US dollar is used as the reference currency throughout the paper to facilitate international comparison, the rate change used is 1CAN$ for 0.67US$.

7.1.2 R&D and innovation in the Canadian construction sector

Little information is available on innovation in the Canadian construction sector. However, existing data show that the construction sector (limited to builders and assemblers) lags significantly behind other sectors in terms of R&D expenditures. In 1997, R&D as a share of the sector added-value was 0.056% in construction, in comparison to 1.53% for all industries (OECD, 2000). While construction represents about 5% of GDP, the industry accounts for only 0.2% of total industrial R&D. Moreover, these firms have decreased their R&D effort by 20% from 1995 to 1999, in comparison of an increase of 19% for all sectors (see Table 7.1).

Table 7.1 Average percentage increase in R&D expenditures from 1993 to 1997, by Canadian industry sector (Source: Statistics Canada – Cat. No. 88-001-XIB, 1999a)

Sector	1995-1999 R&D Change (5 years)	R&D in 1999	
		$US million	% of Total
Mining and oil wells	-33%	92.5	1.4
Manufacturing	24%	4,155	64.5
Construction	-20%	15.4	0.2
Services	15%	1,989	31.1
Total industries	19%	6,403	100

Construction is often portrayed as a mature sector that is slow to innovate and resistant to technical change. It is also said to have little corporate memory, reinventing crucial processes at every project. However, this portrait is not entirely accurate. Because the construction industry recognises the importance of change and adaptation, construction technologies are continuously being improved. New business practices, such as the design-build approach, and new public-private arrangements are profoundly affecting the industry. The view that the sector is conservative and hesitant to change reflects a misunderstanding of the innovative process as it applies to construction, but this is not captured by official statistics.

Other sectors related to construction, such as energy companies (owners of large facilities), consulting engineers and manufacturers of building products, perform about 42% of total construction R&D (see Table 7.2). The 1990s have also witnessed new trends in the funding and performance of R&D in construction.

The federal government is the only sector that has significantly increased its construction R&D expenditures, due in large part to external revenue from joint R&D projects and technical services. Universities have maintained their efforts in construction R&D, while all other sectors have decreased their investments.

Table 7.2 Sector trends among top 50 construction R&D performers
(Source: Canadian Construction R&D Performers and Funders Report, Revay and Assoc., 1999)

Construction R&D Sector	Construction R&D Performed in 1992		Construction R&D Performed in 1998	
	US$ millions	Share %	US$ millions	Share %
Federal government agencies	17.4	18.7	21.4	26.5
Provincial government agencies	6.0	6.5	3.9	4.8
Energy companies	21.6	23.2	11.4	14.1
Manufacturers, Services & Associations	25.5	27.3	22.1	27.4
Universities	19.8	21.3	20.2	25.0
Consulting Engineers & Contractors	2.8	3.1	1.7	2.2
TOTALS	93.1	100.0	80.7	100.0

7.2 PUBLIC INTERVENTIONS

Numerous publicly supported programs to promote R&D and innovation are available to all industries in Canada. Some programs provide funds to private firms for R&D projects, and others promote industrial R&D indirectly through tax credits or employment support for young engineers and scientists. However, construction firms appear not to take advantage of these programs. For example, Technology Partnerships Canada (TPC) provides repayable, long-term risk-sharing capital to private-sector firms for R&D projects. From its inception in 1996 to the end of 1998, TPC invested US $425 million in 67 research projects, leveraging US $1.70 billion in private-sector innovation spending. However, we found no project related to the construction industry that was funded by the program during this period.

We estimate that construction-related projects have been granted less than 2-3% of public investment in R&D in recent years, a small amount compared to the sector's economic impact (13-14% of GDP).

Major programs that are available to support innovation in construction are presented in Table 7.3. The Institute for Research in Construction (IRC) of the National Research Council of Canada (NRCC) is the leader in research, technology and innovation for the Canadian construction industry, performing about two-thirds of the total federal R&D activities in this field (c.f. Table 7.2). IRC also co-ordinates the development of Canada's model construction codes, which enable provincial governments to produce building, fire, plumbing, and energy codes, and provides a national evaluation service for innovative building products. The Industrial Research Assistance Program (IRAP) of NRCC assists small - and medium-sized enterprises (SMEs) and provides some financial assistance for innovative projects; about 4% (or US $1.9 million) of total contributions to firms were allocated to construction in 1997-98.

Table 7.3 Major public programs available to the construction sector to support innovation

Innovation Stage	Name	Resources to Construction	Objectives	Means	Contribution to Innovation
Programs to support R&D	University research	Approx. US$20 M/y	To encourage university-based research and promote collaboration with industry	Funding individual researchers or groups via grants	A few cases have shown industry applications.
	IRC-Research	Approx. US$7 M/y	To act as Canada's construction technology centre.	Performing collaborative R&D projects with the industry, Facilitating access to technology information	Recent case studies have shown significant economic impact.
	IRAP	Approx. US$2 M/y	To support technology development of SMEs	Providing financial support and advice to SMEs for technology development projects	No construction-specific data.

Table 7.3 (Continued) Major public programs available to the construction sector
to support innovation

Innovation Stage	Name	Resources to Construction	Objectives	Means	Contribution to Innovation
	PWGSC	Approx. US$4 M/y	To promote R&D related to property management	Performing and funding R&D in energy efficiency, building control, building longevity, pollution control	No published information
	NRCan – Research	Approx. US$2 M/y	To conduct R&D to help Canadians use their country's resources wisely, efficiently and protect the environment.	Performing R&D projects and providing grants for projects in energy efficiency	No published information.
Programs to support advanced practices and experiment-ation	CMHC – Research	Approx. US$2 M/y	To ensure a competitive mortgage system to help Canadians buy homes and maintain the excellence of their housing.	Performing projects for improving housing, for ensuring a competitive mortgage system and for promoting the export of Canadian housing products and expertise.	A few cases in collaboration with IRC and/or NRCan
Programs to support perform-ance and quality improve-ment	IRC-material evaluation	Approx. US$3.5 M/y	To develop evaluation methods for new building products (and eventually to develop a best practices guide for infrastruc-ture)	Developing and performing evaluation of new building products and publication of evaluation results	A few case studies linked to R&D projects have shown significant impact

Table 7.3 (Continued) Major public programs available to the construction sector
to support innovation

Innovation Stage	Name	Resources to Construction	Objectives	Means	Contribution to Innovation
Programs to support performance and quality improvement	IRC-material evaluation	Approx. US$3.5 M/y	To develop evaluation methods for new building products (and eventually to develop a best practices guide for infrastructure)	Developing and performing evaluation of new building products and publication of evaluation results	A few case studies linked to R&D projects have shown significant impact
	CMHC-Awards	(a few)	To promote quality housing products	Organizing regular contests for awarding and promoting best housing products	(not known)
Programs to support the take-up of systems and procedures	NRCan Commercial Building Incentive Program	Approx. US$5 M/y	To support adoption of energy efficiency building products and systems	(no detail)	(not known)
	IRC – National Building Code	Approx. US$3.5 M/y	To develop national building model codes to foster a safe and sustainable built environment	Developing model codes from best available information and diffusion (publication, information)	A few case studies linked to R&D projects and evaluation have shown significant impact

Two granting councils -- the Natural Sciences and Engineering Research Council (NSERC) and the Social Sciences and Humanities Research Council (SSHRC) -- and the Canadian Foundation for Innovation (CFI) are funding universities. Construction-related R&D performed by universities was estimated at US $20 million in 1998 (Table 7.2).

Natural Resources Canada (NRCan), a federal government department, comprises several research laboratories that focus on specific construction topics. Research topics include advanced environmental building technologies, energy efficiency, building-control systems, integrated Heating Ventilation and Air Conditioning systems (HVAC), reinforced concrete, and geo-technology. Another department, Public Works and Government Services Canada (PWGSC), supports some R&D in property management issues, such as energy efficiency, environmental sustainability, building longevity and maintenance, and productive work environments.

Canada Mortgage and Housing Corporation (CMHC), a federal crown corporation, funds R&D to improve living conditions as well as the technical performance of housing. CMHC also promotes Healthy Housing, a program designed to explore safer and more efficient housing alternatives.

Provincial and territorial governments have constitutional responsibility for building regulations. The IRC provides model building and fire codes, based on the best information collected in an independent, participatory process. These model codes are reviewed, modified if required, and adopted by provincial governments as mandatory minimum requirements. Generally, provincial legislation delegates to municipalities responsibility for the day-to-day enforcement of codes. Model codes are also used or referred to in voluntary standards. Standards certification and quality assurance falls under the National Standards System of Canada, which comprises various accredited standards-writing organisations, certification agencies, accredited laboratories and quality-assurance organisations. Certification has become the norm for certain product areas such as electrical and plumbing.

Unions also serve an important role in the construction industry, with approximately 35-40% of the construction industry labour force claiming union affiliation (Industry Canada, 1998). The construction industry is the single largest participant in apprenticeship training in Canada. In most provinces, union-employer joint training boards or similar organisations administer apprenticeships in the unionised portion of the industry. This co-operative system has avoided the chronic under-training found in other industries. In addition, the provinces and territories, with assistance from the federal government, have established a program of mutual recognition of trade qualifications – the Interprovincial Standards Program. This program, in conjunction with organised apprenticeship training, helps create a highly skilled and mobile work force in Canada.

7.3   EFFECTIVENESS OF PUBLIC INTERVENTIONS

Although Canada's public instruments have had little impact on R&D investments in construction, the overall effectiveness of these programs remains unclear. There are few data on construction industry performance in Canada. The last column of Table 7.3 gives an assessment of each program's contribution to innovation.

The value-added of the industry, limited to builders and installers, slowly decreased from 7% to 5% of the national GDP in the last two decades. Data on manufacturers of building products and on construction services (design, operation and maintenance) are not available. Extrapolating from incomplete import/export data suggesting that economic activity in manufacturing and services related to construction has gradually increased, it may be concluded that construction-related activities have grown at much the same pace as the overall economy during the last two decades (maintaining a share of 14-15% of the GDP).

During this period, governments have reduced their direct intervention in order to encourage firms in construction to initiate R&D projects and to take leadership in developing new standards. However, governments have increased their presence in regulation (at the provincial level) and in the diffusion and acceptance of codes and standards. The programs listed in Table 7.3 have not significantly changed in the last decade. Total investment in R&D has decreased (c.f. Table 7.2), which suggests that the private sector did not fill the R&D gap created by the government's withdrawal.

However, recent case studies have shown some success from public programs that use an integrative approach, facilitating the creation of consortia and the transition of new standards from design to development. This approach involves integration and interaction between R&D, evaluation of innovation and codes development. Construction is an increasingly complex activity, including numerous products, sub-systems and agents, coupled with high requirements for safety and reliability. Therefore, successful innovation processes should be adapted to this business reality. In particular, they should be developed with an integrated approach that takes into account interactions with other building components, as well as safety and reliability issues.

A logical system of building regulations plays an important role in facilitating the adoption of innovations in building products and designs. However, Canada has experienced some delays in the acceptance and enforcement of building regulations based on new model codes.

7.4 TRENDS

Six major international and national issues, presented in Table 7.4, are expected to affect the Canadian construction sector in the medium and long term (3 to 10 years). It is anticipated that these issues will play an increasing role in the development of public policy.

Trade Globalisation – Trade liberalisation under various international agreements (AIT, NAFTA, GATT, and APEC) has increased the export potential of Canadian construction firms and materials suppliers, particularly those operating in shrinking domestic markets. Local regulation and product standards still impede market entry, but the trend is from national to internationally harmonised standards, codes and products. Programs to promote and facilitate export will certainly be in demand, but the process of international harmonisation remains so far incomplete.

Changing Industry Structure – New contracting practices are being adopted, including design-build, build-own-operate-transfer and public-private arrangements to build and operate public facilities/utilities. Instead of the traditional reliance on tight prescriptive specifications and low-bid selection criteria, these new practices allow for "best value" selection criteria. A second major trend relates to the diversification of activities as some builders provide increased services as well as some manufacturers of building sub-systems are also providing more services, including the installation and maintenance of these complex products. An important challenge for public intervention is the development of objective-based building codes, which allows for more creative and integrative solutions to construction. Another challenge in innovation will certainly be the development of an integrative approach for the entire construction sector, including building products, design, assembly and maintenance.

Table 7.4 Developing Trends in Public Policy

Developing Issues	Policy Response
Trade Globalisation	• International agreements and uniform standards/assessment methods • Attracting foreign investment • Promoting/facilitating exports
Changing Industry structure	• Objective-based codes • New contracting practices such as integration of building design, construction and operation • Public-private ownership and increased user fees (direct tax)
Environment/Sustainable Development	• Climate change • Energy, land and primary resources efficiency programs • Land/Resource use • Pollution control
Demographics	• Health/Safety regulations • Quality control regulations
Life Cycle Costs	• Building Longevity • Renovation and Retrofit
Information and Knowledge	• Increasing use of GPS and EDI information systems • Design Monitoring Systems (CAD, scheduling software) improving construction efficiency.

Environment and Sustainable Development – In Canada, homes and other buildings account for more than 30% of the nation's total energy consumption. As such, they are the focus of efforts to improve energy efficiency, which must be achieved without compromising the indoor environment. Pressure will increase to reduce land and primary resource usage and to recycle building waste. In addition, new regulatory measures and environmental assessment requirements are expected to increase the cost, complexity and time required for construction.

Demographics – Large demographic shifts have occurred in the last decades: Canada's population is ageing, while the ratio of children per family has significantly decreased. More specialised facilities to house senior citizens will be needed. Improved building controls and automation will be used to reduce operating and care-giving costs.

Life Cycle Costs – The longevity and durability of buildings is becoming increasingly important, in part to meet environmental pressures for reducing waste and recycling. Consequently, repair and rehabilitation technologies will gain in profile. No major transformations, re-constructions or economic booms are expected, and the pace of new construction will tend to slow.

Information and Knowledge – Information technology systems such as GPS and EDI, computer networks and design-monitoring systems (CAD and cost-schedule planning-monitoring systems) are already beginning to affect construction in a variety of ways, and their impact is expected to continue to grow.

7.5   CONCLUSION

Because construction is still a highly fragmented industry, takes comparatively more time and effort to deploy innovation processes. However, information technologies are facilitating systems integration, and bringing greater flexibility to the adaptation and development of new designs. They will also make national and international harmonisation and product customisation easier to achieve.

A major challenge will certainly be to develop capacity to manage complex information in a strategic manner. Needs and constraints are becoming global and widely diverse. To be effective, innovation programs will have to identify their niches, build on or complement other available capabilities, address complex demographic, environmental and culture issues, and as well as be linked to international networks.

In Canada, some public programs have shown a measure of success. The most promising trends are found among programs that use an integrated approach that facilitates interactions among and transition from R&D, evaluation and codes development. However, no single rule applies; government programs will need to be flexibly applied to each situation.

For instance, for some industry sub-sectors that could be readily integrated, support for innovation would require little more than the circulation of information or some facilitated networking. On the other hand, innovations involving public safety or the environment could require important up-front government investment in R&D or stringent regulations, especially if the private sector cannot capture all the benefits. In all cases, however, an integrative approach that takes into consideration all key players -- from building product manufacturers and designers to builders and end-users -- will be crucial to enhance innovation in construction and to develop effective public instruments.

CHAPTER 8

INNOVATION IN THE CHILEAN CONSTRUCTION INDUSTRY: PUBLIC POLICY INSTRUMENTS

Dr. Alfredo Serpell

8.1   INTRODUCTION

After many years of steady growth, the Chilean construction industry has been passing through difficult times lately due to the recession produced by the so-called "Asian crisis". During the last two years, activity decreased to a very low level, with high interest rates that reduced the investment in construction and delayed the selling of new houses. Thus, during 1998 and 1999 the construction industry suffered a strong decline of output. However, starting the year 2000 a slow recovering is being observed and the construction sector is expected to get into the growing path soon.

These harsh years have brought several consequences for the Chilean construction industry. First, this period of time has revealed several weaknesses that affect the competitiveness of the construction sector. As a result of this, for the first time the construction sector is seriously concerned about its long-term development and several organisations are working to start activities related with the improvement of the construction industry performance.

Second, the scenario has considerably changed with an intensified competition in the different construction markets. The increase of foreign participation in many areas is leaving many local construction companies with limited work opportunities. Business practices and the scale of construction projects have changed reducing the participation of domestic construction companies, particularly in the infrastructure area but also in the big real estate projects.

Third, Chilean construction companies are becoming more proactive regarding their own development. During the difficult years, several healthy companies took advantage of the lack of construction activity to undertake improvement initiatives to prepare them for the new scenario. Research activities, technology developments, quality certification projects and other efforts were carried out during these years. In the foreseeable future it seems that conditions are given to expect more concern about the innovation and technology development of the Chilean construction sector. Several institutions have dedicated their last board meeting to address these issues and a growing concern can be appreciated. However, the challenge is to organise the sector to seriously confront the future and to produce real changes to improve the sector's performance. To achieve this, the construction industry needs the definition of a vision of the sector for the next five or ten years. Also, work should be done at different levels to get the participation of all the actors that are involved in the development of the construction industry, particularly the government, the industry as a whole and the construction companies. The expectations are high at this time as well as the enthusiasm that is perceived in the construction environment.

This chapter will describe the context of the Chilean construction industry, its main characteristics and the current situation with regards to instruments and policies available for innovation and technology development of this sector.

8.2   THE CHILEAN CONTEXT

8.2.1   South American context

Construction is an important productive sector that besides producing the required housing, buildings, and infrastructure needed by any country to develop its productive potential and satisfy its society needs, it also provides jobs and economic activity in a direct and indirect way. This fact makes this sector very important for developing countries (all South American countries fall in this category) where the construction industry is used almost as a tool for governments' economic policies.

The construction industry in the region has been affected by many changes in the last twenty years. The opening of construction markets to foreign competition through the elimination of many trade barriers, the diminishing construction role of public institutions; the incorporation of new construction procurement approaches and the entrance of many large foreign construction companies to the market arena are just few of these changes. Private initiative has taken many areas traditionally in the hands of governments, reducing governments' role to activities where private investment is still insufficient, like: low-income housing, school buildings, government infrastructure, urban infrastructure and infrastructure maintenance, etc. Not all the countries have gone through this process with the same intensity but the trend is there to see.

For a long time, governments in South America were the most important clients of the construction industry and in some countries of the region they still are. However, from the early eighties, the government participation in the economic activity of several countries has been consistently reduced. Governments' activities today are mainly focused on creating the legal framework to regulate the construction industry and to promote its development. For example, during the last two decades, several South American countries have approved regulations related to BOT projects, quality assurance and control, environmental management and control and others. Also, some governments have created instruments and mechanisms to support innovation and research activities in the construction industry.

8.2.2   The Chilean construction industry

The Chilean construction industry is currently facing market and business conditions that are changing at a rapid pace. This situation is forcing construction companies to confront requirements that were unusual a few years ago in the Chilean construction market. The principal and traditional sector concern about the cyclical nature of the construction activity has given step to the recognition of the fact that there are other forces in the market at this time that are equally important for the current and future success of a construction company. These forces are, among others: the growing intensity of competition; the participation of an increasing number of foreign companies; the increasing requirements of clients; and the greater complexity and technology demands of current construction projects.

The current scenario of the Chilean Construction Industry, can be briefly described by the following issues:

- There is still inadequate development of technical and management abilities in Chilean construction companies. Construction companies are reactive in front of new market demands
- contribution margins of projects and total sales have been decreasing consistently during the last ten years
- strong projects bidding competence and negotiation are appreciated now and expected to intensify even more in the future
- construction clients are increasingly more demanding
- increasing number of companies competing in the local market. At least 50 foreign companies within the world largest 300 are working in the country
- quality has become a critical performance parameter
- an accelerated change of construction methods and technology is expected in the near future
- slow reactivation is expected for the year 2000 with an accelerated growth for the next two years.

8.2.2.1   *Structure of the Chilean construction industry*

The structure of the Chilean Construction Industry can be described by the scheme shown in Figure 8.1 (Gazmuri et al., 1993). The industry comprises providers of personnel, materials and supplies for construction projects, including money, contractors and subcontractors, clients and users of the construction services and products.

Figure 8.1 The structure of the Chilean construction industry

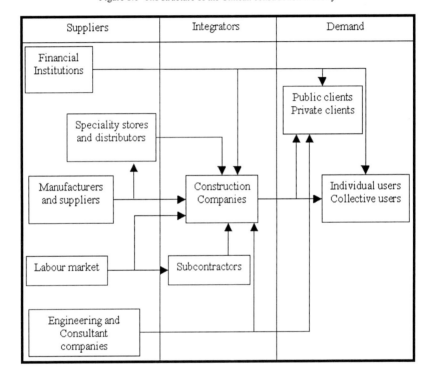

Data about the number of companies that participate in the Chilean market is not reliable. Estimates by the Chilean Construction Chamber indicate an approximate number of 7,500 companies (Pavez, 2000), distributed in the following market sectors:

- Public housing contractors
- private housing and commercial building contractors
- general contractors
- public works contractors
- specialised contractors.

The labour market is varied and involves more than 480,000 workers employed in the construction sector. The unemployment rate is around 10% in average although it has greatly increased during 1999. Some figures to show the size of the Chilean construction industry are shown in Table 8.1 (CChC, 1998; Apertura, 1998).

Table 8.1 Some figures of the Chilean construction industry (in US$ of 1998)

Approved housing construction in 1998	115,000 units
Total Investments in 1998	US$ 9,212 million
Total Investment expected for 1999	US$ 8,263 million
Projected private investment 1998-2000	US$ 24,000 million
GDP Chilean construction/GDP Chile	5.8% to 6%

The areas of major development of the Chilean construction industry can be found in mining construction and electrical projects, due to the fact that Chile has huge mining resources and growing energy requirements. Also, there have been new developments in industrial construction and building construction, especially high-rise buildings due to office demands. These developments have been pushed by the substantial growth of the Chilean economy during the last decade. It can be noted that between 1985 and 1997 the country grew at an average rate of 7.6% per year duplicating the GDP during that period (Pavez, 2000). The last 5 years have seen a considerable growth of the investment on infrastructure development due to the enlargement of the economic activity. Moreover, the infrastructure of the country has become an important bottleneck of the Chilean economic development. According to studies of the Chilean Construction Chamber, the current deficit of infrastructure result in losses of around US$ 2,300 million per year, which is equivalent to 3.4% of the GDP of 1999. It is expected that the country will have to invest almost US$ 30 billion during the next 5 years in this area (Pavez, 2000).

8.2.3   R&D and innovation in the Chilean construction sector

Research, development and innovation expenditure in Chile is well below the levels of expenditure that can be found in developed countries. Figure 8.2 shows the total expenditure in research and development in Chile as a percentage of GDP until 1997.

Figure 8.2 Expenditure in R&D in Chile as percentage of GDP (adapted from *CONICYT, 1998)*

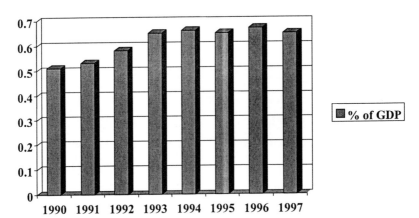

The total expenditure reached US$ 497 million in 1997 in dollars of that year. If this expenditure is subdivided by economic sector, it can be appreciated that in the construction sector this level is very low, reaching no more than one percent of the total, as shown in Figure 8.3.

Figure 8.3 R&D in different Chilean sectors 1997 (adapted from CONICYT, 1998)

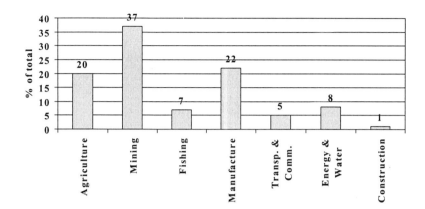

This situation confirms the general image that the construction sector is very conservative and reactive to market demands. In general, the construction industry has been traditionally slower than other sectors to include new advances. Also, a valuable function that should be performed by construction companies is the accumulation and storage of innovation experiences that are accomplished at project and operational levels. Interviews with executives from several Chilean construction companies show that most of them do not register the advances that are achieved in their projects. In this way, successful innovations have a limited impact because they are not shared with other projects that can benefit from them, neither are they used afterwards in new projects. This problem is explained in part because of the lack of conscience of companies about the importance of accumulating their experiences and by the break-up of project teams upon project ending. This situation severely limits the learning capacity of construction organisations (Serpell, 1997).

However it should be considered here that many of the research and development efforts in the construction sector are carried out privately by construction companies and are not registered in national statistics because of limitations in data gathering and information. It should be noted that the information given by the official system (CONICYT) only considers the R&D data generated by public programs or private programs partially supported by public programs. In the author's knowledge, many Chilean construction companies have been investing strongly in R&D activities during the last decade. Also, the Chilean R&D instruments have consistently ignored the construction sector as a priority sector for assigning R&D resources. Still today, one of the most important R&D programs supported by the government does not include the construction sector for research and development financing. It is expected that this program will open a line on infrastructure for the first time during the first months of the year 2000. In this line construction research and innovation will find a space.

Despite these restrictions and the low absolute values of R&D in construction, an increase of the levels of construction R&D registered at the national level can be appreciated. This is clearly shown in Figure 8.4. In average, the R&D activities in the construction sector has grown at a rate of almost 39% during the last eight years.

Figure 8.4 Average annual increase of R&D in the construction sector
(adapted from CONICYT, 1998)

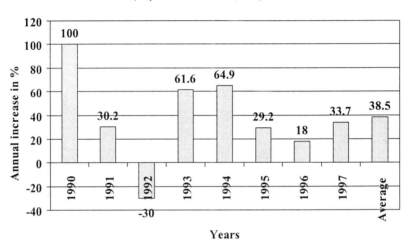

8.3   MAJOR PROGRAMS FOR SUPPORTING R&D AND INNOVATION IN THE CHILEAN CONSTRUCTION INDUSTRY

In Chile there is not a state policy or institution focused on Construction Development like the ones that can be found in other countries, particularly in Europe and Asia. Most of the construction R&D and innovation activities are found in private institutions and public ministries that required construction services, like the Ministry of Public Works and the Ministry of Housing and Urban Development. However the activities of these public organisations are very specific to their needs and do not take into account the construction sector in an integral approach.

Lately, in 1997 was created the Chilean Construction Institute. This organisation was formed through a combined effort of public and private institutions, leading universities and professional associations. The purpose of the Institute is to lead the development of the construction sector by focusing on the most critical issues affecting the construction industry and which require the integrated effort of all the key participants of the construction sector for their solution. The Institute receives public and private funding from its associates. This money will be used to develop applied research projects in the near future. At this time, the Institute is still in its initial forming years. Nevertheless, some minor research programs are under development and funding in the order of US$ 60,000 has been assigned for the year 2000.

8.3.1   General programs

One of the main public instruments to support research and innovation activities that have had a significant impact on research and technology development in Chile is the donation law that was put in operation in December 1987 (Law 18.681, 1987). This law covers donations to universities and professional institutions recognised by the State and can be applied to any field of research and technology development. The law has the purpose of supporting the development of the institutions included, but it considers within the allowed activities, the financing of research projects including equipment, furniture, infrastructure, publications and other research associated activities or resources. The tax effect of this law is to provide the donator a tax credit of 50% of the donation during the corresponding fiscal year.

As well as every other country, Chile also has a National System for Science and Technology. This system involves a network of institutions, organisations and funds that participate in a more or less organised form. This diversification is out of proportion compared to the size of the country and in the author's viewpoint it is an important weakness of the system. There is a high fragmentation of the system, which seems to be the result of the accumulation of many different initiatives pushed by the authorities of the moment instead of a definite policy with a clear purpose.

For this reason, there are many Chilean programs that promote R&D and Innovation. The National Council for Science and Technology (CONICYT) manages several of these programs. The most important of these programs is the National Fund for Science and Technology (FONDECYT). This fund was created in 1983 and it is now totally consolidated. The main purposes of this fund are as follows (FONDECYT, 1999):

- To create an endogenous scientific capacity
- to create new knowledge in science and technology
- to accumulate knowledge to address national problems
- to apply both national and foreign knowledge in the production of goods and services.

In 1997, this fund allocated US$ 37.1 million in currency of the same year to different kinds of research projects in science and technology.

Another program recently created is the National Fund for Scientific and Technology Development (FONDEF). The program started in 1991 with the goal of contributing to the competitiveness of the principle national economic sectors by strengthening the national scientific and technology capacity. This purpose is being achieved by developing strong links between R&D institutions and companies. The program has selected several priority economic areas for support. However, the construction sector has been absent of the selected areas, but can apply for this fund in an indirect way by presenting projects that are related in some way to the priority areas. The fund received US$ 65 million at its beginning. During 1997, US$ 19.2 million in currency of the same year were allocated to several projects. With these funds, the program has financed activities like:

- R&D projects
- scientific and technology infrastructure financing projects
- technology transference projects
- training and workshops for research project formulation.

An important and interesting characteristic of this program is the fact that it requires the project financing to be shared by private companies and R&D institutions. Since the starting of the program, the projects that have been realised have been financed by FONDEF in 45%, R&D institutions in 38% and private companies in 17%.

The Fund for Advanced Studies in Priority Areas (FONDAP) is an additional program recently created. This program was created in 1996 and is focused on the promotion of high productivity research groups. The construction sector is not considered a priority area for this program. No information is available about resources involved in this program.

The Chilean Economic Development Agency (CORFO) manages another group of R&D programs. This agency was founded in 1939 and is responsible for promoting productive activities in Chile. By encouraging competitiveness and investment, contributing to the generation of more and better jobs and equal opportunity for productive modernisation.

CORFO directs its activities to the following areas:

- Innovation and technological development
- modernisation of companies that form alliances to compete
- improvement of business management practices
- financing and development of financial instruments that meet business needs
- productive development in regions and emerging industries.

CORFO makes various financial instruments available to the business community, including long-term credits and co-financing. Co-financing partially covers the cost of business modernisation efforts. They require companies to make a growing contribution over time to ensure that the initiatives that are supported are of real use to the beneficiaries. Some of the most important programs included in this agency are:

- The Development and Innovation Fund (FDI): is granted to companies for activities like: the development and adaptation of new technologies; the spread and transfer of technologies to Chilean companies and institutions; the development of technological abilities necessary to generate and manage technological changes and for improving related markets to develop a national innovation system. There are no restrictions regarding priority areas allowing construction companies to apply for project financing in this case. During 1997, US$ 14.7 million were allocated by this fund.

- National Fund for Production and Technology Development (FONTEC): this fund has 5 lines of activities as follows:

  A. FONTEC - Line 1. This line supports technological innovation projects that consist of research and development of product, process, or service technologies, including models, prototypes and pilot plants for introduction to market.

  B. FONTEC - Line 2. Supports the financing of company technological infrastructure projects for increasing process productivity through technological support services related to production quality assurance. It finances projects for physical infrastructure, installations, scientific or technological equipment, as well as technical training for personnel associated with the infrastructure project.

  C. FONTEC – Line 3. It supports technological transfer projects for strategic alliances. Project goals can be for prospecting, diffusing, transferring or adapting management or production technologies in the allied companies in order to contribute to their productive modernisation.

  D. FONTEC – Line 4. It supports the financing of projects presented jointly by five or more unrelated companies to create Entities or Centres whose goal is the research, development, diffusion, transfer, and adaptation of technologies for company modernisation requirements. Among other things, it finances public works, construction, installation and layout of infrastructure, acquisition of scientific or technological capital goods, equipment for research and development, and expenses involved in linking technological supply and demand.

  E. FONTEC - Line 5. It supports the realisation of pre-investment studies designed to introduce technological innovations at a commercial or industrial scale for products, processes or organisations; or the materialisation of projects with high innovative content that are capable of generating significant economic impacts at a national or regional level.

These studies may be applied to investment projects that result from technological innovation projects previously co-financed by FONTEC, or to projects that haven't been co-financed by FONTEC, but meet the institution's eligibility requirements.

This fund allocated the amount of US$ 14.8 million in 1998 in currency of the same year for several projects.

- Management Support Program for Enterprises (PAG): the purpose of this program is to improve the competitiveness of productive companies, generating higher productivity and quality through consulting on processes.
- Technical Assistance Fund (FAT): through specialised consulting it supports the incorporation of operating management techniques into companies or new technologies to their productive processes that permit them to improve their competitiveness. There are two types of FAT:

    A.   Individual FAT: It is consulting work that one company requires for a specific management area. An intermediary operator must first conduct a diagnostic evaluation of the company before hiring a consultant.
    B.   Collective FAT: It is consulting work that at least three companies in the same industry or facing the same management issue require. An intermediary agent conducts a diagnostic evaluation to determine the pertinence of the project.

- Project Development in Teams (PROFO): this program is oriented to improve the operation of a group of companies that are willing to commit themselves to the materialisation of a shared project. This allows companies to resolve management and commercialisation problems that due to their nature or magnitude are better confronted jointly.
- Provider Development Program (PDP): the objective of this program is to increase competitiveness in production chains and facilitate the establishment of subcontracting relationships between a large company and its smaller sized providers, thus allowing mutually beneficial specialisation and complementation.

It is important to note that in 1998, CORFO awarded almost US$ 50 million in public co-financing and US$ 208.7 million in credits for the above and other programs.

8.3.2   General programs that address R&D and innovation in construction

There are several sources of funds that support R&D and innovation in construction. Information on these programs is scarce. Some of these programs are: Universities funds, R&D tax credits for companies that donate money for R&D to institutions that perform this type of activity, international research and development programs and environmental regulations. No reliable estimates are available about these activities.

8.3.3   Specific programs that address R&D and innovation in construction

One of the most important institutions related to construction in Chile is the Chilean Construction Chamber. This institution is a professional society that has created many organisations to support the construction sector. This institution owns the Corporation for Technology Development that carries out several activities in this regard allocating approximately US$ 1 million per year to these activities. Also the Chilean Construction Chamber itself is very active in R&D and innovation through several internal committees and work teams that are formed by volunteers with expertise in specific areas. However their contribution is limited in general.

Other programs that can be included in this category are the activities carried out by the Construction Industry Institute, which are still very limited at this time, and some potential international research projects that are in the evaluation stage.

Recently, the Ministry of Public Works has created a new innovation fund that will allocate close to US$ 0.5 million for research and innovation programs in the year 2000 dedicated to solve or improve areas that are the responsibility of this government branch.

There are also some other activities that have had indirect impacts on the construction sector, pushing renewed interest for R&D efforts. These activities are the issue of new regulations that affect the construction sector. The R&D and innovation efforts created by new regulations are due to the reactive approach of the construction sector in front of the new scenario. Some of these regulations are:

- Housing Quality Law
- BOT Law
- New Housing Leasing Law
- Environmental Impact Law
- New National Building Codes

Also, additional activities that indirectly affect the construction sector are as follows:

- Construction schools managed by the Chilean Construction Chamber that will prepare well-educated and trained middle level personnel that will be able to apply innovative work practices in the future
- The Construction Training Corporation that performs research related to human resources development in the construction sector
- the training tax credit allowing companies to deduct the expenditures in training from the annual tax statement.

8.4   CHANGE DRIVERS FOR INNOVATION IN THE CHILEAN CONSTRUCTION SECTOR

In general, the Chilean construction sector is very similar to construction sectors in many other countries in the sense that is more reactive than proactive regarding innovation. This is a real fact. However, the discourse can be very different from the real practice. For example, a questionnaire was recently applied to a group of construction managers about reasons that are pushing change in their construction companies. Some of the answers were as follows:

- To successfully compete with local and foreign companies
- to differentiate from competitors and to access specific market niches
- to be prepared to face after-crisis opportunities
- to answer to growing clients' requirements
- to seek other construction markets
- to improve personnel working conditions
- to improve capacity to face increasing complexity and technology requirement of future projects.

As seen from the answers, it seems that the attitude would be highly proactive. Nevertheless, in practice this reasons are more a discourse than real action and for many years, the Chilean construction sector has only reacted to the market forces. Only in very few cases a different approach has been seen.

8.5   BARRIERS TO R&D AND INNOVATION IN THE CHILEAN CONSTRUCTION SECTOR

Unfortunately, the Chilean construction industry faces several barriers to innovation as summarised in Figure 8.5.

Figure 8.5 Barriers to innovation in the Chilean construction industry

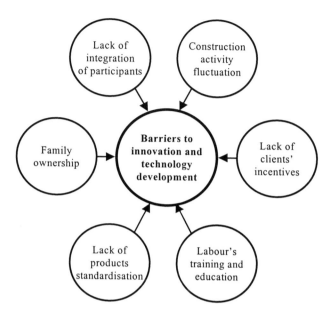

One of the first barriers found in Chilean construction companies is the fact that most of them are family-owned. Also, most of these owners are experienced construction professionals and since the creation of the company they have had an intensive participation in all its activities. For this reason, not enough importance is given to contracting personnel duly qualified for the principle functions within the company. The assignment of responsibilities and the delegation of authority are then extremely limited because of the lacking of confidence on available subordinates. This reality significantly restricts the innovation potential of these companies (Ofori, 1991). Also, the frequent owner-centred administration style of this sort of companies reduces the creativity and motivation of personnel for providing innovative ideas or developing new technologies.

Another barrier is the traditional fluctuation of the activity level of this sector. In practice, the fear to sector fluctuations is an important restraint for innovation.

The lack of standardisation of many Chilean construction products (materials and components) and the inadequate regulations also restrict innovation. Many innovative ideas are rejected due to the lack of suppliers and products capable of complying with the requirements that such initiatives establish.

The deficient qualification of labour is perhaps the principal "bottle neck" of innovation and technology development faced by the Chilean construction industry currently. Many innovation and technology development processes have failed or been limited in scope due to difficulties with labour's training and low education level. Unfortunately, neither the government nor the construction companies have addressed this important problem with effective solutions.

The lack of incentives provided by clients to innovate is also considered a barrier to innovation. It is quite frequent to observe clients that only value the price offered by contractors without considering factors associated with their performance. In spite of this, clients have begun to change, giving increasing importance to performance aspects. Then, more than a limitation this situation should be an incentive for innovation and technology development for construction companies since in this way they can increase their competitiveness.

Finally, the lack of integration among different actors that participate in a construction project is also a barrier. In many cases, innovation requires the participation and co-operation of several parties (e.g: constructability requires the participation of designers; just in time requires the participation and co-operation of suppliers). This is not easy to achieve in construction and reduces the scope of the initiatives that construction companies carry out.

8.6   TRENDS IN THE CHILEAN CONSTRUCTION SECTOR

Several trends can be observed in the Chilean construction sector regarding innovation. Hopefully these trends will have a real positive impact soon.

First, the construction industry might have good opportunities for innovation in the near future. The new government has declared a strong purpose of improving the National System for Science and Technology, by introducing a more integrated approach with a clear set of goals for innovation and technology development. In addition to this, there is an announced program to double the investment on Science and Technology in the next 6 years. There is optimism that this initiative will produce a most needed national plan for R&D and will provide opportunities to every economic sector of the country as well as the construction industry.

Second, the Chilean Construction Institute, although in its initial years have already shown its potential as an engine for innovation in the sector. It is expected that this institution will play a vital role for construction innovation in the long-term and for the first time, an integrated view will take place for construction development.

Third, there is a growing concern about reinforcing the idea of creating strong research centres dedicated to specific areas of research and innovation. New and existing programs have recently invited presentations of proposals for the creation of this type of institutions with the purpose of gathering a group of known researchers together with interested companies in an association that will address specific problems and research needs. The idea is that these centres will provide not only pure research, but more importantly, applied research validated through implementation activities in the real world.

Associated to the last point, it is expected that in the future the public interventions will be more focused to specific areas. These specific areas are those related to public concern, like housing quality, infrastructure development, environmental regulations, health and safety regulations and promotion of construction exports. There is a trend of using research and innovation to solve problems that have an important political component that would be valuable to the government in office.

It is also expected that Chilean Construction Chamber will become more actively involved research and innovation activities in the construction sector. The Chamber has proposed a plan for carrying out research efforts to develop technologies adequate to the reality and idiosyncrasy of the Chilean construction sector through a continuous work with research centres and universities. It considers issues like improving the management capacity of construction companies, the use of new contract forms, the development of more complex and sophisticated mechanism of financial engineering, the application of INTERNET capabilities and the creation of new businesses. However it should be noted that the Chamber has not been very effective in the achievement of these goals because of the many different visions that are found inside this institution.

8.7 CONCLUSIONS AND RESEARCH AGENDA

This chapter has briefly described the situation of research and innovation in construction in Chile and the public instruments available for this purpose. The current situation is clearly not positive and many changes need to be introduced in the future on this regard. However, the situation and the attitude of the most important actors related to the construction sector are receptive to these changes and so, there is a fertile ground to improve what the country is currently doing with respect to research and innovation in construction.

The main conclusion of the review presented in the previous sections is that public policy instruments for research and innovation in Chile have to date been very limited for construction. In general it can be said that there are not a clear concern about the importance of innovation for this sector and that the main public actions have been associated only with regulating the construction activity so to avoid negative effects for the society. Although an almost 39% of annual growth of R&D from the private sector as shown previously looks quite impressive, the fact is that the total amount for these activities is still very reduced if considered in absolute terms. Also, most of the private investment on R&D activities has been done by construction companies trying to adapt their organisations to the new conditions and not as a direct effect of public instruments.

There is a clear lack of strategy for the development of the construction sector in Chile. This fact is not particular to construction only. However in this case it is apparently that in addition, the construction sector is not considered a priority sector for the economic development of the country. Fortunately, opportunities do exist today to change this situation. Particularly important is the activity that can be realised by the private sector through the Chilean Construction Chamber and the Construction Institute. These institutions can become key actors for the innovation of the Chilean construction industry in the future, particularly the last one that has within its associates, the participation of the two government branches that are most related to the construction industry.

What is necessary to achieve a successful innovation activity in the future is the statement of a vision for the Chilean construction industry that can direct the future efforts to develop this important sector within an integrated plan. While this vision is not available, the many individual efforts carried out by universities, companies and other institution will remain ineffective to produce concrete changes and will be lost very soon. The increasing competition that is visualised today can become a very strong drive for innovation in the sector. In fact today much more concern is perceived everywhere in the construction industry given the growing participation of foreign companies that are reducing the participation of domestic construction companies. However, the problem here is that if action is not taken soon, it might be late in the future.

Many research areas are to be addressed soon regarding innovation in construction and the role of public policy instruments. Some of these areas are as follows:

- What is the current situation of innovation activities in the Chilean construction sector? As for today, there is not enough information about this situation in Chile.
- Barriers and drivers for innovation in the construction sector. Particularly important is to know the real importance that is assigned by the construction industry to research and innovation activities in the development of the sector.
- Mechanisms for innovation in the construction industry. How innovation can be carried out in an industry that is so fragmented? What kinds of co-operative schemes can be practical for construction? How can technology be transferred between different sectors and countries?
- Strategic planning for innovation. How to set goals, plans and activities for innovation? What innovation layers should exist and how to provide an integrated approach? What are the roles of the government, the national innovation system, the construction industry, the companies and the associated activities like education and others? How can these roles be institutionalised?
- Information about research and innovation. How can data and information be gathered and structured so that comparisons between national sectors and countries can be made?
- Implementation issues. How can the research and innovation results be successfully transferred into the construction industry?

CHAPTER 9

INNOVATION IN THE DANISH CONSTRUCTION SECTOR: THE ROLE OF PUBLIC POLICY INSTRUMENTS

Henrik L. Bang, Sten Bonke, and Lennie Clausen

9.1 THE OBJECTIVES OF THE STUDY AND THE ISSUE OF INNOVATION SYSTEMS

In this first section of the paper the objectives of the study relative to the issue of innovation systems in construction will be outlined. Subsequently, the issues of innovation and innovation systems conceptually are discussed. Finally, the outline of the paper is presented.

9.1.1 Objectives of the study

In this paper we are concerned with the issue of innovation systems in construction. The paper is written as a Danish response to the following thematic question raised in CIB Task Group 35 on innovation systems in construction[1]: *Which current public policy instruments facilitate innovation in construction?* Thus, in our analysis of innovation systems we focus in the following on public policy instruments, how they are used and to which degree they enhance innovativeness in the Danish construction sector.

One of our main arguments is that different public policy instruments may be superior at facilitating different kinds of innovations. Our effort is therefore aimed at identifying and classifying different kinds of public policy instruments which have been used in the Danish context and to evaluate their effectiveness with respect to promoting different kinds of innovative activities. Another main argument is that frequently several public policy instruments may be identified that simultaneously influence innovative developments. The degree, of which different instruments are co-ordinated or, conversely, are in conflict with other public policy instruments or with industry initiatives, is bound to influence the overall effectiveness with respect to specific innovative activities.

---

[1]   CIB Task Group 35 identified two other important questions: What is the role of knowledge brokers in the innovation process? How does the client-industry interface influence innovation? These two questions will not be our primary focus in this paper, however.

Table 9.1 Taxonomy of public policy policies and instruments, which have an effect
on construction innovation (adapted from Winch, 1999).

	Direct	**Indirect**
Construction Specific	Instruments explicitly aimed at construction firms (developing their innovative capabilities) or promotion of certain construction technologies.	Public policies, which have an incentive/disincentive effect on innovation (institutional framework guiding the behaviour of the firms in the construction industry – without explicit reference to innovation).
General	Instruments developed for a number of sectors, which are available to construction firms.	Public policies and governance structures directed towards the economy as a whole.

We concentrate on instruments, which are construction-specific and have either direct or indirect impact on innovation in the Danish construction sector (the top part of Table 9.1). An example of a construction-specific instrument with a direct effect on innovation may be a programme supporting new uses of IT on construction sites through making the equipment freely available for contractors. An instrument that is construction-specific with an indirect effect may for instance be a change in building regulations requiring more energy efficient windows in new buildings.

Our main focus will be on policies and instruments brought forward by the Danish Ministry of Housing and Urban Affairs[1]. Since its establishment just after the end of the Second World War the Ministry has played a pivotal role in promoting innovation in the Danish construction sector. Consequently, due to the established area of activities of the Ministry, our focus will be biased towards housing and building at the expense of civil engineering activities. Also activities initiated by the Ministry of Business and Industry and the Ministry of Energy and the Environment will be given some consideration, however.

The general, non-specific, instruments we do not discuss specifically in this paper. Since there is a separate Ministry aiming at promoting innovation in the construction sector it has been natural for us to focus on the more powerful, specific instruments. In general, however, innovation in Danish business firms is considered an important issue, which has been debated frequently in recent years. Especially the Ministry of

---

[1]   We thank Mr. Ib Steen Olsen, Chief of Advisory Division, Ministry of Housing and Urban Affairs, for useful comments during the process of writing this paper.

Business and Industry has been active in initiating investigations and analyses through the Agency for Development of Business and Industry. We refer interested readers to the numerous reports published by the Agency (e.g. Lundvall, 1999, a summary report concerning strength and weaknesses in the Danish innovation system).

9.1.2 The concept of innovation

Initially, during the establishment of CIB Task Group 35, it was decided that the following common definition of the concept of innovation should be employed:

Innovation is the successful exploitation of new ideas.

This definition reflects the view taken by many authors that it is not so much the invention that is important as its practical development for commercial use by business firms which means that it must be both practically useful and economical. The innovation process pertains to '… the work after the invention but before the wider diffusion' (Sundbo, 1995, p. 19).

Following the above definition it is natural that the situation of construction firms take central place in our analysis. This is in stark contrast to the prevailing view in earlier decades, namely, that investments in research and development (R&D) automatically would lead to increased levels of innovativeness (the so-called 'linear model'). There is a growing understanding that the macroeconomic indicators of investments in R&D are not so important when getting to terms with a nation's innovative ability. Rather, getting to understand the strategic capabilities of firms, their ability to interpret markets and how to organise the innovation process are much more pressing issues.

Since the 1980s many governments have been active in promoting innovation. The general trend is from a narrow focus on stimulating R&D expenditure to a much broader spectrum of programmes using a wide variety of public policy instruments The conviction driving government involvement is apparently that innovation leads to improved productivity, increased competitiveness and speeds up economic growth.

9.1.3 The concept of innovation systems

In this section we attempt to justify the idea that innovations – in particular construction innovations – are generated not only by individuals, organisations and institutions, but also to a great extent by their complex patterns of interactions. Thus, organisational and institutional configurations are important determinants of innovation activity. This line of reasoning follows from the developments that have called the simple cause-effect relationship between R&D spending and innovation into question and which has also challenged the notion that consideration of national systems of knowledge diffusion is sufficient (ibid., p. 73).

In recent years there have been numerous attempts at defining the concept of (National) Systems of Innovation (e.g. Lundvall, 1992, and Edquist, 1997) and related concepts such Technological Systems (Carlsson, 1995). On this basis – and since we focus specifically on construction as an economic subsector of the national economy – we shall define a national sectorial system of innovation for construction as:

> *A network of interacting organisations in the built environment involved in the generation, diffusion and utilisation of new technologies in a specific country and under a particular institutional infrastructure.*

In this definition we have made a distinction between institutions and organisations. *Institutions* can be regarded as 'things that pattern behaviour' or 'guide-posts for action', i.e. norms, rules, routines, laws, etc., while *organisations* are 'formal structures with an explicit purpose' (Edquist, 1997: 26).

Organisations include (1) business firms; (2) universities and similar organisations providing basic and applied research and related training; (3) Private and public organisations providing general education and vocational training; (4) 'Knowledge brokers', providing information; (5) Financial organisations; (6) government, financing and performing a variety of activities that regulate and promote innovation (see Figure 9.1). According to this listing of organisations our main emphasis will be on the business firms, where innovations take place, and on government activities, which also include regulation and maintaining various institutions. As already mentioned, the Ministry of Housing will be our main focus when considering the activities of government with relation to construction innovation systems.

When considering what makes up the infrastructure in construction some other types of organisations such as trade unions, trade associations and professional bodies, can be identified in addition to those already mentioned. Carlsson emphasises that organisations should not be seen as distinct entities, however. Rather, an institutional set-up should be characterised by its linkages. What should be considered, as already observed, are the networks of actors that generate, diffuses and utilise knowledge. We should thus concentrate on flows of knowledge and competence in the system. These flows can under conditions of entrepreneur-ship and sufficient critical mass translate into synergistic clusters of firms and technologies sometimes referred to as 'development blocks' (Carlsson, 1995, p. 7) – a kind of dynamic specialisation within specific sectors based on knowledge sharing. The question is whether Danish construction can be considered a development block or, alternatively, what changes are necessary for realising its potential as one. In the following sections we attempt discussing in general terms the role of the public sector as a policy maker in the national systems of innovation for the Danish construction sector.

Figure 9.1 Innovation in the Danish Construction Sector

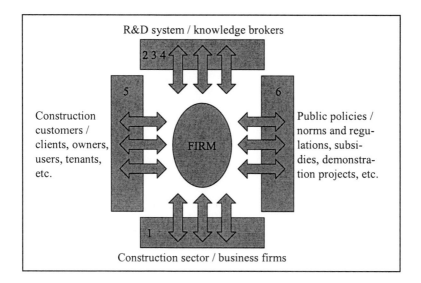

R&D system / knowledge brokers

Construction customers / clients, owners, users, tenants, etc.

FIRM

Public policies / norms and regulations, subsidies, demonstration projects, etc.

Construction sector / business firms

9.1.4   Outline of paper

This introductory section is followed by more detailed discussions (in section 9.2) of the context for Danish construction innovation and presentations of Danish government instruments and their effects in a historical context.

Section 9.2 begins with a short general introduction to the Danish society, followed by background information and key figures of the Danish construction industry. Moreover, section 9.2 is presenting an overview of the historical developments in construction since the Second World War concluding with a consideration of current societal changes and policy challenges. Section 9.2 is round off with a presentation of key characteristics of the Danish R&D infrastructure with a special emphasis on the system of production and distribution of knowledge (the R&D system).

Then, in section 9.3 we return to the role of public policies with special emphasis on classifying and describing the specific public policy instruments used in the last 10-15 years, whereas in section 9.4 we briefly evaluate their effect on construction innovation. Section 9.5 contains an overview of some of the most prominent current trends in public policy strategies regarding innovation in the Danish construction industry, and at last in section 9.6 we sum up our main findings and recommendations.

9.2   CONTEXT OF INNOVATION

In this section we provide some general information on the Danish society. In addition to the more factual information we describe some of the most important contextual preconditions for the Danish public policies. Main innovations in construction are identified in a short historical review of the post-war period in order to establish a background for the discussion of current agendas in public policies and the future development of the sector.

9.2.1   General context[1]

Denmark is a small country situated in Northern Europe. It has a population of 5.2 Mio. Denmark is a constitutional monarchy and a member of the EU, NATO and UN.

The neighbouring countries of Denmark are Norway, Sweden, Poland and Germany. Denmark consists of one large peninsula and a number of smaller islands (see Figure 9.2). Thus, only the border to Germany is on land (south of Jutland). Denmark is surrounded by the North Sea, Skagerak and Kattegat and the Baltic sea. The geographical area of Denmark is 43,100-sq. km. and the coastline is 7,300 km. In general, distances are short in Denmark and the country is highly urbanised.

Denmark is quite cold in winter, whereas summers are mild. Average temperature in January is 0 degrees Celsius. In July the average temperature is 16 degrees Celsius.

Figure 9.2 Map of Denmark

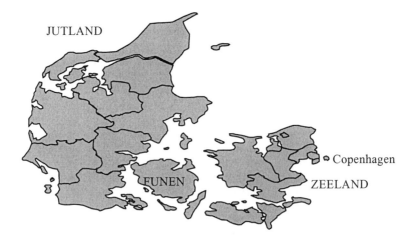

---

[1]   Most of the factual information presented here is from the Danish Energy Agency (1998).

The Danish currency is 1 krone (DKK) = 100 øre. The exchange rate is 1 USD = 8.26 kroner (DKK).

The distribution of employment in Denmark reflects the fact that the economy is highly developed with a large public sector due to its 'welfare state' orientation:

Employment in agriculture and fishery	1.9%
Employment in manufacturing	18.4%
Employment in market services	34.9%
Employment in non-market services	34.8%

9.2.2 Construction industry context

The construction sector is one of the largest in the Danish economy, but also one which is frequently referred to as problematic due to low productivity, poor quality and lack of consumer responsiveness. Increasingly, the sector and the problems associated with it are attracting the attention of both media and politicians. It is open for discussion how severe the productivity problem is for building construction. For instance during the last decade the average Danish household has increased its spending on housing (rent, mortgage, etc.) by several percentage points. Spending increased from 17 to 22% in the 1976-87 period and in 1997/98 it was around 20% – measured as a proportion of total spending (Danmarks Statistik, 1996, 1997, 1998).

Similarly, services in most countries account for increasing shares of the total value of production (and consumption) despite the poor productivity performance of services when compared to manufacturing. Thus, the willingness to pay increasing shares of total spending may be an indication that in wealthy nations housing is increasingly becoming a luxury good associated with consumer lifestyles and leisure occupations. Nevertheless, firms, which succeed in improving productivity, should be able to gain in competitiveness. In 1995 the total volume of the Danish market for construction was DKK 106.2 billion with 168.800 persons working in the Danish construction sector, that is, 6.7% of the total working population, or more than three times the employment in agriculture and fishery. Construction activity contributed some DKK 42.7 billion (5.6%) to the gross domestic product.

The Danish market for construction is traditionally divided into three main segments: (1) Housing, (2) other building and (3) civil engineering. In 1995 investment in new housing was approximately DKK 31 billion, in new buildings other than housing about DKK 23 billion, and in civil engineering some DKK 24 billion (See Figure 9.3). Maintenance and repair activities (in building) constitute the remaining DKK 28 billion. Especially in recent years, the proportion of maintenance and repair has been growing significantly.

Figure 9.3 Investment in new construction in Denmark 1995. The three main segments are housing, other building (commercial and public etc.) and civil engineering. The figure for civil engineering activities (DKK 24.4 billion) includes maintenance & repair, whereas figures for housing (DKK 31.0 billion) and other building (DKK 23.2 billion) include only new construction.

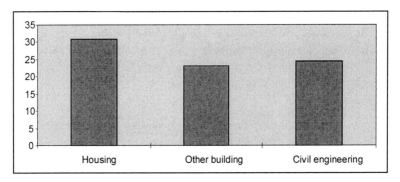

In 1994 construction activities involved approximately 31,300 firms. Almost 5,400 were contracting firms, with a total turnover of DKK 42.4 billion. Only around 100 of these contracting firms had turnovers larger than DKK 100 million, and less than twenty had turnovers of more than DKK 500 million. Evidently, general contracting firms are by far the largest category with respect to total turnover. Their size undoubtedly reflects extensive use of other categories of firms as subcontractors. Joinery is the largest category with respect to number of firms. An example of a category with relatively few firms and a relatively high total turnover is electrical contractors. Other groups of firms in construction are in the trades of bricklaying, electrical contracting, plumbing, joinery, and painting/glazing or in other construction. Of these trades joinery is the most significant group with around 7,300 firms and a total turnover of DKK 14.3 billion.

When taking a broader view, including the whole value chain contributing to bringing about the services buildings deliver (but excluding civil engineering) the 'building/housing resource area' is among the largest with respect to both employment and value added – 'foodstuffs' being the other main 'resource area' (Danmarks Statistik, 1996). Around 381,000 are involved in the building/housing area, contributing DKK 122 billion of value added. Employment in the 'building/housing resource area' (construction industry in its broadest sense) is some 15% of total employment – almost of the same size as the entire manufacturing sector of the economy.

With respect to exports and foreign trade, the building/housing area contributed DKK 51 billion, most of which accounted for by building materials and components and consulting services.[4] However, general contracting firms' export value amounted to DKK 3.9 billion, or approximately 8% of building/housing area total exports (1994 figures – Danmarks Statistik, 1996).

In addition to the firms directly engaged in the execution of construction activities on construction sites a number of other types of firms

---

[4]   In 1995 45% of the export value represented building materials and components - and 23% consulting services (Boligministeriet 1997).

are active in the construction process. Following value chain thinking we can include in our consideration all firms that are involved in the process of bringing about construction services to users. For the building/housing area the following kinds of activities are included (Danmarks Statistik, 1996):

- Extraction of raw materials, aggregates etc.
- manufacturing of building materials and components
- building materials trading and wholesale
- general building contracting
- specialist and trade contracting
- building consulting
- other activities (cleaning, support activities and building administration).

For each of these activities certain types of firms can be identified. Some basic information on each group of firms is provided in Table 9.2 below (the firms in the figure follow the wider 'building/housing resource area' definition, whereby more firms are included than in construction in the traditional sense).

Table 9.2 Number of firms, employment (no. of persons), turnover and value added, or contribution to GDP, (in million DKK) for the population of each type of firms constituting the building/housing resource area. The total number of firms in the 'building/housing resource area' is approximately 62,000 (1994 figures).

	Number of firms	Employment	Turnover	Value added
Extraction	3,016	4,757	2,838	-
Manufacturing	3,783	85,999	68,811	33,402
Trading	6,908	47,303	85,616	21,764
General contracting	4,521	49,445	40,224	14,699
Specialist contracting	23,836	112,550	53,010	26,655
Consulting	11,702	37,818	23,329	17,161
Other	7,959	21,465	10,730	7,933
Total	*61,725*	*381,266*	*284,558*	*121,615*

A total of 4.9 million m^2 of buildings was completed in Denmark in 1995. Housing counted for approximately 1.4 million m^2 (27.8%), commercial building 2.7 million m^2 (54.8%) and other kinds of building 0.9 million m^2 (17.4%). Commercial building can be broken further down, cf. Figure 9.4. Of the total housing area completed in 1995, 73.7% (1,014,000 m^2) were by private clients, 17.7% (243,000 m^2) by housing societies and 8.6% (118,000 m^2) by public clients. In 1995 around 5,000 new buildings (encompassing some 12,000 dwellings) were completed. The number of buildings completed at the same time indicates the total number of clients engaged in new building in one year.

Figure 9.4 Main types of building construction completed in 1995 (1000 m^2). The three most important categories of commercial buildings are farm buildings (1,085,000 m^2), industrial buildings (778,000 m^2) and office buildings (643,000). The category of 'other buildings' encompasses buildings for cultural and institutional use, weekend cabins and other kinds of buildings.

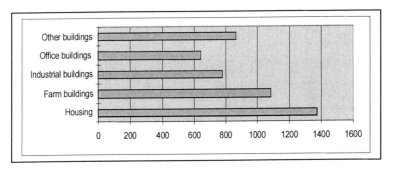

The existing stock of buildings (and other constructed facilities) in Denmark is considerable. Actually, the proportion of new buildings produced in one year to the stock of existing buildings is 4.1% measured by value and only 0.8 measured by m^2. The total value of real property is estimated at DKK 1,750 billion, of which land value is DKK 430 billion. Total building value is thus DKK 1,320 bill or 1.6 times total Danish GDP. Total building stock in Denmark is about 624 million m^2. Around 315 million m^2 (50.5%) are housing buildings, 253 million m^2 (40.5%) are commercial buildings and 56 million m^2 (9.0%) are classified as 'other buildings'. The total number of housing dwellings is about 2.3-2.4 million of which approximately 41% are single family houses, 39% are multi-level buildings (flats/condominiums) and 13% are multi-family single-level buildings (semi-detached etc.)

More than half of today's total housing stock has been constructed after 1960 (Bonke & Levring, 1996: 2). The main share of these buildings was constructed in the 1960s and 1970s. In two decades 824.000 housing units were constructed, that is, an average of more than 41,000 units per year (see Figure 9.5).

Figure 9.5 Average number of housing units being constructed per year for each time period (in 1000s). The period covered is 1900-94 with 1900-19 as the period with the lowest level of annual housing construction (14,000 units) and 1970-the highest (52,000 units).

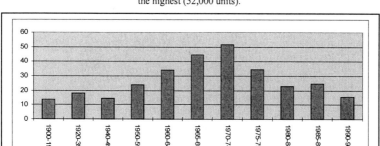

Other points to be made about the Danish construction industry are for example the high level of union membership and the extensive regulation and subsidies of the entire housing sector – both public and private (through the tax policy). Furthermore, the public sector housing has traditionally been the primary driver of innovation in Danish construction – with the Ministry of Housing at the steering wheel. Technologically, the modular concrete element technology constitutes a dominant trajectory in the Danish built environment since the (first) industrialisation process took off for real in the 1950s and 1960s.

9.2.3   The historical context for public policy strategies

As earlier mentioned, Denmark belongs to the so-called Scandinavian 'welfare states'. These are constituted on a Lutheran alliance between state, church and society that originated in the 16th century Reformation (which was a state-driven project). Seen in this framework modern time social democratic welfare politics could be redefined as products of 'secularised Lutheranism'. In this perspective it can be argued (Selle, 1996) that the fusion of state, church and society has been instrumental in creation of a specific Nordic 'organisation society', permeated by social organisation in which explicitly formulated 'democratic' rules and obligations regulate peoples' lives. The state, although being the highest level in this process of organisation, is merely a 'membership association' guaranteeing equal treatment and fulfilment of its members' needs. Providing for the needs for housing is obviously of some importance as well in welfare states such as the Danish one.

The establishment of the Danish Ministry of Housing right after the Second World War (1947) - and in particular the Ministry's initial objectives – can very well be described within the context of the 'welfare state' discussion. The need for dwellings was urgent due to high birth rates, urbanisation and of course low production outputs during a long period of wartime resource shortages. The social democratic party in charge considered the situation a threat against its emerging notion of the welfare state (Bertelsen, 1997). And the building industry itself, like in most other countries, was dominated by craft-based companies and methods and therefore incapable of upgrading its products radically in terms of quantity and quality. Thus the young Ministry could define its assignment as a project of innovation and restructuring, encompassed by the term 'industrialisation'[5]

9.2.3.1   Instruments of the industrialisation process

The basic idea of the industrialisation concept was to transfer a major part of the work, traditionally done on site by craftsmen, into factories in order to change it towards a mechanised production process.   This strategy was politically and socially acceptable because it could increase the output from the construction industry in accordance with the demands from society and at the same time activate the large reserve of unemployed, unskilled labour (Bonke and Levring, 1996:   1).   In 1953 the Ministry of Housing implemented the instrumental launch of this development rather dramatically.   In a departmental order it was required that no more than 15% of labour time for the production of a basic building structure should be consumed by skilled labour, if government financial support for housing was to be obtained.   The Ministry did not specify how this target was to be achieved by the actors of the construction process – only the accumulated result was defined.

By setting such financial restrictions the Ministry paved the way for new technologies and actors entering the construction sector.   From traditionally being dominated by architects and craftsmen as key actors in a site oriented production process it now became obvious that contractors and consulting engineers gained importance in connection with the introduction of reinforced concrete as the new basic building material.   The use of concrete reduced the need for energy, which had been necessary for the traditional production of bricks, *and* at the same time eliminated the bottleneck represented by the scarcity of skilled bricklayers.   The large contractors were in an advantageous position to exploit this situation, having accumulated substantial stocks of machinery and capital in connection with civil works during wartime

The next major step forward towards industrialisation was taken in another legislative action – through the issue of a departmental order for the assembly of prefabricated building in 1960 [*Montagecirkulæret*].   This legislation defined the entire building process as being part of an industrial production system in order to make further reductions in the consumption of labour and materials, and it differed from previous attempts at industrialisation in two distinct ways:

1.   Planning and design were now conceptualised as an integral part of the industrialised construction process

---

[5]   The early policymakers in the Building Ministry obviously operated upon a conception of industrialisation, which was inspired by the success of new mass-producing consumer oriented industries within plastics, electronics and mechanics. However, they could also implement into their policy-modelling the ideals of functionalism, developed within the *Bauhaus* school, which in the inter-war period were demonstrated in practice by outstanding Danish designers and architects.

2.   The building industry was guaranteed substantial long-term markets in non-profit housing when working within the technological provisions of this legislation (so-called 'big plans' of multi-storey buildings).

During preparations for *Montagecirkulæret* the Ministry had earlier introduced certain practices for the design of buildings. These practices included the 'modulus-grid system', aiming for a standardisation of dimensions of components for buildings. This system is presumably considered the most crucial precondition for the production of building materials on a large-scale industrial basis. And indeed for the development of the most important national characteristic of Danish construction technology – the pre-cast concrete element building.[6]

From the 1960s concrete element building systems formed the technological core of the Danish construction sector. As already indicated the structural contractor was in a key position of technological development, thus, the contractor could exploit the organisational consequences for the building process, for instance by taking over the responsibility for crucial parts of planning, design and construction. Historically these companies were used to working with concrete from activities in civil engineering. They employed a large proportion of unskilled labourers but also the necessary white-collar (engineering) staff for planning purposes. And they had access to the necessary equipment for the assembly of the larger concrete panels. Finally the larger contracting companies had a financial background making it possible for them to diversify into production of pre-cast concrete panels, and a number of pre-cast concrete factories were established by these companies during the early 1960s.

Moreover, the 'component concept' of the Danish innovation policy radically influenced also other areas of building production and crafts. Traditionally, each building part, be it for instance a window, door or cupboard, had hitherto been produced individually by joiners in workshops or on site according to specification. Now, however, it became possible due to the modulus-grid standard sizes to exploit economies of scale by applying factory-based methods and machinery to the production of building components.

Not surprisingly, these developments created technical problems in the interfaces between the different building components/materials/trades, by some characterised as the very basic challenge for a prefabrication-oriented innovation policy. And indeed, much effort since then has been allocated to establishing certain standards and guidelines for connections and joints between various components and materials – as well as to assessing the effects of the combinations in question.

But it stands beyond any doubt that the initial measures of technical co-ordination provided by the legislative framework led to the foundation of a remarkable strong building component industry in Denmark, giving many Danish producers a relative advantage over competitors in other countries.

---

[6]   A few, well-known consultants and contractors played a very crucial role in developing and testing these new design practises, in particular Malmstrøm a/s, Jespersen & Søn a/s and Larsen & Nielsen a/s. Today, only Malmstrøm remains.

Positive figures in import-export relationships indicate the continuous positive effects of the technology policy.

9.2.3.2   *Innovation and productivity*

The consequences of the comprehensive state policy of innovation were manifold. Concerning work productivity the effects were impressive. The amount of manpower spent producing one 'standard' dwelling unit reduced sharply – from 1950 to 1980 by approximately 50%. These developments in fact enabled the building industry to supply the quantity of housing originally aimed for in political and social objectives underlying the innovation strategy.

The time spent on site reduced by approximately 65% while white collars' office work more than doubled, obviously due to increased need of planning. The increased prefabrication also led to an absolute increase in time consumption in factories. However this increase was relatively minor mainly due to the fact that standard sizes for building components had been introduced, thus promoting investments in effective automation technology.

The developments in labour productivity (time consumption for production of one standard flat of 80 m²) from 1950 to 1980 is illustrated in the figure below.

Figure 9.6 Time consumption producing one standard flat, 1950 and 1980

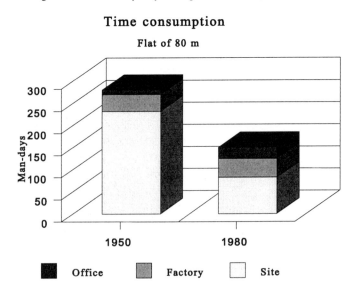

Time consumption

Flat of 80 m

9.2.3.3 Planning legislation and other indirect innovation instruments

In this brief historical context could also be mentioned that the Danish (social democratic) government as a means to boost the private investments in buildings during the 1960s introduced very attractive tax rules for depreciation of buildings. Also tax deduction for mortgage interest in combination with the rather high inflation rates of the 1960s made it extremely lucrative not least for families to buy single family houses. A single-family housing production industry virtually came into existence during that decade.

Also the legislative basis for planning and approval of new buildings was changed during this period. The perhaps most signification change was represented by the National Building Regulation [*Landsbyggeloven*] which from 1960 unified design conditions for construction all over the country.

The physical planning also plays an important role as a precondition to the long period of growth in construction output. In the Copenhagen region for instance this planning started during the early 1940s. The developments were planned in order to provide a sound environment giving tenants in heavily populated areas easy access to open space and green belts in conjunction with the living areas. In 1947 the so-called 'Finger Plan' [*Fingerplanen*] was conceptualised, very much consistent with functionalistic ideals. The plan defined guidelines for the localisation of new construction for housing and industry and inserted wedges of green areas into the very centres of residential areas.

The emphasis on the social well being of the tenants was obviously considered by the planners. The approach aimed at reducing potential social and environmental problems resulting from increased density of population in existing residential areas. Similar plans were developed for other larger towns all over Denmark.

From the beginning the idea of the industrialised building technology was associated with high-rise multi storey buildings of 15 - 20 floors. This approach was completely new compared to the traditional Danish way of building. Earlier building structures were based on traditional materials like bricks and mortar, which limited the height of the buildings to a maximum of 6-8 storeys. The utilisation of reinforced concrete allowed far taller buildings allowing the architects to experiment with designs which had not been possible before.

Despite the planning intentions – and indeed many successful estates of good functionalistic architecture and well-functioning social structure - such high rise-areas have occasionally developed into problematic housing environments with an unbalanced proportion of socially challenged people.

It can be considered a reaction to such social-political problems that some of the 'traditional qualities' of the Danish way of housing were restored through the developing of the industrialised semi-detached house. This model was implemented from the early 1970s and was soon widely adapted by the non-profit housing sector in Denmark.

Through all the developments of the different concepts for dwellings the idea of prefabrication has maintained its importance. Though looking quite individual the semi-detached low-house in Denmark consists of mainly standard size components put together in accordance with the overall modulus-grid system. The number of units in the individual housing project has been reduced drastically, hence, 40-50 units per project has become the norm.

9.2.3.4   *Regulating energy consumption – raising innovation by indirect measures*

Due to the increase in oil prices following the first oil-crisis in 1973 strong political focus was put on energy savings in buildings. For example, new standards for insulation were set leading to technological changes in the construction of exterior walls and windows. The cavity wall with 150 MM. insulation material or more was introduced and double-glazing became standard as well.

Historically the U-values for different parts of the building have developed in the following way according to revisions of the Building Regulations (see Table 9.3).

Table 9.3 Energy lost per square meter for building parts. All values are in W/m^{2}K

	1972	1977	1982	1995
Light exterior wall	0.60	0.30	0.30	0.20
Heavy exterior wall	1.00	0.40	0.40	0.30
Floor	0.45	0.30	0.30	0.20
Roof	0.45	0.20	0.20	0.15
Window	3.60	2.90	2.90	1.80

The most recent revision of the Danish Building Regulations stipulates 200 mm of insulation in the walls and 3-layer glazing as standard. As a result of these ongoing developments the loss of energy from buildings have reduced significantly. For a 3-storey building with a basement the loss has been reduced over the years as illustrated below.

Looking back on 25 years of intensive and impressive energy saving innovations a particular consequence has to be mentioned. The increased amounts of insulation in walls and roofs as well as the extensive use of air-tight sealing materials have radically altered the humidity balance in the constructions. In many instances, the indoor-climate in the buildings deteriorated as a result. A detailed understanding of humidity transport mechanisms has therefore become essential, just as the ventilation and air-change are now much more important design and management aspects of any building.

Figure 9.7 Energy loss from 3-storey building

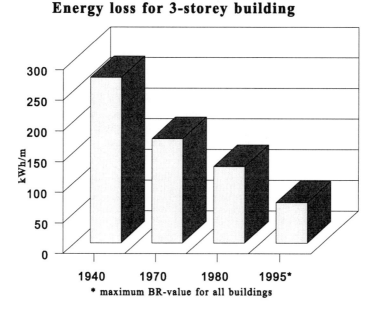

9.2.3.5   *Concluding remarks*

This sporadic short-story of the post-war historical innovation context in Danish construction takes its point of departure in the political framework of the Scandinavian societal model, where scopes for actions are fixed through political and social negotiations. The state initiated innovation policies, which soon led to the emergence of industrialised productions systems in construction. It is also a story showing the efficiency of the systemic alliance between institutional and industrial actors in achieving defined goals. Such developments, however, often tend to loose their impressive momentum after some time unless constantly refuelled by new authoritative goal setting. In the next sections we will be looking further into the more recent public innovation policies and their instruments aiming at refining the concepts of industrialised construction technology.

9.2.4   Innovation and the Danish construction R&D system

Reports on Danish building sector R&D indicate that about 200 organisations (institutes, institutions, councils, associations, etc.) are involved in various aspects of building R&D.[2] From the reports we estimate that building R&D activities currently amount to approximately DKK 400-500 million annually (which is less than 0.1% of total production value for building)[7]. A special network (called SOFUS-BYG) was established in 1971 to promote co-operation and co-ordination among the primary actors among building R&D organisations. SOFUS-BYG counts a number of the most important building research organisations in Denmark (see Table 9.5).

Among the main producers of new knowledge are different kinds of private firms, universities, a sectoral research institute and several (approved) technological institutes. It was estimated by FRI (1990) that research institutes (sectoral and technological) spent about 57% of their turnover (DKK 130 million); consultants about 3% (DKK 95 million); general contractors 0.2% (DKK 50 million); and manufacturers of components and materials 0.5% (DKK 150 million) on building R&D (see Figure 9.8).

Figure 9.8 Spending on R&D by research institutes and private firms. Components and materials producers are the biggest spenders (35%) closely followed by the research institutes. Then follows the consultants (primarily consulting engineers) with 22% and general contractors with 12 % . Total spending for these four groups was DKK 425 million (FRI, 1990).

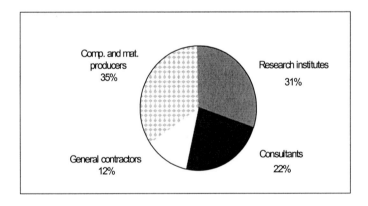

The FRI survey also showed how research institutes and firms financed their building R&D with respect to funds generated internally and externally from private firms and from public sources (1990, pp. 26-27) – see Table 9.4.

Table 9.4 Funding of R&D activities for research institutes and firms.
Firms rely mainly on internal funds - research institutes on public (and private) sources

	Internal	**Private**	**Public**
Research institutes	15%	32%	53%
Firms	80%	9%	11%

Table 9.5 The research organisations participating in the SOFUS-BYG network. SOFUS-BYG (network on research, development and services for the building sector) brings together a number of public and other (non-profit) organisations to ensure that their total technological knowledge is accessible for the Danish building sector. The 24 organisations can be divided into three main groups: University research, sectoral research institutes and technological service institutes. One of the technological service institutes, the Geo-technical Institute, is not an approved institute, and does not receive subsidies. The Danish Building Centre is not a research organisation, but is active in distribution of research based knowledge.

Group	**Organisation**
1. University Research:	Aalborg University:
	• Institute of Building Technology
	Aarhus School of Architecture
	Copenhagen Academy of Arts - School of Architecture:
	• Institute of Building Technology
	Technical University of Denmark:
	• Institute for Structures and Materials
	• Institute for Applied Building and Environmental Technology
	• Institute for Buildings and Energy
	• Institute for Acoustical Technology
	• Institute for Energy Technology.
2. Sectoral research institutes:	Danish Building Research Institute:
	Materials & Structures
	Energy & Indoor Climate
	Town & Building Planning.

Table 9.5 (Continued) The research organisations participating in the SOFUS-BYG network.
SOFUS-BYG (network on research, development and services for the building sector)
brings together a number of public and other (non-profit) organisations to ensure that
their total technological knowledge is accessible for the Danish building sector. The 24
organisations can be divided into three main groups: University research, sectoral research
institutes and technological service institutes. One of the technological service institutes,
the Geo-technical Institute, is not an approved institute, and does not receive subsidies.
The Danish Building Centre is not a research organisation, but is active in distribution of
research based knowledge.

Group	Organisation
3. Technological service institutes:	Danish Technological Institute:
	BPS Centre (Building Planning System)
	Productivity Centre
	Building Component Centre
	Brickwork and Masonry Centre
	Energy
	Wood technology
	DELTA:
	Acoustics and Vibrations - Lyngby Department
	Acoustics and Vibrations - Aarhus Department
	FORCE Institute:
	• Instrument and Sensor Technology
	• Materials and Maintenance
	DK-Teknik
	Danish Geo-technical Institute
	Danish Maritime Institute (Department of Wind Technology)
	Danish Institute of Fire Technology
Other organisations:	Danish Building Centre

9.3   AN OVERVIEW OF PUBLIC POLICY INSTRUMENTS

Using the classification of public policy instruments presented in the
introduction as a starting point the following overview will concentrate on
construction specific policy instruments mentioned in Table 9.6.

Table 9.6 Examples of direct and indirect policy instruments administered by the Danish Ministry of Housing and Urban Affairs, which affect the innovative behaviour of construction firms

	Direct	Indirect
Construction Specific	Demonstration Projects Programme Development Programmes • Process and Product Development in the Building Industry • The Urban Renewal Project • ECO HOUSE '99	Technical regulations, which affect the constructed product • The Building Regulations Procurement regulations, which affect the construction process • Procurement and tender regulations The Quality Assurance and Liability Reform

9.3.1   Direct construction specific policy instruments

This category of public policy instruments has been dominated by two types of innovation programmes, both run by the Ministry of Housing and Urban Affairs: The Demonstration Projects Programme, and a series of development programmes targeting specific technological themes. Basically, demonstration projects should be viewed more as a generic type of instrument than as a development programmes as such, since demonstration projects often form part of development programmes.

Development programmes are generally top down and politically defined (mission driven) whereas demonstration projects are more bottom up with ideas coming from industry entrepreneurs. Development programmes may have some advantages over demonstration projects in terms of continuity and learning over multiple projects and awareness creation. Using a biological analogy we may say that the use of demonstration projects is strengthening the 'natural' mechanism of mutations in the sector. Employing development programmes on the other hand is potentially a more drastic way, which may either be regarded as strengthening the forces of 'natural' selection or as plain transplantation through surgery (with inherent risks of rejection!).

9.3.1.1   The Demonstration Projects Programme

This is a continuous arrangement first established in 1977 and governed by the Danish Ministry of Housing and Urban Affairs. Demonstration projects are development projects, in which new ideas are tested and demonstrated in affiliation to state subsidised non-profit housing projects. The housing projects receive additional funding from the Ministry as to cover the additional development and evaluation activities, but the principal rule is that the firms themselves should finance the main part of the development costs. Demonstration projects receive exemption from procurement regulations, but apart from that they normally have to perform to the same level of construction costs, standard, and legislation as other non-profit housing projects. The main results are commonly made public via reports. The Demonstration projects programme can be seen as a 'brokering institution' (or facilitating) offered by the regulators of the industry.

In the majority of cases the project coalitions split up after completion of the demonstration projects. However, in recent years a still larger proportion of the demonstration projects has been allocated to specific development programmes (e.g. the development programmes presented below). At the present time, all major development programmes involves demonstration projects either as a development tool or as a diffusion mechanism.

Demonstration projects are supposed to meet some more or less explicitly stated requirements: They should involve the 'system integrators' of the building industry, i.e. the project-based firms such as architects, contractors and consulting engineers. Focus is on 'systemic' innovations and the innovation idea has to be of general relevance to the development of the built environment. Thematically, the demonstration projects in the 1980s mainly focused on the development of new building systems and product innovations, but in the 1990s there has been an increasingly interest in organisational developments and process innovations, for example in the form of building logistics projects and exploring partnering arrangements.

9.3.1.2   Development programmes

The Danish Ministry of Housing and Urban Affairs is actively promoting technological change by setting up research and development programmes targeting specific areas of the built environment. These programmes usually follow the same recipe: The Ministry defines the scope and objectives of the programmes; sets the requirements regarding evaluation, documentation and dissemination of results; and presents the overall themes – usually based on recommendations from expert committees involving industry representatives. It is the construction firms, though, that generates the concrete proposals and innovation ideas.

Traditionally a development programme includes a competition among invited (pre-qualified) groups of firms organised in temporary project organisations for the programme period. The winners then carry out development activities including demonstration projects based on their innovation ideas and conceptual designs. The first of these development programmes was the programme for development of multi-storey housing in Denmark. This was the first major development programme that specifically included the integral use of demonstration projects as a diffusion instrument. 6 consortia pre-qualified, and the winning designs have been tested and used in 12 experimental building projects (mid 1980s till early 1990s). This programme has subsequently acted as role model for later programmes.

Current examples of major development programmes:

- Process and Product development in the Building industry (PPB programme)
- The Urban Renewal Project
- ECO-HOUSE '99

9.3.1.3   Process and product development in the building industry – PPB Programme

Initiated, administrated and funded by The Ministry of Business and Industries in collaboration with the Ministry of Housing and Urban Affairs. The programme is managed by a group of monitors, working on behalf of the ministries. Programme period: 1994-1997 (originally), now extended to 2001. The programme has a total budget of DKK 100 million of which the Danish State finances half the sum and half by the four consortia. To this should be added the budget of the related demonstration building projects (app. 1500 dwellings equivalent to DKK 1,5 billion).

The main objective of the programme is to improve productivity and the international competitiveness of the construction industry. This is to be achieved firstly by developing new building systems that meet the demands for healthy, environmentally friendly and energy efficient housing, secondly by means of a further industrialisation of the production process, and finally by initiating a reorganisation of the building process. Private firms, working in long-term, collaborative networks are expected to represent the appropriate innovative environment for such goals. Thus the PPB programme is intended to be a synthesis of the contemporary construction industry developments.

From the outset process and organisational developments were given high priority in the programme. Particular focus has been put on the increased complexity and disintegration of the organisation of the building process as this is generally perceived as a major hindrance to further industrial development and to productivity improvements. The need for vertical integration of the different actors and their functions in the construction process was strongly emphasised in the programme.

Consequently, the innovation activities of the programme are carried out by four consortia, representing the whole value adding chain of construction. The selection of these consortia was arranged as a tendering process including a pre-qualification round and subsequent invited competition. At the end of 1994 four consortia were nominated to carry out their proposals for process and product developments and to test their ideas in a series of demonstration projects. This model secures a long perspective on the innovation activities and promotes common learning processes among the parties in the consortia.

9.3.1.4   The Urban Renewal Project

This is the first major programme directed towards urban renewal and refurbishment. The programme was initiated, administrated and funded by the Danish Ministry of Housing and Urban Affairs. The programme period was originally 1995-1997 but it was prolonged into 1999. The government funded budget amounted to a total of app. DKK 170 million. The main objectives of the programme can be summarised in the following keywords: Increased industrialisation, productivity improvements, product development, and internationalisation of the urban renewal industry. The programme included three types of projects: (1) Demonstration projects (applying for up to 90% of total funding), (2) analyses, (3) international (demonstration) projects. In total approximately 150 projects have been initiated within the programme. Extended efforts with regard to evaluation and documentation were handled by a programme secretariat operated by the Ministry. The formal dissemination of results takes place mainly through reports and seminars. So far the effects of the programme are difficult to measure, since many demonstration projects and evaluation reports have not yet been completed (mid 2000). However, the programme administration body believe that *implementation* of the results will turn out to be a major challenge to the construction industry.[8] With this problem in mind the Ministry is planning different initiatives in order to institutionalise the proposed changes.

9.3.1.5   ECO-HOUSE 99

This programme was initiated, administrated and funded by the Danish Ministry of Housing and Urban Affairs in collaboration with KAB, a large nation-wide non-profit housing group, and has a much narrower objective than the two programmes described above. A competition took place in 1996 with two winners appointed (project coalitions comprising architects, contractors and consulting engineers). The competition's main themes were:

*   Integration of ecological and architectural solutions
*   integration of clients in the development activities
*   focus on ecological aspects on the basis of existing technology
*   assessment of life cycle costs of the buildings.

Three demonstration projects were completed in 1998-1999 by the two winning teams. It is expected that the ECO-HOUSE demonstration projects will function as 'exemplary' or 'paradigmatic' cases for the future developments with regard to ecological building technology.

---

[8]   Judgement by head of the Urban Renewal Project secretariat at the programme closing conference in 1999.

9.3.2 Indirect construction specific policy instruments

This category of policy instruments can be divided into two main groups:

- Technical Regulations, which principally affect the constructed product (e.g. the Building Regulations);
- Procurement Regulations, which affect the construction process (e.g. procurement and tender regulations, standard forms of contract, and the Quality Assurance and Liability Reform).

Furthermore, the Ministry of Environment and Energy handles planning regulations, which affect the constructed product, and the Ministry of Labour administrates labour market regulations, which affect the construction process. There is no systematic understanding regarding in what way the regulatory framework affects innovation processes in the Danish construction industry.

9.3.2.1   *Building Regulations*

The Building Regulations [*Bygningsreglementet*] is a central instrument in the Ministry of Housing and Urban Affair's actions in relation to technological regulation and development in the building industry. The aim of the Building Regulations is to ensure the basic quality of the building, and a certain standard with regard to fire resistance, health and safety aspects in general, etc. The Building Regulations also promotes environmentally friendly construction methods, and regulates the interaction of new buildings with the existing built environment. The Building Regulations applies to all building activities in Denmark (a simplified version exists for work on smaller buildings, i.e. single-family houses and semidetached housing), and is primarily based on functional demands rather than detailed technical requirements or prescriptive regulations. The role of the Building Regulations in a historical context is further elaborated in Bonke & Levring, 1996.

It is generally perceived that the Building Regulations have a positive as well as a negative impact on the development of new building technologies. Some parts of the industry claim that the Building Regulations indirectly favour some materials and building methods over others, and thereby support the evolution of existing (dominant) technological trajectories, most notably the industrialised building concept based on prefabricated concrete panels. For example, until recently the fire resistance regulations blocked the introduction of wood-based multi-storey buildings in Denmark. This was changed as a consequence of development activities within the 'Process and Product Development in the Building Industry' programme.

9.3.2.2   Procurement and tender regulations

Following the strong interest in new forms of co-operation and re-engineering the building process in the second half of the 1990's there have also been a strong focus on the institutional framework, which is guiding the distribution of roles and activities in the building process. Traditionally, procurement and tender regulations in the Danish building process have supported the separation of design and production (construction) into two rather distinct phases. This fundamental characteristic has gradually changed and today a broader variety of different procurement routes, organisational arrangements and tender methods are influencing the distribution of work functions and activities. However, what is known as the 'main contract' procurement route with its sequential approach to design and production activities is still dominating, and the well-recognised problems of disintegration is yet to be overcome.

The traditional procurement and tender regulations are generally perceived as barriers to a more widespread implementation of partnering-like arrangements and other types of re-organising and re-engineering the building process. However, whether this regulation is a real barrier to new and more integrated forms of co-operation in the building process is not clear. It may be that the limited use of new forms of co-operation up till now is misinterpreted as a result of too tight procurement regulations, while in fact the problem is rather a case of clients' institutionalised conception of construction industry behaviour? For example, some (public) clients express a general mistrust in the ability of construction firms to maintain a high level of architectural integrity and quality under a different procurement regime than the dominating traditional 'main contract' procurement route.

9.3.2.3   The Quality Assurance and Liability Reform

Studies in the early 1980s revealed a considerable amount of technical defaults in new buildings. This led to the introduction of the Quality Assurance and Liability Reform by the Ministry of Housing and Urban Affairs in 1986. The basic idea behind the Reform was to urge the actors of the construction process to establish procedures to improve the quality of the constructed product. The Reform applied to public financed and subsidised building activities, but it was the Ministry's clear ambition that the concepts and procedures should diffuse into private building activities (in 1989 a similar reform directed towards public financed renovation activities followed).

The main elements of the Reform:

- Formal procedures for documentation of quality in design and execution. Additional activities in the design phase have been introduced emphasising the need for design reviews among other things with the aim to include production experience in the design activities.
- Unification of periods of liability for all parties involved in the building project. The aim was to establish a more coherent attitude towards the quality aspects among the different parties in the building process.

- The establishment of the Building Defects Fund [*Byggeskadefonden*]. The Fund works as an insurance pool against defects. If defects are discovered the responsible firms are charged for the repair costs, yet the Fund covers the expenses if a responsible party cannot be found or if the firms are unable to pay. The Fund also employs consultants to conduct the mandatory 5-year inspection of buildings, which was another main element of the reform.

As a direct effect of the Reform a number of construction firms during the 1990s have developed new management strategies and administrative procedures (administrative innovations), partly based on the ISO-9000 standards framework. Industry associations as well as technological service institutions to a large extent supported these developments. A revision of the Reform is currently under consideration, since it is acknowledged that there has been no substantial decline in the number of defects in new buildings.

9.4 EFFECTIVENESS OF PUBLIC POLICY INSTRUMENTS

As indicated in section 9.2 the effectiveness of public interventions behind the industrialisation process in Danish construction during the 50s and 60s appears to be uncontested - at least what the mere technical objects and simple output measures are concerned. The 'Open Danish Building System' - a prefabricated and modular co-ordinated concrete slab technology - actually did succeed in providing high quality housing at low cost (Bertelsen & Nielsen 1999). Recent times' innovation programmes, however, certainly constitute more complicated developments, when it comes to the interpretation of their outcomes.

1950 – 1973	1973 - mid 1990s
<ul><li>Expansive economy (continuous growth periods)</li><li>Keynesianism, subvention economy</li><li>Active, radical state intervention</li><li>National regulation frameworks</li><li>Scarcity of skilled manpower resources</li><li>Fulfilment of basic social and material needs</li><li>Industrialisation as mass production</li><li>Collectivism, conformity</li></ul>	<ul><li>Contractive economy (frequent crises)</li><li>Neo-liberalism, market as driving force</li><li>Re-active, adaptive state intervention</li><li>National de-regulation, globalisation</li><li>Scarcity of natural resources</li><li>Fulfilment of spiritual demands</li><li>Consumer oriented production, service society</li><li>Individualism, flexibility</li></ul>

Now initially, before moving into discussing the specific aspects of public instruments' effectiveness, a few societal premises should be listed to indicate the different conditions for innovative developments within these two distinct periods in Danish construction.

Although this tentative general labelling - nor the periodic division - does not permeate every socio-economic and technological segment of construction, it does however provide a perspective to understanding the essence and scope of public instruments under certain background conditions.

Thus, it seems obvious that setting up the agenda for innovation in construction as well as choosing the means to achieve the goals was a relatively less complicated assignment during the first of these two post-war eras. A very central outcome of this first 'top down' development period in which, as indicated, the Danish state played a crucial role, was in fact the establishment of the modern contracting system as a national institutionalisation of the specific production technology, implemented through the innovation policy (Bonke & Levring, 1996: 1). Then in the second period - primarily dealt with in this paper - as public policy shifted bearing from radical towards adaptive, the institutionalised system would gradually constitute a factor of increasing resistance against system alterations as such and thus tend to neutralise the effect of transverse public policy instruments. Hence, it is a quite frequent impression when observing the last 10-15 years' public debate about future strategies for the construction sector that there is a growing distance between the vigorousness of viewpoints explicitly formulated by opinion leaders and the weakness of means operated by policy makers.

Although this argument of a mature production sector and its institutions as a non-innovative system - having the game all to itself - has not been analysed within this context, there are however considerable specific signs of lacking effectiveness in the Danish innovation programmes. For instance in the *demonstration projects programme*, which as mentioned has been given a central instrumental role in most if not all public innovation schemes, a number of problematic issues have been identified:

- *The intended diffusion effect does not seem to be achieved.*
  Insufficient evaluation and documentation of results from the projects could be the obvious cause; lack of general business incentives for innovation among potential users might be a more basic barrier against diffusion.
- *The development activities tend to be overrun by regular production problems.*
  This indicates that the long-term, strategic competence development goals in firms are subordinated to short-term, economic project events.
- *Fear of competitors' appropriation of innovations may prevent leading edge firms from engaging in experimental building projects.*
  The Ministry's demand for openness and immediate publication of results obviously could have a selective effect on potential firms and types of innovations.
- *Firms are critical of what they perceive as a large administrative burden in experimental building projects.*

It is however, also observed that firms' capability of handling the regular planning and managerial tasks in R&D projects is generally poor.

These types of 'practical' problems are of course contributing negatively to the effectiveness of the different innovation schemes in which demonstration projects are applied as a physical testing and diffusion instrument. However, also a few, more fundamental imperfections characterising the set-up of public programmes should be mentioned here:

- *Lack of performance data.*
  There are severe problems of measuring performance in construction firms. In particular it appears difficult to provide reliable data for the large quantity of production work on site, which is sub-contracted to small firms. Lack of comparable data impedes the documentation of new technology's effectiveness.
- *Imbalanced involvement of actors*
  This problem is particularly evident in programmes/projects focusing on vertical integration along the value chain, and in innovations aiming at reorganisation of functional roles. It displays the institutionalised division of labour in the Danish contracting system where projects are moulded upon transactions between autonomous actors/firms. In this system large contractors and consulting engineers tend to monopolise superior decision processes whereas the influence of clients, architects, suppliers and subcontractors (SMEs) become lined out within their specialised fields of competence. The public instruments of innovation have not hitherto managed to dismount such traditional, power based organisation patterns.
- *Insufficient activation of knowledge institutions*
  There is generally a low awareness of the potential effect of activating knowledge institutions more directly as change agents in innovation programmes; R&D institutions should be more consistently employed in analysing and evaluating assignments linked to public programmes. Likewise, education and training institutions should be regarded as obvious objects to - and thus participants in - any innovation process concerning the implementation of new technology in construction.

Consequently, when assessing the overall impression of public policy instruments' effectiveness, these critical issues do deteriorate the possibility of successfully achieving the planned goals. During ten to fifteen years of innovation programmes ministries and state agencies have managed to put the need for changes on the agenda as a broadly accepted, rhetoric object. The lack of implementation of results in business practise is however a disturbing fact that should bring policy makers to rethink their instruments.

9.5   TRENDS IN PUBLIC POLICY STRATEGIES AND INSTRUMENTS

Present time construction industry policy trends represent a wide variety of issues and initiatives, as will be demonstrated in this section. Overall, these trends in public policy initiatives form part of a wider process towards *industrial reengineering and restructuring* of the Danish construction industry, and the policy makers are apparently extending their relationships with industry and research institutions for an even stronger involvement in reshaping an internationally competitive construction industry. These and other trends will be further elaborated in the following section.

9.5.1   Strategic action agenda

In 1998 the Ministry of Housing and Urban Affairs released a strategic action agenda including 'guide-posts for action' and a prioritised list of subjects for further construction developments in the years to come. The action agenda is the result of an intensive dialogue between the Ministry, research organisations, professional organisations, industry associations, and companies from the construction industry. It is yet to be seen whether this action agenda will usher in a new epoch of long-term commitment to the solution of the fundamental construction innovation problems. The strategic action agenda includes – among others – the following initiatives: [9]

- Focus on *quality of the built environment,* e.g. by educational initiatives aimed at strengthening the role of the client, introducing benchmarking activities, and further use of life cycle assessments in the both design and production processes.
- Focus on *productivity and new forms of co-operation* in the building process.   This stream of activities includes programmes for introduction and development of partnering principles and policies to support implementation of information technologies in the building process.
- Focus on *construction R&D and the building sector's use of knowledge.* Under this headline a number of initiatives regarding R&D in the building industry has been initiated in recent years, with a specific attention to the role of knowledge brokers and the technological service institutions in the diffusion of new knowledge and construction technologies throughout the building industry.

---

[9]   A full list of initiatives in the Strategic Action Agenda is available on the Ministry of Housing and Urban Affairs homepage: www.bm.dk, and the publication: Ministry of Housing and Urban Affairs (1998).

9.5.2   New policy-makers and inter-ministerial partnerships

Although the Ministry of Housing and Urban Affairs is still central to Danish construction innovation, new actors have entered the construction innovation scene and gained influence. For instance, the Ministry of Energy was instrumental in promoting energy efficiency as a key objective in the 1970s. Later the Ministry of the Environment has had an impact with regard initiating programmes for recycling of building materials and introducing environmental assessments in the design process. There is now a joint Ministry for the two areas, the Ministry of Environment and Energy, which is quite influential. More recently, the Ministry of Business and Industry has increased its involvement in the construction sector substantially through its Agency for Development of Business and Industry. Primarily through financing and administrating R&D programmes in close collaboration with the Ministry of Housing and Urban Affairs, but also by carrying out benchmarking analyses etc. Spring 2000 witnessed the establishment of an inter-ministerial 'task force', casting representatives from the two ministries. The aim is to develop and co-ordinate the two ministries' policy strategies in the construction area.

9.5.3   Large programmes – 'project building'

A very noticeable feature of the public policy strategies and instruments with regards to Danish construction innovation has been the reliance on demonstration projects administered by the Ministry of Housing and Urban Affairs. Demonstration projects are still the 'backbone' of the Danish construction innovation policy. But a most important trend in the Danish construction sector over the last couple of decades has been an increasing preference for 'large programmes' in the form of initiatives that are well co-ordinated and with a long term perspective. Thus, the preference of the Ministry of Housing and Urban Affairs has moved from an emphasis on individual demonstration projects where the ideas came from the industry to using demonstration projects mainly as integral parts of large development programmes with politically defined objectives - although usually harmonised with industry priorities.

These changes have increased the time horizon for initiatives considerably, because each development programme typically consists of several demonstration projects. Increasingly, learning from one demonstration project to the next is integrated into the design of programmes. The increasing time horizon is somewhat paradoxical to the extent that it goes beyond the political re-election period, which is defined by a maximum of four years for general elections. This would seem to indicate that government officials (the civil service) are gaining influence at the expense of the politicians.

As an illustration of the above mentioned observations the Danish Ministry of Housing and Urban Affairs and the Agency for Development of Business and Industry have recently launched the ambitious ten-year 'Project Building' programme [*Projekt Hus*][10] in co-operation. The aim is to improve the productivity and quality of building construction. The overall objective of

---

[10]   The programme covers not only housing but all types of buildings.

this major new initiative has been stated as 'twice the value at half the price', thus, a synthesis of consumer and producer oriented productivity improvements is envisaged. This ambitious objective will be achieved through improvements in the competence of building clients (to take account of consumer preferences) and increased use of industrial products and methods.[11]

The programme is still in the conceptual phase, and for the moment ten core themes have been identified:

- Increased consumer satisfaction
- tendering and client's selection of partners
- the designers' services
- industrialised processes
- new building components
- co-operation between firms and on site
- architectural aspects
- external conditions (e.g. national building regulations and EU directives)
- the knowledge basis
- quality assurance and management of the development projects as well as the demonstration projects under 'Project Building'.

For each theme a group of industry representatives and management consultants have entered a 'theme group', which proposes innovation ideas and develops new concepts for a renewed construction industry. In co-operation with clients the new concepts will subsequently be tested in real building projects and/or demonstration projects. Key performance indicators will be measured to document the effects of the new concepts.

9.5.4   Increased client and consumer orientation

The 'Project Building' programme is an example of an innovation strategy aiming for both consumer and producer oriented productivity development in the building sector, thus attempting to balance the driving forces for essential productivity improvement in Danish construction. It is worth noticing that the focus on consumers and clients is very strong in the current public debate in Denmark (e.g. the Danish Building Development Council initiative).

In addition to the 'Project Building' programme a few other examples of initiatives supporting an increased user orientation in the building sector should be mentioned. For instance, the strong focus on clients is a general trend which was initiated by the Danish Building Development Council [*Byggeriets Udviklingsråd, BUR*] and supported by the Ministry of Housing and Urban Affairs. The objective has been to mobilise especially the large building clients in order to establish a market-driven pressure for better performance in the construction sector. A related effort is concerned with upgrading clients in terms of competence building through education. Also, the developments towards increased reliance on life cycle costing of buildings

---

[11]   For further information refer to the 'Project Building' home page: www.projekt-hus.dk.

is bound to make the role of the client more crucial for improving construction sector performance. In other words, the client has now been placed in focus as an important actor of change.

9.5.5 Process innovation, organisational change and innovative capabilities

It could be argued that traditionally there has been a preoccupation with products (buildings, components, etc.) in the public policies aimed at supporting innovation in construction. In recent years, however, the focus tends to be shifting from product to process and organisational change, emphasising such issues as productivity, new organisational models, new forms of collaboration etc. It is likely that for implementation of such process innovations there is a distinct need for longer-term perspectives as observed in the trend towards using development programmes as vehicles for change.

Parallel to this shift innovation strategies are now also explicitly reflecting questions of industry politics and business economics, that is, taking a point of departure in construction as private firm business activities when trying to understand the innovation problems of the sector. Obviously, the strengthening of this perspective is an effect of the growing involvement of the Ministry of Business and Industry. The business enterprise approach to innovation probably implies future initiatives regarding development of innovative capabilities and innovative cultures in firms.

9.5.6 Diffusion and implementation: developing the learning capacity

Recent years have furthermore demonstrated an increased orientation towards the implementation of results from research, development and demonstration projects in construction, i.e. getting from invention to innovation through commercialisation. Especially the mechanisms of dissemination have been targeted as a major problem area – and several reports from the Ministry of Housing and Urban Affairs and the Danish Building Development Council have pinpointed the problems, e.g. Bang (1997) and Dræbye (1997). In particular the role of 'knowledge brokers' in the construction innovation process appears to be of great importance to the dissemination effect.

Increasingly, the emphasis on dissemination and implementation is now finding its way into government instruments as means of promoting innovation. At a conference held in 99 as a conclusion of the Urban Renewal Programme, it was announced that progress in the implementation of results should be evaluated after one and five years (in so-called 'implementation audits') in order to press for real commercialisation of development results. At the same occasion specific actors responsible for implementation, such as public clients and trade associations, were identified and nominated.

Furthermore, as a result of the critical reports mentioned above, the Ministry of Housing and Urban Affairs has recently introduced a strategy projects (also known as the '*Kollekolle*' initiative) for a more comprehensive pre-screening of project managers and the teams, applying for participation in subsidised demonstration projects. In the application phase such teams will have to document dissemination expertise and present plans concerning the implementation of results into industrial practise. The pre-screening will be supplemented by a procedure for evaluation of business economic relevance of the results, once the projects are near completion.

9.6   CONCLUSION

During the last two decades the spirits of public policy in general have gradually shifted from Keynesian policies towards more market-oriented and neo-liberal opinions. This obviously constitutes an instrumental problem in innovation policy – not least if the policy makers define their goals in terms of industrial reengineering.   As earlier shown the mechanism of the first industrialisation phase in the 50s and 60s were in fact a quite efficient combination of instruments, including a comprehensive set of construction specific, direct technological guidelines and instructions.   These were, however, subsequently very efficiently implemented in the business processes of the actors of the sector though more indirect instruments of influence attached to the innovative technologies.

Present time politics are clearly based on faith in the market forces. This also applies to the Social Democratic coalition government, which succeeded the conservatives from the early 1990s. Generally this shift did lead to a certain reluctance in public policy towards the use of very specific and active types of instruments, and of course also to an intentional phasing out of market subsidies as an element of policy-making. Consequently, since the early 1990s the role of the modern state has very much been limited to that of the critical buyer: the private sectors are better off left alone and expected to solve developmental problems more efficiently under the forces of competition.

Seen under this perspective we might be facing a paradox in relation to the innovative challenges, imposed upon the construction sector. On one hand innovation programmes are now defining inter-firm organisational structures and production processes to be crucial areas of developments. On the other hand market forces do not seem to urge firms to reorganise, neither do public policy instruments at hand interfere at that level. Seen under this perspective Danish public innovation policy may find itself in a deadlock which is not only determined by macro economic and political forces but perhaps also reinforced by micro level barriers to innovation.

Thus as demonstrated, throughout the post-war period Danish construction firms did establish a capability of implementing solutions to particular technological problems, no matter if originated in daily business practise or defined and orchestrated by public bodies. Confronted with the recent need for more transverse and radical innovations in construction, however, this capability of micro organisations does appear to have developed into a 'competency trap' (March, 1991). This means that an innovation competence in firms, although refined to high efficiency in relation to certain well-known types of problems and solution, does at the same time neglect the existence of innovation challenges of different nature. Consequently, the building up of alternative innovative competencies (and technologies) is likely to b obstructed by firms themselves.

Presumably, only a policy of long-term and co-ordinated innovation instruments applied to all levels, institutions and actors in the national contracting system will be able to bring innovative impact back to the construction process. As indicated, the 'Project Building' programme is representing elements of this recognition. However, time is still to show whether a targeted public policy intervention will take place during the implementation of the programme objectives.

Appendix: Framework for positioning public policy instruments[1] in the Danish Construction industry

Name of Programme	Multi-storey Housing Programme	Project Urban Renewal	Process and Product Development in the the Building industry	ECO House Programme
Category of programme: 1,2,3,4[2]	Mainly 2 and 3 (small element of 1).	Mainly 2 and 3	Mainly 2 and 3 (and a small element of 1)	Mainly 2
Resources (Public / private) funding in total (app.), DKK[3] ([4])	20 million DKK.	500 million DKK. (?)	100 million DKK.	1.5 mill DKK.
Objectives	Flexibility of building systems	Process and product development and implementation in renovation	Process and product development and implementation	
Means	Consortia competition (6 consortia) Demonstration projects	Projects by single firms, organisations and (mainly) consortia Demonstration projects	Consortia competition (4 consortia) Demonstration projects (+ PhD studies)	Consortia (Architect driven) competition (2 consortia) Demonstration projects
Contribution to innovation (results)	A new ready for use design concept, tested in a number of demonstration projects	Product solutions for renovation, (prefab bathrooms etc.). New labour org. 'multi-skilled gangs'	New design concept for timber-based multi storey buildings. Reorganised design processes.	New technical solutions for sustainable construction. Energy saving design. New building materials.
Strategies for diffusion of results etc.	Reports, workshops	Reports, workshops, conferences, knowledge-brokers	Reports, workshops	Reports, seminars, newspapers
Special features?	Ideas generated by consortia	Ideas generated by firms, organisations etc.	Ideas generated by consortia	Ideas generated by consortia

[1] Public policy instruments in category: construction specific and direct (cf. To taxonomy presented by Winch, 1999), and only instruments and programs which are promoted by the Ministry of Housing and Building.

[2] Four categories are used to characterise the public policy instruments: (1) Programs to support R&D. (2) Programs to support advanced practices and experimentation, (3) Programs to support performance and quality improvement, (4) Programs to support taking up of systems and procedures.

[3] 1 UK£ ~ DKK 12,14; 1 US$ ~ DKK 8,26

[4] Additional public funding in relation to demonstration projects is not included.

Appendix: (Continued) Framework for positioning public policy instruments in the Danish Construction industry

Name	Project New Forms of Collaboration	Project House
Category of programme: 1,2,3,4	2, 3 and 4	1, 2, 3 and 4
Resources **(Public and private funding in total (app.), DKK**	Only in affiliation to demonstration projects. Public funding app. DKK 1 million	Phase 1: App. DKK 10 million + Demonstration projects (DKK 1-3 billion)
Objectives	To reduce costs (5-20%) and improve quality - new rules for actors' interaction	'Twice the value at half the price' Reorganisation of interaction between actors; Industrialisation
Means	Agreement on collaboration and teambuilding Demonstration projects	Conceptualisation groups Demonstration projects
Contribution to innovation	Highlighting the role of clients. New forms of interaction. Increased awareness and use of partnering in non-profit housing	Still intentional: Bringing the client, the component industry and suppliers into focus as innovation drivers. De-regulation.
Strategies for diffusion of results etc.	Demonstration in practice Reports, seminars	Demonstration Creating a 'Tsunami': making key actors from all areas of influence commit themselves to the overall objective of the programme
Special features?		More than 150 VIPs involved in conceptualisation groups. Scheduled for duration of 10 years

CHAPTER 10

INNOVATION IN THE FINNISH CONSTRUCTION AND REAL ESTATE INDUSTRY - THE ROLE OF PUBLIC POLICY INSTRUMENTS

Tapio Koivu and Kaj Mäntylä

10.1   INTRODUCTION

This paper draws mainly from findings and preliminary results of a study "Developing Innovation in the Real Estate and Construction Sector – Background and Experiments" (Koivu et al, 2000).   The study has been initiated and performed by VTT Building Technology in 1999 and is continuing until 2000. The aim of the project is to find out new methodologies for handling the innovation chain to benefit the industry.

   Innovation is the result of a complex interaction between various actors and institutions. Technical change does occur through feedback loops within this system. In the centre of the innovation system are the firms, the way they organise production and innovation and the channels by which they gain access to external sources of knowledge. These sources might be other firms, public and private research institutes, universities or transfer institutions – either regional, national or international. Here, the innovative firm is seen as operating within a complex network of co-operating and competing firms and other institutions, building on a range of joint ventures and close linkages with suppliers and customers.

   The effectiveness of the innovation system requires co-operation in the network of companies, RTD performers and financing organisations, companies specialised in technology transfer and commercialisation.   This networking needs to work on both regional and international levels. Especially, networking needs to be beneficial for the small and medium sized enterprises.

   This paper discusses the role of public policy instruments and the national innovation system from the viewpoint of the construction industry and facilities management. The instruments are analysed in terms of a model of the innovation and diffusion process in construction. Table 10.1 presents the main objectives, means and contribution to innovation of the public policy instruments.

Table 10.1 Framework for positioning public policy instruments

	Programme to support R&D	Programme to support advanced practices and experimentation	Programme to support performance and quality improvement	Programme to support taking up of systems and procedures
Name / Type	Various national programmes National support to enhance participation in the 5th Framework Programme (EU) Innovation Relay Centres, multinational	Technology Clinic activities Pilot Building –programme Technology centres Regional knowledge centre - programme	Centres for employment and economic development: aid for company specific improvement (education etc..) Agenda of the National Quality Centre	Rakli: "Progress" –pilot programme Motiva: Energy saving promotion programmes TULI –activities Sitra: innovative projects
Resources	Gov. funding by Tekes and other agencies	Gov. funding by Tekes, expertise provided by RTD performers	Gov. funding, networks of companies	Networks, Gov. and other funding, expertise provided
Objectives	Science and Technology development	Technology transfer and testing to/in companies	Varying depending on agenda	Technology transfer, commercialisation
Means	RTD-projects specific to the programme themes & company needs & ideas Exchange of ideas	Consulting of companies to help transfer	Exchange of information, education and training programmes	Evaluation of ideas, feasibility and market studies, creating co-operation networks, seeking of risk capital, etc..
Contribution to innovation	Sources of ideas, R&D funding and performing knowledge & information, Exchange of knowledge	Transfer of technology, lowering the threshold of taking new technology into use	Infrastructure for exchange of ideas, information and experience Support to training & use of new technology	Systematic help in commercialisation of ideas

10.2   THE FINNISH CONTEXT

10.2.1   General context

Finland has a population of approx. 5.1 million people. The country has been a member of the European Union since 1995 and a member of the European Monetary Union since 1999.

Finland can be ranked in the 15 richest countries in the world in terms of GDP. At the moment, the country is experiencing a very favourable economic cycle. The growth in GDP has been between 4-6% per year since 1994 and is estimated at 4-5% for the next 2 years. The level of taxation is among the highest in the world, but reciprocally the welfare state provides a very large variety of services for the citizens and companies.

The institutional context of the country is very mature with regard to innovation systems. One of the key tasks outlined in the Finnish Technology Policy is to improve the national innovation system as a whole. The share of RTD expenditure (totally FIM 22 billion) has risen to 3.1% of the GDP by 1999. Parallel to this development, the utilisation of RTD results has been one of the main focuses.

The Finnish innovation system as a whole has been ranked by OECD comparison as one of the five most effective ones in the world. During the previous decade, the chosen national science and technology policy has efficiently played a role in producing a new, strong industrial cluster – information technology and communications. To some extent, the same managerial approach applies to the construction industry as well. Naturally, some peculiarities of the industry need to be taken into account, such as project based business and conservative management culture.

The cultural environment in terms of innovation can be regarded as pragmatic and favourable for entrepreneurship despite the taxation.

Finland is quite large (total area is 338,000 square km) with a relatively small population. Concerning the regional development there are major differences between different areas. There are about 10 larger growth centres in Finland (major cities and their surroundings), where most of the new jobs in the industrial and service sectors are created, and to which people are moving from other parts of the country. The rest of the country relies more on traditional sectors. The strong urbanisation process will probably continue for several years in the future.

10.2.2   Construction industry context

The built environment represents 75% of Finland's real national wealth. Construction investments account for 65% of all fixed investments. Upkeep of infrastructure, construction, building products industry, real estate and their related service make up the real estate and construction cluster. The cluster is presented in Figure 10.1.

Figure 10.1 The real estate and construction cluster

The real estate and construction cluster

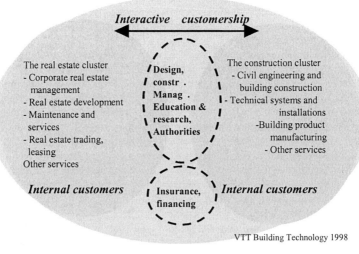

VTT Building Technology 1998

In the 1990s, there has been a trend of transferring ownership of real estate to enterprises established for that purpose. New technical features of buildings and the challenges of business and production of public services, have differentiated property management from the core activities of enterprises and institutions. Hence, the real estate industry is now divided into three main fields: ownership of facilities, use of facilities and production of related services.

The value of the real estate and construction cluster's domestic demand in 1998 was FIM 150 billion. The cluster is an active exporter and has production abroad. The value of operations in Finland and abroad in 1998 was over FIM 200 billion (Table 10.2). Compared to the national GDP, the value of operations in the real estate and construction sector (excl. turn-over of affiliated companies abroad) was 26-28% in years 1997-1998. (VTT Building Technology, 1999)

Table 10.2 The value of operations in the real estate and construction cluster
in 1997-1999 (VTT Building Technology, 1998, 1999 and 2000).

Operations in the Real Estate and Construction Cluster	Value (billion FIM) in years		
	1997	1998	1999
Maintenance of buildings	65	64	
Renovation and modernisation	24	27	30
Total new construction	31	39	45
Civil engineering	19	19	20
Exports of building materials and products, building services and municipal engineering products	20	31	33
Turn-over of affiliated companies abroad	20	25	
Total	**180**	**205**	

In 1998, there were about 45,000 firms (a rough estimate) operating in the construction sector (construction industry, building materials and products industry, building services). The construction sector employed directly about 200,000 persons in 1997 (about 10% of total employment). The real estate sector employed about 200,000 persons in 1997 (VTT Building Technology, 1998)

Construction investment volumes in Finland have fluctuated significantly following the fluctuations in GDP volume. In the late 1980s and 1990s, the annual change in construction investment volumes have varied between –20% and +16%, while the annual change in GDP has varied between –7% and +6%. The favourable development of the Finnish economy in the late 1990s has created confidence in the future, which is indicated by the continuing growth of the demand for construction in the next coming years. Renovation and modernisation and maintenance of infrastructure are not tied to the national economy in the same way: their volume will increase as buildings and constructions grow older. In 2000-2002, investment in renovation and modernisation of the Finnish building stock is predicted to increase 3-5% annually (VTT Building Technology, 2000).

Tapio Koivu and Kaj Mäntylä

Table 10.3 New building starts (million m³) in Finland in 1998 - 2000
(VTT Building Technology, 2000).

New-building starts (million m³)	1998	1999	2000
All Buildings	36.4	38.2	38.2
Residential Buildings	11.4	12.9	14.0
Commercial and Office Buildings	5.3	6.7	6.2
Public Buildings	2.6	2.8	2.7
Public Buildings	9.8	8.5	9.8
Other Buildings	7.3	7.3	7.3

10.2.3 Innovation in the construction industry

In 1998, close to FIM 1,200 million was invested in the research and development of the real estate and construction cluster, which accounted for 6% of Finland's total R&D expenditure. Public R&D financing was FIM 500 million and private sector financing was FIM 680 million. The main public financiers where the Technology Agency of Finland (FIM 250 million), the road administration (FIM 100 million), the Technical Research Centre of Finland (FIM 60 million) and the universities (FIM 30 million). Of the private sector financing came FIM 525 million from the building materials and products industry, FIM 95 million from the construction industry and FIM 60 million from the building services sector (VTT Building Technology, 1999).

Innovativeness is a key factor of competitiveness of the construction industry. The ability to generate new knowledge and new products and applications to the markets at the right time determines the success of an organisation in a competitive environment. However, the construction sector uses less than 1% of its turn-over to RTD-activities. One consequence of this is that companies do not have capabilities focusing on an excessive amount of successful innovations yet the surrounding environment needs to be followed. Especially in the SMEs of the construction sector RTD-activities are often so minor, that the SMEs have difficulties in following even the technical development in construction.

The amount of financial resources used to RTD-activities tells very little of the effectiveness of the innovation process and RTD. A single construction process, for example a novel building, can be an innovation by definition. The key issue is how to find a more effective framework for innovations in the processes of construction (Atkin, 1999).

Some characteristics of the innovation processes of the facilities management and construction industry have been recognised:

- The sector is very fragmented.
- The sector as a whole has a very low potential for added value. Competition is based on price, and this do not encourage firms to take risks in innovating.

- The demand fluctuates considerably and does not always follow the cycles of the rest of the economy.
- The business is project based, it is difficult to verify the results and impacts of innovations.
- Product or process improvement cycles are slow.
- The industry's ability to manage the innovation process is not at the same level as in other industries.
- There are limitation to the risk taking in developing new solutions or products.
- The sector lacks educated personnel in companies.
- The culture in the sector is conservative and in some countries, not end-user oriented.
- Governmental control plays a significantly stronger role in innovation process than in most other sectors.

However, the potential for significantly better processes and competitiveness is recognised. If this potential can be used, a far better productivity, end-user satisfaction and cost-effectiveness can be obtained.

10.3   PUBLIC POLICY INSTRUMENTS

The basic elements of a national innovation system can be condensed into four elements: organisations, institutions, learning and interaction. Organisations refer to those active in RTD and in using new knowledge or supporting this activity. By institutions one refers to different support and relay systems, such as patent institutions. This includes values and normative systems attached to the innovation system. Interaction helps in the process of learning, understanding and applying new knowledge or solutions.

This paper mainly describes those Finnish public organisations, which take an active part in supporting the companies in the innovation process or take a role in creating new knowledge or generating new potential for innovations.

Various methodologies intervene in the way the companies and the construction sectors are assisted in their efforts to produce effective innovations. Some of the main functions are described in this chapter. Public interventions both focus on creating new business and on the other hand, creating new standards, regulations and practices for the use of the sector as a whole. Special emphasis has been put into speeding up the process of commercialisation of ideas.

10.3.1 Public policy instruments in the role of identifying sources of innovation or ideas

In Finland, few organisations are actively looking at the changes in the environment, societies and values at a time perspective of 15-30 years. These organisations, like The Society of Future Studies or The Committee for the Future of the Parliament of Finland, have a profound effect when national policies are written. This has an effect on the technology policy and thus to the innovation system as a whole. The direct effect on the real estate and construction sector, however, is not very profound.

The Technical Research Centre of Finland (VTT) performs anticipation-studies of different technological realms. These studies often serve the needs of customers, but they can also be connected to the development of VTTs own activities. There is also a special group for Technology Studies in VTT, which performs studies on the innovation system as a whole. Technology foresight and assessment forms a central part of its work. As an example of the work, the VTT Group for Technology Studies is presently developing a reliable tool, based on numerous rounds of expert polls, for measuring innovativeness and identifying most important innovations in Finland.

10.3.2 Public policy instruments in the role of evaluating and upgrading ideas

The network of technology centres in major cities offers different kind of services connected to the innovation process. One of the most interesting service provided is the TULI-activity (From Research to Business). This process aims at finding and evaluating ideas and upgrading results of RTD work into commercial use (see Figure 10.2), including the following:

- Licensing or selling of a technology
- starting of a spin-off or a new company and
- further RTD or co-operative projects with companies or in groups of companies.

Foundation for Finnish Inventions supports and helps private individuals and entrepreneurs to develop inventions and their further upgrading and use. The basic functions of the foundation are consultation, evaluation and protection of inventions and financing of marketing and activities leading to market success. The Foundation is working on more than a thousand projects in different areas of technology every year. The Foundation serves as a link between inventors, innovators, consumers, businesses and industries in Finland or other parts of the world be it a matter of setting up production, licensing or any other means of exploiting an invention. (Foundation for Finnish Inventions, 2000).

Figure 10.2 The TULI process (From Research to Business)

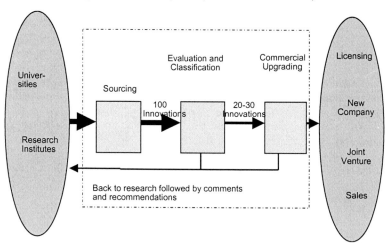

The Foundation provides funding for developing inventions by private inventors and small entrepreneurs. Approx. 20-25% of projects funded are commercialised either in the form of the inventor-entrepreneur's own production or under a licence agreement. The industrial and commercial implementation of invention projects is promoted by various methods of marketing and marketing communication. The Foundation's Inventions Market also provides entrepreneurs with new business and product ideas via the Internet. The Foundation receives majority of its funding from the Ministry of Trade and Industry. (Foundation for Finnish Inventions, 2000).

10.3.3 Research and development activities

National public research centres have a key role among the public policy instruments in Finland. The Technical Research Centre of Finland (VTT) is the central public institution concerning (technical) innovation in the construction industry. The division for building technology in VTT has a market share of 15% of the approx. 200 MEURO RTD market volume in the Finnish FM & construction industry. The role and profile of VTT is to perform applied research and technical development projects. VTT is controlled by the Ministry of Trade and Industry and receives national funding to upkeep infrastructure for RTD. (Technical Research Centre of Finland, 1999)

VTT was once a large public-sector body, but now the institute has become more focused on the needs of private industry. There has been a cultural shift from being a scientific research community to a more customer-oriented and flexible commercial organisation. Long-term survival and prosperity mean developing commercial interests and seeking out virgin areas of RTD that are linked to industry.

Currently, VTT is carrying out a specific project which evaluates industrial innovations in Finland. According to the results of the Sfinno survey of 1,000 companies and nearly 1,500 innovations between 1984-1998, the product market conditions combined with technology push are the most important factors for companies in Finland to innovate. Market niche recognition, customer feedback and collaboration, consumer knowledge and commercialisation are highly critical factors on technology and marketing innovations. Commercialisation of core technology is deemed to be quite important especially for small firms, which have a narrow market niche and a modest R&D budget. (Palmberg et al, 2000)

Nearly 90% of all reported innovations have involved collaborative work. About 2/3 of all innovations received public support during their development. Tekes Technology Programmes helped 25% of the innovations and enabled traditional industries to modernise. In most cases, local universities provide basic scientific knowledge which big firms find very helpful, while VTT's applied research is particularly useful to small firms (Technology Agency of Finland, 2000a).

The Technology Agency of Finland (Tekes) is the most important actor in the Finnish innovation system. Tekes was established in 1980s and has been growing in influence ever since. Today, its funding to RTD and innovation is about 2.1-2.4 billion FIM per year in loans, direct and confidential support to companies and public support to RTD performers. In 1999, the share of construction- and wood product-technology of the total Tekes R&D funding was 13%. Tekes funding directed to construction in 1998 was FIM 220 million covering public and confidential projects.

One of the main forms of action of Tekes is the extensive national technology programmes. Technology programmes are an essential part of the Finnish innovation chain. In co-operation with companies, universities and research institutes and sector authorities, the technology programmes aim at gaining new technology expertise and product development options in the important business areas of the future. The programmes also offer good frameworks for international R&D co-operation. In 1999, Tekes provided FIM 1.1 billion to financing technology programmes. Tekes support is usually about half of the total costs of programmes. Half comes from participating companies. In 2000, about 60 extensive national technology programmes are under way (Technology Agency of Finland, 2000b), and over 20 of these are related to construction (listed in Appendix 1).

Tekes technology programmes are intended to improve the activity, industry and services of the real estate and construction cluster. The technology programmes of the late 1990s and the next few years, e.g. Rembrandt, ProBuild and the nascent Infra programme, focus on the development of a certain core area or part of it. The aim of the future technology programmes is to increase co-operation between core areas and sectors serving them. Such co-operation is presently promoted by, e.g., the VERA and the Healthy House programmes.

Other highlights include activities like the Pilot Building programme supported by Tekes, ministry of Environment and Housing Funding agency.

The purpose of this activity is to help in the risk taking involved in testing new solutions in housing projects.

Tekes, together with expert organisations and specialists in different sectors, provides services for technology transfer as well. These <u>Technology Clinic Activities</u> focus on small and medium sized companies. The projects are normally very small and aim at transferring a particular technology to the use of the SME. The overall aim of this activity is to lower the threshold of SMEs to enter technology programmes, to help in technology transfer and to develop facilities for technology-based services. Tekes finances about half of the costs of the projects, and the second half comes from companies. In 2000, there are 15 different Technology Clinic operating (Technology Agency of Finland, 2000c), and 8 of these are related directly or indirectly to the construction and real estate sector (listed in Appendix 1).

For example, the Technology Strategy Clinic activity aims at filling the gap between the business aims of SMEs and RTD projects. This service comprises of a review of the company strategy from viewpoint of finding new technological possibilities. The technology and market reviews are customised according to the need of the companies. The possibilities available are further processed in workshops facilitated by the RTD performers in co-operation with the company key personnel and outside experts. The process itself develops the company's internal processes and increases its competence in the marketplace. A normal process for formulating a technology strategy takes 3 to 6 months. So far, over 70 companies of the real estate and construction industry have used this service.

The Finnish National Fund for Research and Development (Sitra) is an independent public foundation operating under the responsibility of the Finnish Parliament. Sitra finances and implements research, training and innovative projects of significant national importance.

The purpose of Sitra's research operations is to identify the pressures of change faced by the Finnish society, as well as to provide a basis of information and operative models for national success strategies (Sitra, 2000). Sitra has launched a research project to develop the innovation system. This program will be completed in 2001, and the following are among the focus themes of the programme:

- Competitiveness in "skill-based"-industries: Developing tacit knowledge, co-operation and marketing in handicraft, culture and personal service industries
- improving the utilisation, diffusion and transfer of knowledge in the Finnish economy: a policy perspective
- the role of knowledge-intensive business services in the Finnish innovation system
- innovation networks and "learning communities": their nature, operation, evolution, and framework conditions
- regional clustering of innovation: Implications for policy makers
- customised innovation: Improving market-orientation and market-knowledge in Finnish companies
- barriers to (innovation) competition in Finnish markets

- managing the interface between design and manufacturing: benchmarking international best practice
- the internationalisation of the Finnish innovation system: current state and future challenges.

The aim of Sitra's innovative projects is to develop competitive new business activities and social innovations which promote the prosperity of the Finnish economy. The objective of the projects is to create new innovative co-operation networks. The cost of setting up such networks is often prohibitive for an individual company or organisation. On the other hand, once functioning, the benefits of networks may exceed their operating costs considerably. By creating new forms of co-operation, Sitra complements the networks that already exist within the national innovation system, and thus promotes the development of both the traditional and new sectors of the Finnish economy. (Sitra, 2000)

The Academy of Finland is an expert organisation for research funding. The main function of the Academy is to enhance the quality of basic research in Finland by long-term selective research funding allocated on a competitive basis, by systematic evaluation and by influencing science policy. The Academy of Finland grants over 150 million Euros (over FIM 900 million) for the funding of research in 2000. This represents about 12% of the total Finnish government research funding. The Academy's operations cover all scientific disciplines. The wide range of high-level basic research funded provides a sound basis for an innovative applied research and the exploitation of new knowledge. The Academy operates within the administrative sector of the Ministry of Education. (Academy of Finland, 2000)

Finland has five major Universities of Technology in Helsinki, Tampere, Lappeenranta, Oulu and Åbo. Three of these, Tampere, Oulu and Helsinki, are both active in giving highest education in construction and perform basic research. Finland has a number of technical colleges, which are very active at regional level in testing and in RTD as well. All universities and colleges are operating under The Ministry of Education and Culture.

The Finnish Trade Associations in the area of FM and construction have been active in promoting innovation and RTD activities. The Finnish Association of Business Facility Owners (RAKLI) has become one of the most active associations in regards to the restructuring of the industry. Its policy is to enhance the role of professional owners and users of facilities as customers of the industry's value adding chain. At the moment it manages one of the most important national technology programmes (Rembrand) developing service businesses in the real estate area.

The Union on Design and Consultancy Firms focuses mainly on promoting quality of design and productivity by developing trade specific quality management models and instructions. Its member organisations are not very active in RTD activities although the impact on construction is evident. The Finnish Contractors' Union (RTK) focuses mainly on labour policies, but is active in RTD as well. RTK co-ordinates 3 – 4 major RTD projects each year.

10.3.4   Commercialisation of Innovations

The Finnish National Fund for Research and Development (Sitra) invests in technology firms and venture capital funds, both in Finland and abroad. Sitra provides funding for promising technology enterprises by entering as a minority shareholder. The aim is to internationalise and grow the companies in co-operation with their management. Sitra invests in companies, which are based on technological innovations or other unique expertise. Sitra evaluates the following points when preparing for an investment decision (Sitra, 2000):

- Market potential of the products of the company
- the unique nature of technology and possibilities for protecting it
- growth potential of the company
- strengths and weaknesses of the management
- competitiveness (which is evaluated in detail).

Innovation Relay Centres (IRC) form a European wide network of 54 IRC-units with their partner organisations. Altogether, there are approximately 200 organisations in the network. The Finnish Innovation Relay Centres-network covers 13 units, which operate under the Technology and Economic Development Centres. They are all active in developing or transferring technology. The network offers assistance in acquiring new technology from universities or RTD performers or even from other industries. The funding of most EU-programmes requires co-operation with foreign partners. The IRC-network offers channels for this purpose as well.

Finland has a unique network of 13 technology centres in major cities. These centres are mostly jointly owned by research centres, universities, cities and companies. The technology centres aim at helping the application of new technology and give aid in commercialising innovations in enterprises. In addition, the centres aim at establishing new enterprises and supporting them by, for instance, means of knowledge and financing. The technology centres are a subgroup of the knowledge centre network. Several technology centres are active in co-operating with the real estate and construction industry.

Centres for Regional employment and economic development are a joint form of services to companies by the Ministries of Trade and Industry, Employment and Agriculture. The services are offered in 15 cities, covering all 19 provinces of Finland. Their tasks are to:

- Support and advise small and medium sized companies in all phases of their life cycle
- promote the technological development of their client companies
- assist companies in exports and internationalisation
- execute regional employment policy
- plan and perform training and educational programmes for the workforce
- take part in regional development.

The Regional knowledge centre –programme is a joint activity of 5 Ministries (Internal Affairs, Trade and Industry, Employment, Education and

Agriculture), industry and commerce, technology centres, universities, research centres, cities and municipalities. The knowledge centre programme promotes the use of high level international knowledge in companies and in creating jobs. The knowledge centre programmes are connected to technology, culture, design and media. Several knowledge centres have relevance to the real estate and construction industry.

There are 30 centres for new business operating in different parts of the country. Together these centres handle about 15,000 new business ideas and help about 5,000 enterprise getting a start every year. They receive a part of their funding from the state. Together with their networks of advisers (other entrepreneurs, experts in industry and commerce, authorities), the centres for new business help persons intending to be entrepreneurs and starting stage firms in e.g. following issues: evaluation of the business idea, establishment of a firm, creating contacts and a network of experts for the firm, product marketing (Suomen Jobs & Society, 2000).

Cities and municipalities support the utilisation of knowledge in entrepreneurship, and improve professional education to fit the needs of the local economy. Municipalities are often partners in local development companies etc., which systematically produce and direct operational and economical support into firms evolving and commercialising innovations.

10.4   EFFECTIVENESS OF PUBLIC INTERVENTIONS

Public interventions can never replace the role of the entrepreneur. Apparently, construction sector is not in the lack of services or economic support. Despite this only a small portion of companies make use of any sort of outside help. The services provided often seem too far from the everyday problems of the companies and practitioners. The best results have been obtained by interventions, which have had a close contact with practice.

How the different public policy instruments are used is revealed partly by two studies published by the Association of Finnish Entrepreneurs (1999 and 2000). According to the latest study, only 17% of the SMEs of the construction sector had acquired outside funding during the previous year. More than 80% of the SMEs were not interested in outside funding (over two thirds in 1999) for developing their enterprise. The majority of the companies feel that they cope on their own or do not wish to expand their operations. None of the SMEs are aiming at finding risk funding from the outside (4% in 1999). The main reason for this is the willingness to use own capital and the fear of losing power. Answers of the SMEs from construction sector to the question "Are you going to acquire outside funding of any kind in the near future" were the following in 2000 and in 1999:

	5/2000	**5/1999**
No intentions	84%	68%
Bank loan	10%	14%
Loan from a finance company	1%	10%
Outside risk capital	0%	4%
Other types of funding	5%	4%

The commercialisation of ideas or inventions often coincide with the start of an enterprise. It is normal, that only 50 – 60% of companies pass the "economical death valley" (the time it takes the net earnings of the company to turn to positive) during their first years in business. However, the share of surviving companies can rise even to 90% if economic and educational support is given to the entrepreneurs. There may even be a greater potential for benefits, if the process of innovation is better known and better controlled (Paasio, 1999).

Effective results have been recorded in the evaluations of the technology programmes, where the key success factor has been an adequate level of investment and risk taking in a specific focus.

Operation and production processes in the construction sector are based more and more on utililization of information technology. Today, construction sector companies have at their disposal advanced information technology systems aimed to support their internal processes. There has been a shift towards development of project-specific data management and transfer between the parties of a project. The goal is also to utilise design and production data in the management and maintenance of properties.

Life-cycle economics of buildings are becoming an important competition factor in construction projects. Up to now, decisions on investments in property and use-phase services have generally been made separately and independently. Acquisition costs of the investment have a central role in decision making, and often decisions are made only on grounds of acquisition costs. However, also the use costs of buildings should be included as a criterion in the decision making process. In recent years, methods and tools of life-cycle economics serving this purpose have been developed in R&D-programmes. (Technology Agency of Finland, 1998)

Tangible effects can be seen in the rise of interest of the companies towards sustainable construction and in the demand of tools to help companies evaluate their impact on the environment. The Construction Sector Environmental Programme has created criteria and recommendations for the assessment of the environmental impacts of building products, methods of calculating environmental loads, assessment of the usability of by-products and recyclable materials and the reduction of construction wastes. Effective results have also been recorded in several technology programmes, where the aim have been to find ways to decrease energy consumption connected to buildings.

Another effect has been the improvements in the area of quality management. Due to systematic development work from the beginning of 1990s, close to 300 companies of the Finnish construction sector now have quality system certificates and systems, which improve productivity and save in terms of costs of non-conformances.

10.5   TRENDS AND CONCLUSIONS

The Finnish construction sector is facing a new challenge: the merging of real estate and construction clusters is changing the business logic and processes. The paradigm shift is from production orientation to end-user orientation. This shift may take place very slowly, but definite indications of its effects on the sector are very much visible today already.

The real estate and construction sector is not known for its innovative nature. However, looking merely at the share of RTD investments of the average turnover, the comparison with other industries does not seem fair. Construction industry bases its innovativeness to different types of breakthroughs compared to high-tech areas like IT and communications. The success is very often based on combining different areas of knowledge. Due to this, successful innovations need co-operation between parties involved in the value chain. How to divide risks and benefits of RTD work becomes crucial.

Most of the SMEs of the industry face obstacles difficult to overcome when deciding on investing in RTD. In practice, it is not possible to find enough resources to invest in real breakthroughs, which actually would be justified with anything other than reducing production costs. Adding value and rising the price at the same time is virtually unknown in the sector. This seems to be the very reason for the unwillingness of the companies to take more actively part in any type of RTD work.

In terms of innovation systems and support, it seems very evident that there is no shortage of services for the companies of construction sector when it comes to providing assistance, consultancy and funding of innovation in Finland. However, the processes and practices should be tailored to fit the construction industry.   The following points should be taken into consideration:

- New procedures and practices should be considered in providing assistance to companies
- new procedures should be considered in ways of sharing risks and benefits of development work on building project level
- find out new methods for managing technology development work to take into account the aims of the businesses as well as end-users of building products
- measures to give more and transparent information on the effectiveness of RTD work should be developed and taken into use by companies as well as public agencies
- benchmark these activities among European companies in order to share knowledge of success.

In future, RTD performers need to shape their operations to fit the company needs in a more effective way. In particular, this applies to research institutes and centres. Interaction between basic and applied research and their connection to business processes is in need of redefinition. New approaches enabling the concurrent innovation process should be adapted to the industry.

Appendix 1 Ongoing Tekes Technology Programmes and Clinic Activities Related to the Real Estate and Construction Sector in 2000.

Ongoing Tekes Technology Programmes in 2000 (Technology Agency of Finland, 2000b):

Business concepts for industries 2000-2004
Business Process Re-engineering 1997-2000
Competitive Reliability 1995-2000
Control of Vibration and Sound - VÄRE 1999-2002
Environmental Cluster Research Programme 1997-2000
Environmental Technology in Construction 1994-1999
Healthy Building 1998-2002
Information Networking in the Construction Process - VERA 1997-2002
Infrastructure of sustainable society – EKO-INFRA 2000-2002
Lightweight Panels 1998-2002 - KENNO
Model Factory Concept 1996 - 2000
Progressive building process - ProBuild 1997-2001
Quality in Business Networks 1998-2001
Rapid Product Development – RAPID 1996-1999
Real Estate Management and Services 1999-2003 - Rembrand
Sensus® 1998 - 2003
Smart Machines and Systems 2010
Technology and development programme for the furniture sector DIVAN 1999-2002
Technology and development programme for stone industry 1999-2002 - STONE
Transport Chain Development Programme KETJU 1998-2002
Value Added Wood Chain 1998-2003
Wood Energy 1999-2003
Wood in Construction 1995-1999

Tekes financed Technology Clinic Activities in 2000 (Technology Agency of Finland, 2000c):

LonWorks-Clinic
Manufacturing and Applicating of Functional Materials Clinic
Product Adjustment Clinic for Building Materials and Products
Operation Strength Clinic
Real Estate Life Cycle Clinic
Technology Strategy Clinic
Technical Wood Clinic
Wood-fuel Clinic

CHAPTER 11

INNOVATION IN THE FRENCH CONSTRUCTION INDUSTRY: THE ROLE OF PUBLIC POLICY INSTRUMENTS

Elisabeth Campagnac and Jean Luc Salagnac

ABSTRACT

This paper provides an analysis of the innovation and public policy context in France, and its influence on the construction sector. French policy instruments for promoting innovation in construction reflect the general R&D and public policy context. The organisation of R&D activity in France has the following peculiarities:

- The innovation system is strongly regulated by public actors, with and important public research sector which also offers numerous incentives for promoting public-private co-operative R&D. The share of public funding is the highest in comparison to other OECD's countries
- the role of the State is shifting from direct funding towards a more incentive mode
- R&D activity is concentrated in research centres, which are rather independent of universities and other educational organisms. This situation has been changing during the last ten years with the development of tighter partnerships between research centres and universities
- a general trend is the development of networking between public and private actors, both at a national and at an international level
- private R&D is a combination of centralised structures linked to the main industries (for instance construction materials) and of a great number of small structures close to SMEs.

These features shape R&D activities in construction, even though R&D efforts in this industry, in spite of its economic importance, are far beyond other sectors.

Different approaches and programs are being developed to promote innovation in construction in the recent years, but results and effectiveness of these new efforts are still to be assessed.

11.1   INTRODUCTION

Current public policy instruments inherit features from a long tradition born after World War II.

The role of the State was then predominant. After fifty years, there is now a renewal of these instruments which is characterised by a shift towards a closer co-operation between private and public actors. This trend is also true for construction, though not yet so significant because of the small part of the total R&D budget dedicated to this sector of activity.

We first give data explaining the way these instruments have been built.

Then, we present some information concerning the construction sector, that confirm that R&D activity of this sector is hardly visible compared to other sectors, in spite of its economical importance.

Synthetic information on active policy instruments in construction is then given to complete this presentation.

11.2   GENERAL DATA ABOUT PUBLIC POLICY INSTRUMENTS

11.2.1   R&D budgets and human resources

Available statistical data allow to rank French R&D efforts among other European countries.

In 1994, the French R&D budget was 21% of the total R&D budget of the European Union[1]. The 28.4 billions Euros budget represented 2.4% of the French GNP. The rate of increase of this budget on the 1985-1994 period was 9%.

The origin of this budget is as follows:

State funding (%)	43
Private companies funding (%)	49
Non national funding (%)	8

Non national sources include European programmes (EC Framework Programmes, ESA) for European countries, R&D budgets allocated by multinational corporations to their subsidiaries.

Compared to other countries, France is the country where the share of state funding is the most important, but companies funding is now more important than public funding.

This latter point is the result of a long term evolution: over the 1959-1995 period, the share of State and private companies funding (including civil and military R&D) respectively shifted from 70% to 50% and from 30% to 50%.

---

[1]   Source: OECD and OST. (1998)

Over this period, the national effort of R&D was multiplied by a factor of 7.

Other indicators help characterising the French research activity, which is part of the European scientific production:

- 5.1% of the world scientific production is made by France (compared to 8.5% by United Kingdom and 6.3% by Germany)
- globally, 33% of world scientific production is made by the European Union members, compared to 34% by United States 2
- patents registered by the European office of patents represent 43% of the world activity (34% for USA and 15% for Japan).

About 400,000 people work for R&D activities (including researchers, research engineers, Ph.D. students, technicians, workmen and employees) both in the public (250,000) and in the private sectors.

Public R&D works in France are carried out:

- By public research centres (multidisciplinary like CNRS, or dedicated to specific fields like INSERM (medical research), INRA (agriculture), CSTB (building industry) or LCPC (civil engineering))
- by universities and schools of engineer (the so-called "grandes écoles")
- 100,000 full-time researchers or "part-time" researchers (mainly teachers)
- 110,000 technicians and administrative employees
- 42,000 PhD students.

The R&D centres of about 5,000 companies are in charge of the efforts of the private sector.

11.2.2 Main features of public civil research organisation

In France, 61% of the total public funding is dedicated to civil R&D, and 39% to military programmes. This situation is close to the UK situation (respectively 63% and 37%), but differs from the average European situation (80% and 20%), and from the German figures (91% and 9%).

The civil R&D works are carried out as follows:

- 65% by public research organisms, with various juridical structures allowing or encouraging commercial activity
- 30.9% by universities and schools of engineers
- 4.1% by non profit associations.

---

[2]   Mustar Ph. (editor) « Les chiffres clés de la science et de la technologie », Observatoire des Sciences et des Techniques (OST), Economica, édition 1998-1999.

The main part of the research is then carried out by organisms which are not linked to universities or other educational structures. Education is not among the main missions of these research institutions (like INSERM, INRA, CSTB), although information dissemination is a clearly assigned activity.

This situation is changing. More and more partnerships are being concluded between educational structures and research centres. An example is the development of associations between the CNRS and laboratories which are integrated in schools of engineers (this is the case of the LATTS at the ENPC).

11.2.3 Main features of private civil research organisation

In 1995, at least 4,600 companies and 50 professional bodies (technical centres) had a permanent and organised research activity. Such an activity is defined by the employment of the equivalent of a full-time researcher a year.

All together, the private sector employs 162,000 scientists and technicians, including 66,600 researchers and engineers, and 95,400 technicians and related workers.

This increasing number of companies involved in R&D is one of the most important changes of the French system of R&D during the last ten years. Most of the new comers are SME with less than 500 employees. They often take advantage of fiscal incentive mechanisms (tax credits).

The volume of research carried out by these companies increased of 61% during the last ten years (in other words, of 4.9% a year, compared with GNP increasing of 2.2% a year).

As in most of other countries, this R&D effort is concentrated on a limited number of private companies. Among the 4,650 involved in a research activity, 75% of the total R&D expenditure is carried out by 185 companies, which receive more than 90% of the public funding.

Industrial research is also highly concentrated on a few numbers of industrial sectors: 60% of the national R&D efforts are concentrated on five industrial domains[3]:

Sector of activity	Share of total R&D effort (%)
Aeronautics and space activities:	13.2
Car industry	12.9
Pharmaceutical industry	12
Electronics industry (radio, TV, communication)	11
Process control industry	10.3
Agri-business	1.8
Construction	0.8

---

[3]   Full report available, in French, at :www.finances.gouv.fr/innovation/guillaume/

Some of these sectors have significant expenses in R&D, compared with their turn over (for instance, the aerospace industry). On the opposite, some sectors with large national turnover have small R&D budgets. This is for instance the case of agri-business and food industry (1.8%) and construction (0.8%). These sectors take of course advantage of the results of R&D results obtained in other sectors and incorporated in the machines, processes and materials they use.

11.2.4 Links Between Companies and Public Research

In 1995, one third of the R&D budget of the companies was transferred to public laboratories (CNRS, INRA, INRETS, INSERM). This is ten times more than at the beginning of the 80s.

This illustrates the general trend that results from the will to build tighter links between public and private research.

The effects of this policy are for instance the creation of spin off companies by researchers (especially in high technology domains).

Other public tools are also available for private companies, for instance:

*   ANVAR (National Agency under the umbrella of the Ministry of Industry) that supports SMEs in the development of new products or processes (in 1998, 3,244 actions (feasibility and projects developments, recruitment of young researchers, etc.) have been engaged to support innovation in all sectors of activities with a total amount of 1.38 billions Francs (210 millions Euro)
*   incentive budgets to promote co-operation between public laboratories and industries
*   specific budgets dedicated to a limited number of projects presenting a "technological gaps".

11.3 THE CASE OF THE CONSTRUCTION INDUSTRY

11.3.1 Introduction

France is a 60 millions people country. Construction activity employs about 11% of workers and represents 9% of the GNP. In spite of this important economic position, R&D activity is far beyond the level of other sectors.

State funding for construction (buildings and civil engineering) has been very important (directly or indirectly through fiscal or other incentive means) in France during the second half of the 20th century. This has shaped the way the construction activity is regulated. Regulation of the relationships between the actors, especially within the frame of public markets has been strongly determined by the State.

Table 11.1 Main data about the French construction market (1997)[4]

Unit: Billion French Franc	Turnover	Clients			New constructions		Maintenance/refurbishment	
		Public	Private corporations	Private houses	Houses	Commercial buildings	Houses	Commercial buildings
Buildings	444	113	101	230	109	80	143	107
Civil Eng.	134	110	24		50		84	

4 D.A.E.I., 1997 (French Ministry of Construction): "Enquêtes annuelles d'entreprises", Comptes de la Nation, Direction de la Comptabilité publique

For more than ten years, the trend has been a decrease of the share of new construction compared to the activity on existing constructions. Individual houses market represents about 60% of the market of new houses. Export activity is 84 billion French francs.

Table 11.2 Turnover of the different types of companies on the construction market

Size of company	Number of companies	Turnover (billion F)
<10 employees	247,000	191
11 to 50 employees	14,000	128
51 to 200 employees	1,150	53
>200 employees	170	185
Major companies	5	32

Construction industry is directly bound to the general economic context and reflects its variations. Globalisation of economy and the decrease of the role of the State also affect the construction activity.

Significant changes also occur from the development of European regulations, concerning products performances evaluation and normalisation, as well as the relationships between actors. Globalisation of the economy and European integration are likely to deeply change the way construction industry is regulated.

To cope with these changes, construction industry has to be innovative concerning the development of new services associated to the design of product, the reengineering of the whole production process and the adaptation of its management to face the increase of fast changing and diverse markets.

11.3.2 Innovation in construction

General incentive mechanisms presented in section 11.2 are available for construction industry, but the share of the R&D budget is far beyond the share of the total activity.

For instance, only 4% of the annual budget of ANVAR are dedicated to construction related projects (including development for the site activity and for industry related projects such as product manufacturing or site machine developments).

The following table presents the main French organisms involved in construction R&D. They support the policy instruments dedicated to construction. For instance, CSTB is asked by the ministry of construction to conclude research contracts with industrial partners in order to develop innovative products for the French building industry.

Table 11.3 Framework for Positioning French Public Policy Instruments

	Name Web site (EVA = English version available)	Resources in Construction (Million Francs 1999)	Objectives	Means	Contribution to Innovation
Programs to support R&D	CSTB (Centre Scientifique et Technique du B) www.cstb.fr (EVA)	100	Promote the development of innovative products and methods for the building industry	Annual programme approved by the Ministry of construction	Development of innovative products and of assessment methods
	LCPC (Laboratoire central de Ponts et chaussées) www.lcpc.fr (EVA)	280	Promote the development of innovative products and methods for civil engineering	Annual programme approved by the Ministry of construction	Development of innovative products and methods
	ADEME (Agence de l'Environnement et de la Maîtrise de l'Energie) www.ademe.fr (EVA)	20 + 45 (dedicated to non renewable energy) +190 (for information diffusion)	Promote the development of environmental friendly technologies for buildings.	Research and prospective studies Expertise and advisory services Information and action to influence the behaviour of actors in the economy	Support to the development of design methods, of new products. Dissemination of results.
Programs to support advanced practices and experimentation	PUCA (Plan Urbanisme Construction et Architecture) www.equipement.gouv.fr/reche rche/default.htm	65	Stimulate innovation in the building sector (urbanism, architecture and construction technology and organisation)	A permanent team of 40 people in charge of programmes launching, projects selection, networking and programmes assessment. One of the main tools is experimentation on site.	Generation of new ideas and dissemination of results of experimentation and work groups.

Table 11.3 (Continued) Framework for Positioning French Public Policy Instruments

	Name Web site (EVA = English version available)	Resources in Construction (Million Francs 1999)	Objectives	Means	Contribution to Innovation
Programs to support advanced practices and experimentation (continued)	RGC&U (Réseau Génie Civil et Urbain) www.equipement.gouv.fr/recherche/default.htm	NC	Stimulate co-operation between construction-related industries and public laboratories.	The networks promotes the presentation of innovative projects and selects (gives a label) to some projects which are not financed by the network.	Co-operation between construction industry and public laboratories/
	RERAU (Réhabilitation des Réseaux d'Assainissement Urbains)	NC	Promote innovative technologies within the field of sewers, water distribution networks.	Working groups. Experimentation.	Promote innovative technologies and the development of assessment methods.
Programs to support performance and quality improvement	AQC (Agence Qualité Construction) www.qualitéconstruction.fr (EVA)	20	Promotes quality insurance philosophy in construction. -Reduction of the number and size of building activity defects -The improvement of quality in construction both by taking preventative action and by promoting the co-operation	10 persons. Collection of data on insurance declarations. Production of statistical analysis. Organisation of working groups related to sticky points.	Assessment of innovative technologies as used currently on sites.
	MFQ branche construction (Mouvement Français pour la Qualité)	NC	Promotes quality insurance philosophy in construction	Organisation of working groups related to stocky points. Production and dissemination of recommendations	Browsing of ideas. Formalising of procedures.

Table 11.3 (Continued) Framework for Positioning French Public Policy Instruments

	Name Web site (EVA = English version available)	Resources in Construction (Million Francs 1999)	Objectives	Means	Contribution to Innovation
	CTI (Centres Techniques Industriels) (concrete products, bricks and tiles, wood, etc...)	Depends on construction industrial branches (financing from a tax based on the activity)	Solve production problems of the industrialists which are affiliated to the technological centre. Develop innovative production methods	Each industrial branch has his own centre.	Patents. Development of production methods.
	MIQCP (Mission Interministérielle pour la Qualité des Constructions Publiques)	NC	Develop and assess procurement procedures within the frame of public markets.	Capacity to organise the discussion between all the actors involved in public construction markets	Browse ideas. Assessment of quality procedures.
Programs to support taking up of systems and procedures	ANVAR (Agence Françqie de l'Innovation) www.anvar.fr (EVA)	1 380 for all the industrial sectors (Construction = about 4% of the budget)	Facilitate the development of innovative products in SMEs in all the sectors of activity.	Expertise. Innovation projects evaluation. Grants and loans.	Capacity to select projects and to support the development of selected projects. Helps for the dissemination of innovations. Construction products (2/3), engine (1:3)
	ANRT / CIFRE (Agence Nationale de la Recherche et de la Technologie) www.anrt.asso.fr	NC	Facilitate the employment of young researchers within the frame of co-operation with public laboratories.	Incentive fiscal mechanisms.	Promote R&D activity in SME companies.

11.4 CONCLUSION

The order of magnitude of the share of R&D in construction (including building and civil engineering) in France is about 1 to 2% of the total French R&D effort, when the share of this activity in the national economy is about 8 to 10%. This situation is not exceptional and is rather similar in most of other countries.

Nevertheless, the overview of some of the French public research policy instruments shows that construction is identified as a specific area. These instruments do exist, even if the means which are dedicated are far beyond the means dedicated for more strategic industrial sectors.

For historical and sociological reasons, these instruments are characterised by a great importance of public funds, which has an impact on:

- The choice of research themes: public policy is driven by contextual factors which mirror the main subjects of interest on a given period. The key issues have for instance been productivity during the 60s and 70s, production management during the 80s, environmental management and IT during the 90s
- the access to these funding: public funded research programs can more easily be accessed by big companies than by SMEs
- the dissemination of results: in spite of efforts to disseminate information related to public funded programs, a great part of the construction community is not aware of these results. This situation is also due to the structure of this community, which consists of a very great number of small companies.

Most of the efforts made during the last few years tend to improve the management of these public funds.

Different ways are being developed:

- Participation in European programs: a part of the national budget is dedicated to projects involving other countries. A cross fertilisation of ideas and a better dissemination of results is expected
- promotion of a closer co-operation between private actors (especially SMEs) and public funded bodies: a faster development of close to market products is expected
- development of innovation networks: in order to promote exchange of experiences between different regions and different actors.

One of the problems to be solved is to find an equilibrium between the use of public funds and a fair return in terms of benefits for the different actors. This may be a direct financial return for the laboratory that puts money in a development project. It may also be a non-financial return such as a free licence to use a patented product or process, the right to use a specific equipment developed jointly with a private company, or any other agreement between the partners.

Table 11.4 R&D budget in Europe: position of France in European Union

Country	1994 R&D budget (billion Euros)	1994 Share of R&D budgets in GNP (%)	Evolution of R&D budgets aver the 1985-1994 period (%)
Germany	40.1	2.3	+ 15
France	**28.4**	**2.4**	**+ 9**
United Kingdom	24.2	2.2	+ 9
Italy	13.3	1.2	+ 2
Netherlands	6.3	2.0	+ 12
Sweden	5.2	3.3	+ 17
Spain	4.7	0.8	+ 10
Belgium	3.1	1.6	+ 9
Austria	2.7	1.5	+ 34
Finland	2.1	2.3	+ 24
Denmark	1.9	1.8	+ 27
Portugal	0.8	0.6	+ 38
Ireland	0.8	1.4	+ 120
Greece	0.6	0.5	+ 58
Total E.U	134	1.9	+ 12

Table 11.5 French R&D funding in 1994: public, private and non national sources
(source: OECD and OST. (1998)

Country	State funding %	Companies funding %	Non national funding %	Total funding (billions Euro)
U.S.A	41	59	0	181
Japan	27	73	0	75
E.U	41	53	6	134
France	**43**	**49**	**8**	**28**
Germany	37	61	2	40
U.K	37	50	13	24

Table 11.6 Evolution of the expenses of R&D in France (1959-1995)
(source: MENRT-DGRST, OST)

	1959	1967	1973	1979	1985	1991	1995
R&D budgets (billions of current Francs)	3.1	12.2	19.8	44.0	106.3	164.0	179.4
R&D budgets (billions of 1995 Francs)	**26**	**75**	**85**	**102**	**138**	**177**	**179.4**
R&D budgets / GNP (%)	1.15	2.16	1.76	1.78	2.25	2.40	2.35
Share of state funding (%)	70	71	63	56	57	54	50
Share of companies funding (%)	30	29	27	44	43	46	50
Share of R&D works carried out by public centres	55	49	42	40	41	39	38
Share of R&D works carried out by private companies	45	51	58	60	59	61	62

Table 11.7 Participation of French private companies and public centres in the
European programmes (source: CD-ROM Cordis, OST (PCRD 1990-1996)

	Total number of projects	Industrial	Public
Scientific and technical co-operation	1257	41	1216
Energy	666	301	365
Environment	807	113	694
IT	2044	1422	622
Industrial technology and material	1264	888	376
Life science	998	237	761
Total	7036	3002	4034

Table11.8 Employment of researchers and R&D engineers (source: OCDE - OST)

	Total number (equivalent to full-time job)	For 1,000 active people
USA	926,700	7.3
Japan	551,990	8.3
European Union	798,611	4.8
Germany	229,389	5.8
France	**149,193**	**5.9**
UK	146,000	5.1

Table 11.9 Number of employees in some of the public research centres
(source: projet de loi de Finance pour 1998, OST)

Name of the research centre	Scientific domain	Total employment
CNRS	Multidisciplinary	26,277
CEA	Nuclear energy	11,406
INRA	Agriculture	8,515
INSERM	Health	4,960
CNES	Space	2,416
CIRAD	International co-operation in agriculture	1,790
ORSTOM	Co-operation multidisciplinary	1,609
IFREMER	Sea	1,330
INRIA	IT	737
CEMAGREF	Agriculture	609
CSTB	Building	560
LCPC	Civil engineering	550
INRETS	Transport	411

Table 11.10 The key technologies: international comparisons on technological
position (based on world share) 1996 (source: Mustar Ph. (editor)
« Les chiffres clés de la science et de la technologie »,
Observatoire des Sciences et des Techniques (OST), Economica, édition 1998-1999)

	USA %	Japan %	E.U %	France %	Germany %	U K %
Electronic components	43.1	22.5	28.6	**5.0**	12.0	3.5
TV and telecom.	33.1	29.5	31.9	**5.3**	7.3	8.0
Computers	67.4	9.1	19.5	**4.9**	4.8	5.1
Instrumentation	47.1	9.9	36.4	**7.2**	13.5	6.3
Chemical and pharmacy	59.8	5.4	25.8	**6.6**	5.5	5.8
Biotechnology	57.1	5.3	29.7	**6.4**	6.4	8.2
Material	34.0	18.4	39.3	**7.8**	15.1	6.6
Industrial process	36.1	15.7	41.6	**6.5**	20.7	4.6
Environment	24.9	6.8	59.1	**12.2**	25.9	6.2
Transportation	20.0	18.1	53.1	**12.1**	27.1	3.3
Construction	48.4	5.6	33.7	**5.8**	12.5	6.3
Total	43.8	13.0	36.0	**7.2**	13.6	5.8

CHAPTER 12

INNOVATION AND INNOVATION POLICY IN THE GERMAN CONSTRUCTION SECTOR[1]

Thomas Cleff and Annette Rudolph-Cleff

12.1   INTRODUCTION

The German building and construction industry has been experiencing difficult times for a long time now. For four years in the old Länder and for three years in the new Länder, construction output has really been on the decrease. The building and construction industry in the new Länder reveals particular weaknesses. Here, the downward trend is really dramatic. For 1998, total construction output reached a value of 369.73 billion DM (1995 Prices) and was therefore around 20 billion DM above the value for 1991. This increase in value corresponds to an actual increase of 5.9% in construction output. In the same period of time, the gross national product increased from 2,853.6 billion DM to 3,186.7 billion DM, meaning that the relative growth rate of the GNP lay at 11.7%. From that, one can deduce a decrease in the share of the national economic net product represented by construction output, falling from 12.2% in 1991 to 11.6% in 1998. For the old Federal territory, this negative change turned out to be much more emphatic: in this case, the share of the GNP represented by construction output actually fell from 11.3% (1991) to 8.3% (1998).

The tendency to reduce the average size of a business is an incessant trend in the German building and construction industry.   This statement is especially true for the East German building and construction sector. In the new Länder the medium-sized business categories (20-49, 50-99 employees) are represented more frequently than in the old Federal territory, meaning that an altogether higher value for the average size of business arises for the East.

The building and construction sector is structured in a medium-sized manner, as already demonstrated by the average size of business, and in view of this fact, vast differences between East and West can no longer be determined. Also signalling this is the high proportion of workshops among the total amount of businesses belonging to the building and construction sector. In Germany, 72% of businesses are classified as workshops (71.4% in the West and 74.1% in the East) and 76.1% (76.2% in the West and 76.0% in the East) of all workers from the

---

[1]   This paper is dedicated to our friend and assistant Brendan Schumacher. During the work on this paper Brendan was killed in a tragic car accident. You are no longer there where you were before, but you are everywhere where we are now.

building and construction sector are employed in these businesses. In 1998, around 1,180,000 people were employed in 81,301 businesses belonging to the building and construction sector.

Overall, due to the negative economic trend and the decrease of expenditure for public works one must assume that even in the near future construction and construction-related activities will develop more slowly than the economy as a whole. And this will not come without affecting employment in this sector. In a construction market which is becoming narrower and narrower, technological innovations through the improvement of processes and products are becoming the "sine qua non" of a competitive enterprise. In order to maintain or improve the competitiveness of a construction company, the technological surroundings of a construction company, along with its integration into networks and flows of knowledge, is subsequently of particular significance.

In a first chapter, this paper will deal with the flow of knowledge and new ideas among different actors and institutions and might lead to a better understanding of innovation in the German construction sector. Questions concerning the innovation behaviour of construction firms are to be answered on the basis of results from the Mannheim Innovation Panel[2]. In 1997, 163 companies of the German construction sector[3] took part in the survey. The results are weighted corresponding to the distribution of the population reported by the German Federal Statistics Office with regard to stratification of company size and region (new and old Länder).

From the point of view of an evolutionary approach, the national system of innovation is path-dependent in a way that traditions and historical issues lay a basis from where the system can move on. It is not necessary to deeply analyse the structure of German research history to find that the federal structure of society forms the basis of the institutional pattern of the German innovation system. The funding for education, research, technological transfer and innovation performance are shared between the Federal Government, the Länder's governments, and the private business sector. Out of this a variety of innovation-supporting programmes in order to improve the capability to innovate were developed.

These programmes are principally aimed at new technologies and new branches such as the IT-sector. This funding often overlooks traditional branches, although users or even lead-users of new technologies often don't participate. In a

---

[2]   On behalf of the Federal Ministry of Education, Science, Research and Technology (BMBF), the ZEW in co-operation with infas has been conducting annual systematic surveys on the dynamics of innovation in the German economy since 1993 (The Mannheim Innovation Panel). Approximately 2,500 companies from the manufacturing, mining and construction sector took part in the survey. The survey focuses on the development and diffusion of innovative activities, the development of innovation success, the significance and structure of factors hampering innovation as well as the extension and results of public promotion of innovative activities. The target group of the survey is companies with at least five employees.

[3]   The construction sector consists of all firms with activities in the NACE-subgroup "(45) construction": (451) Site preparation, (452) Building of complete constructions or parts thereof, civil engineering, (453) Building installation, (454) Building completion, (455) Renting of construction or demolition equipment with operator.

second part to this paper, the extent to which the construction sector is integrated into the innovation funding of German politic will be examined.

12.2　RESEARCH AND INNOVATION ACTIVITY IN THE GERMAN CONSTRUCTION SECTOR

12.2.1　Innovation Activity in the German Construction Sector

After a marked decrease at the beginning of the 1990s, 1995 shows an initial stabilisation and 1996 an increase in **innovation activity** of the manufacturing sector.[4] Companies extended their budgets for innovation and the share of innovative firms has clearly risen up to 60%, a trend that continued in 1997 and 1998. In some industries, the level of 1992 was reached for the first time. The development of the innovation activities is a time-lag picture of the business cycle. Correspondingly, the share of innovators in the construction sector in 1996 again increases slightly. Above all, this can be led back to the continuous increase of companies with product innovations. However, with a share of 33% of innovating companies (see Figure 12.1), the construction sector is one of the industries with below-average innovation activities. The share of innovators rises with the company's size: the highest share of innovators is in large companies with 200 or more employees (59%). For companies with between 100 and 200 employees, this share lies at 49% and for those with up to 100 employees, only at 30%.

Innovative construction companies most often report innovation expenditures for "machinery and equipment" (see Figure 12.2). After that come expenditures for "training of employees", "intramural R&D" and expenditure for the acquisition of "external know-how". The frequency of mentioning expenditures for R&D and "design of products" rises with the company size. On the other hand, it falls for expenditures on patents and licenses. However, the frequency of mentioning different types of expenditures for innovation in the construction sector vanishes behind this frequency in the manufacturing industry. In the new Länder such structured connections cannot be developed: still, expenditures for "machinery and equipment" are mentioned most often, but the individual expenditure positions are mentioned less frequently than in the old Länder. Innovators in the new Länder therefore rarely use the whole spectrum of innovation activities. What is striking in comparison to the West, is the significance of "training of employees", which is high and rises in accordance with company size. Here there seems to be a need to catch up for the East German construction industry.

---

[4]　This paper will accordingly use the term "innovation activity" in its wider sense as construed in the "Oslo Manual". (OECD, 1997a).

Figure 12.1 Innovation behaviour in the construction sector
(source: ZEW, 1998)

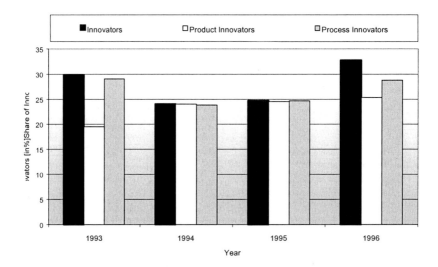

12.2.2   The importance of R&D and other sources of innovation

At first glance, the construction industry seems to be neither an R&D-intensive industry nor very innovative and intense user of technology. Altogether, only 35% of innovative construction firms carry out their own R&D activities. The total R&D expenditure of the construction sector amounts to 176 million DM. The R&D intensity in terms of R&D expenditure as a share of sales or construction volume in 1995 is 0.3% (Stifterverband, 1997). That is far below the R&D intensity of almost all manufacturing and service industries.

As important as R&D, is the number of various information sources that firms use to come in touch with fruitful ideas. Some sources are closer to the market, e.g. suppliers, customers or competitions, than others, which are more related to the science fields of universities or private or government R&D labs. The competitors have the highest significance as an external source of information, followed by suppliers and clients. Information from these groups can be gathered in conferences and meetings as well as in journals. The importance of fairs or exhibitions is a clear hint towards the role that informal communication plays in a national system of innovation. The demand-pull seems to be far more important than the technology push both to small and large companies of the construction sector. Universities are the most important information source used by construction firms to get research-based information. Research institutes (Fraunhofer institutions or Max-Planck) and patent disclosure fall behind universities and technical colleges. Altogether, the significance of individual information sources rises with company size.   Particularly for research-based information, e.g. from universities, this relation is distinctive.   This indicates an absent ability to absorb technical know-how in small companies (SMEs).   Large companies can frequently fall back on staff divisions, engineer divisions etc. which – having own university education – can integrate external knowledge into the company.   In small companies these interfaces are often missing.

Figure 12.2   Structure of innovation expenditure in Germany
(source: ZEW (1998):   Mannheim Innovation Panel)

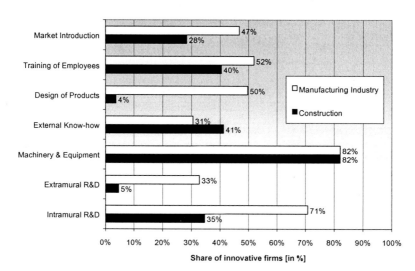

Details weighted according to the number of innovative companies.

An intensive form of knowledge-flow frequently goes beyond the simple use of different information sources.   Vertical and horizontal co-operation enables a construction firm to gain from the knowledge of the particular partner. Of course, at the same time there is an outflow of knowledge to the partner.   Here the observation is confirmed that companies that are concerned about a know-how drain to the competitor will seek to implement their innovation development work preferably without involving other institutions as partners.

Figure 12.3 Share of sources of information that have great significance for innovators
(source: ZEW, 1998)

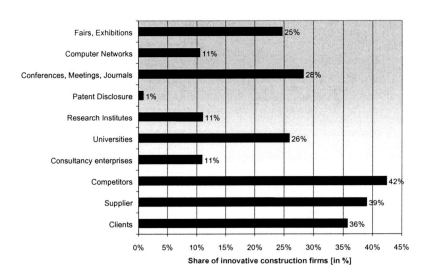

Details weighted according to the number of innovative companies.

Co-operative projects both between the science sector and the industry and inter-firm collaborations are the most effective form of knowledge and technology transfer.   The co-operation partner can always contribute complementary knowledge only. As was the case for the frequency of mentioning of different types of expenditures for innovation, it becomes clear for co-operation that the probability to co-operate is much higher in the manufacturing industry than it is in the construction sector. In the construction sector the inclination to co-operate in the new and old Länder, as well as between the different size groups of companies, is developed in different ways. Whereas in the old Länder 10% of the innovative construction firms co-operate, the respective number for the new Länder is 23%. This is an indication that a large proportion of East German construction firms has to obtain their know-how from West Germany. Particularly distinctive is the co-operation behaviour in firms with between 50 and 99 employees. For one, this "critical" firm size allows for the adoption of greater construction plans, but on the other side of the coin, such projects require specialist knowledge which, particularly among new technologies, goes beyond internal company know-how and must correspondingly be bought.

Figure 12.4 Share of co-operating innovators (source: ZEW, 1998)

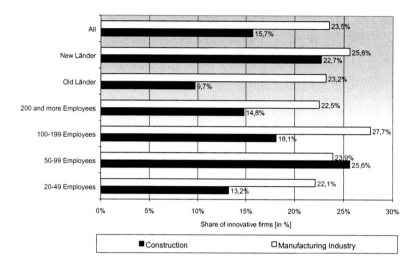

Details weighted according to the number of innovative companies.

Clients are the most important co-operation partners of construction firms followed by competitors and suppliers. Universities and research institutes (Fraunhofer institutions or Max-Planck) fall significantly behind these co-operation partners.

Large construction firms co-operate more with partners in the own company group as well as with customers and consultancy firms. For innovative small and medium sized enterprises, competitors or suppliers are the most preferable partners. It is obvious that SMEs co-operate on a lower level with institutions in the science sector such as universities and research institutions. The ability to adapt external technical knowledge from research institutions is conditional on an in-house corporate innovation potential. In-house innovative activities, expressed in R&D or innovation intensity, increase the likelihood of companies co-operating with universities.   Empirical analysis for the manufacturing sector likewise reveals that the probability of small and medium sized companies co-operating with a university rises.

Details weighted according to the number of innovative companies:

- the larger the company is in terms of payroll
- the higher the company's R&D personnel intensity is[5]
- the higher the significance of universities as sources of information is
- the less the company sees its innovations as easy to imitate
- when the CEO has a doctorate.[6]

Figure 12.5 Share of co-operation partners of innovators
(source: ZEW, 1998)

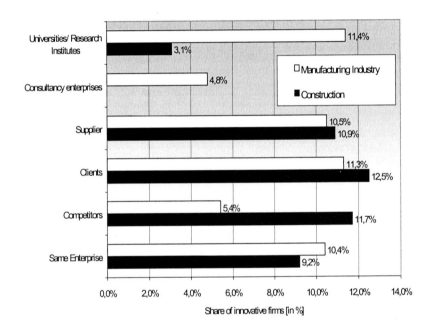

In summary, it can be observed that the inclination to co-operate in the construction sector is – compared to the manufacturing sector - developed below average. Still, as can be seen in Figure 12.3, very different sources of information are used to improve the quality of innovations. Size-related effect is highly significant in the case of science-based information partners. When it comes to co-operative projects with research institutions, size effect is less marked. The greater affinity of large companies with universities is also transferred to their small and

---

5   The R&D personnel intensity is the quotient of the number of R&D employees and the
        total payroll.
6   Beise et al. (1995).

medium-size subsidiaries: subsidiaries of large companies are more likely to co-operate with universities than are independent medium-size companies of equal size.

Universities and research institutes are a vital source of information for many companies' innovation activities. Furthermore, the transfer of personnel is a particularly essential component in the transfer of know-how. Other firms only introduce a new technology when it has developed out of a stage of R&D into a standardised technology. The use of standardised technologies requires a smaller amount of collaboration with universities, but nonetheless makes it possible for the firms to improve products and processes and consequently improve competitiveness. Which structural factors exert an influence on the strategic decision for the moment of introducing technologies in the construction sector, are to be examined in the following paragraph.

12.2.3 Technological life cycle: a conceptual framework

Every technology goes through a life cycle, the phases of which have various strategic competitive approaches. In the realms of a technology management, one tries to recognise the development from one phase into another and correspondingly control the use of resources and the required flow of information. Here firms must be able to drawn lines between the various phases. Here Arthur D. Little (1993) mentions qualitative differentiation characteristics, whose operationalisation remains open. In the framework of this paper, variables from the Mannheim Innovation Panel concerning the differentiation of individual technology phases will be consulted upon.

Figure 12.6 clarifies this, in that it differentiates between the use of innovation expenditure which, in turn, can be categorised into four phases of the technology life cycle. In this way, the 4 phases conceptually depends on a technology's degree of maturity (see Ford and Ryan, 1981; Arthur D. Little, 1993). This model illustrates a linear approach of innovation from R&D to standard practice. Many authors consider innovation processes as more complex and non-linear. However, differentiating the phases by means of a linear model makes a categorisation of the individual firms into the phases of research, advanced practice, best practice and standard practice possible.

Some companies make use of the opportunity for developing new knowledge through carrying out their own R&D or through assigning R&D duties to another party. A firm is ordered to the **research** phase if it has carried out intramural or extramural R&D. Within this phase, new product or process ideas develop, which are made available through spin-off or spill-over effects of **advanced practice**. Due to their degree of diffusion, which is still small, the new technologies still have a high development potential. Here it is a matter of potential key technologies that are able to fundamentally change the competition. Products and processes are made marketable in this phase and, through further refinements and targeted marketing instruments, they are used for the improvement of company's competitiveness. In the framework of operationalisation, an

advanced practice comes in action when a company uses universities or other
higher education institutions, research institutions or a patent disclosure as sources
of information for their own innovations, or co-operates with universities or
research institutions in the application of R&D results, instead of carry out
intramural R&D or extramural R&D.   The share of companies in the research-
based and the advanced practice phase of the technology life cycle lies at 49% of
innovative construction companies.

Figure 12.6 The technology life cycle (source: ZEW)

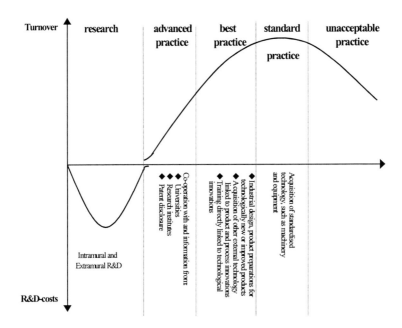

During the growth phase of a new technology (**best practice**),
application-related knowledge is further matured. The technological risk from the
individual company's point of view is smaller than in the formation phase of an
idea.   Through improvement innovations and by adapting to the customers'
specific wishes, more and more application possibilities of the new technologies
gradually open up. At the same time, the remaining market potential slowly
decreases (Pleschak and Sabisch, 1996). Firms belonging to best practice show
themselves in the Mannheim Innovation Panel by way of expenditure for the
industrial design, product preparations for technologically new or improved
products, for the acquisition of other external technology linked to product and

technological innovations. The proportion of best practice innovators lies at twelve per cent of all innovative construction firms.

In the maturing phase (**standard practice**) of a technology, measures of standardisation are to the fore. The differentiation potential of the technology is very small, meaning that through this technology, companies rarely stand out from their competitors any longer. The technology degenerates more and more into the basic technology (Specht and Beckmann, 1996). In the Mannheim Innovation Panel, as far as their innovations are concerned, innovators of standard practice exclusively restrict themselves to the acquisition of standardised technology, such as machinery and equipment. The share of these companies amounts to 34% of all innovative construction firms.

The descriptive analysis has identified innovators acting in different phases of the technology life cycle. The aim of a multivariate analysis now is to study the influence of independent structural factors - such as the company's size, labour productivity and the geographical origin of the company (old or new Länder) - in order to quantify the specific contribution made by each individual factor to the decision of a construction firm to engage in one phase of the technology life cycle. The dependent variable (phase) has an ordinal structure, whose value becomes smaller the closer a company is to being active at the beginning of the technology life cycle (research). The robust ordered logit model in Figure 12.7 shows an influence of company size which is particularly significant. The larger a company is, the more likely it is to be operative in the early stages of a technology life cycle (research, advanced practice). Regional aspects and labour productivity, however, do not play a significant role. Only in the decision for or against R&D are we able to identify regional influences with a weak significance. The results of a maximum-likelihood logit model show a more frequent R&D activity for West Germany than for East Germany.

Figure 12.7 Ordered logit model to identify different types of innovators
(source: ZEW, 1998)

Ordered logit estimates				Number of jobs	=	58	
				Wald chi2 (3)	=	15.23	
				Prob > chi2	=	0.0016	
Log likelihood	=	-55.776865		Pseudo R2	=	0.1056	

	Robust						
Aspect	**Correlation**	**Std. Err.**	**z**	**P> z**	**[95% Confidence Interval]**	
Company size	-.4694885	1.753277	-2.616	0.009	-8212259	-1177512
New laender	-.4694885	1.753277	-2.616	0.164	-3399014	2.007654
Labour productivity	1.753277	3.535548	0.496	0.620	-5.176271	8.682825

The political instruments of innovation funding are supposed to support the dynamic process of acquisition of new knowledge in the various phases. Companies are to be put in the position where they can substitute for old products and processes through the help of new technologies, so as to maintain a long-term competitiveness and thus not find themselves in a situation of "unacceptable practice". The efficient employment of public policy instruments thus requires the knowledge of difficulties regarding the companies' abilities to generate or adapt new knowledge in the four phases of the technology life cycle. Above all for the construction sector, there have been no empirical investigations in this area up until now.

For this purpose, Winch (1999) theoretically systematised public policy instruments into "construction specific" and "general". In a further dimension he subdivided public policy instruments, with respect to their policy objectives, into "direct promotion of innovation objectives" and "other policy objectives with consequences of stimulating innovation".

Figure 12.8 Public policy instruments (source: Winch (1999))

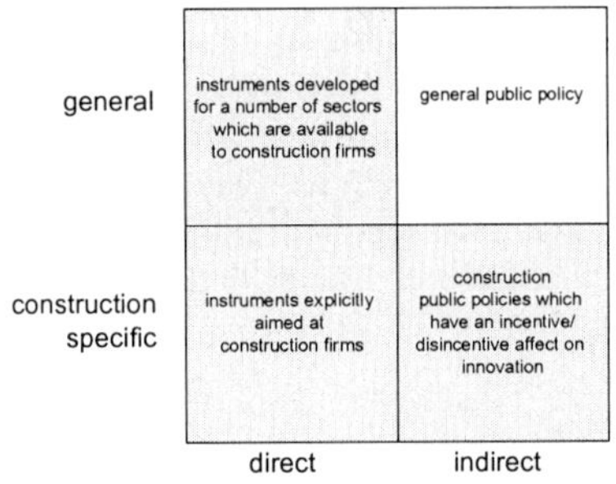

There is absolutely no doubt that general indirect public policies such as fiscal policy exert an important impact on the innovation abilities of firms. Nevertheless, influences on sectors outside the construction sector are to be ignored at this point. In the following, only those instruments that directly or indirectly improve the capability of construction firms to innovate, will be dealt with.

Due to this reason, to start with, the institutional mapping of the System of Innovation is described in the following. Spill-over and spin-off effects from institutions such as universities, research institutes etc. directly support the whole industry's ability to innovate (see section 12.3.2).

Furthermore, the different national supporting programmes are examined as to the extent in which they directly or indirectly support innovations in construction (see section 12.3.3). One question in particular arises here, namely which construction firms are involved to what extent in innovation supporting programmes (see section 12.3.4). The following sections show the development of Federal and Länder's expenditure on innovation.

12.3   INNOVATION POLICY IN THE GERMAN CONSTRUCTION SECTOR

12.3.1   Public expenditure on research

German expenditure on "Research" – comprising of all funds for research and experimental development, resources for scientific education and training and other scientific and technological activities – amounted in 1997 to approximately 109.2 billion DM. Between 1991 and 1997, a continuous rise of 13.6% can be seen. During this time the share of the Federal Government declined from 20.9% to 18.4%. In contrast, the share of Länder and local government – mainly due to the development of expenditure in the new Länder – increased from 27.9% to 31.9%. Although the share of all public authorities increased from 50.5% to 51.8%, the quotient of research expenditure to the total public budget decreased from 5.0% to 3.0%. Over the same period, the share of the business enterprises sector decreased from 49.5% in 1991 to the minimum of 47.3% in 1996. With the recovery of the economy in 1997 we discover an expansion up to 48.2%.[7] Research and Development (R&D), with 83.7 billion DM for 1997, forms the main part of the research expenditure. This amount also includes the costs for R&D performed abroad (2.7 billion DM in 1995). The more appropriate "gross domestic expenditure on R&D" (GERD) for evaluating the expenditure for R&D performed in Germany continually increases by 11.1% from 74.5 billion DM in 1991 to 82.8 billion DM in 1997.[8]

During the recession between 1991 and 1994, the enterprise sector lowered their expenditure on R&D to intensify R&D activities from 1995 onwards. Between 1994 and 1997 we observe a reversed trend with a continuous rise of 8.4% of R&D expenditure. The development of the higher educational, governmental and non-profit R&D expenditure is a mirror image of the development in the private industry: whereas the R&D expenditure growth was rapid between 1991 and 1994[9], we see a stagnation from 1995 onwards. Nevertheless, the important share of the business enterprise sector at 67% of the gross domestic expenditure on R&D is decisive for the development of the total R&D expenditure: between 1991 and 1994 the total gross domestic expenditure on R&D rose by only 3.6% whereas during 1994 and 1997 the increase was double that, at 7.5%. Furthermore, the rise of GERD in the business enterprise sector and the decrease in other sectors led to a shift in the sectoral structure of GERD in

---

[7]   BmbF (1998), p. 9
[8]   BmbF (1998), p. 11.
[9]   8.7% Government and private non-profit sector; 18.7% higher education sector. The growth rate in the higher education sector due to expenditures to improve the system of research institutes and universities in the new Länder. cf. BmbF (1998), p. 15.

1997. The share of the business sector increased between 1996 and 1997, from 66.3% to 67.0%.[10]

Figure 12.9 Gross domestic expenditure on R&D (GERD) in Germany
(source: BMBF, 1998, p. 13)

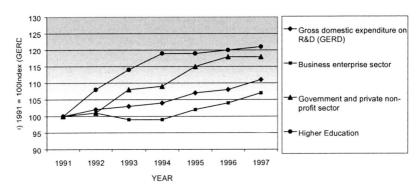

Change in the methods applied to the determination of R&D expenditure in the higher education sector between 1994 and 1995.

12.3.1.1   Federal expenditure on R&D

In 1981 the Federal Government spent 10.5 billion DM on R&D. With an average annual growth rate of 4.7%, the federal R&D expenditure increased to a total of 17.4 billion DM in 1992. Since 1992, we state an inconsistent decline with an annual average of –1.1%. A total of 66% (in 1996) of the federal R&D expenditure was financed by the "Bundesministerium für Bildung und Forschung (BMBF)"[11], 17.5% by the "Bundesministerium der Verteidigung"[12], "Bundesministerium für Wirtschaft"[13], and 11.1% by other ministries including the General Fiscal Administration.[14] Federal research funding is compartmentalised in the **funding types** of project funding, basic funding for institutions (statutory funding), university-related funding and international co-operation. Their share of the total expenditure on research over the long-term period between 1981 and 1998 reveals the different evolution of these funding types: the share of project funding declined from 51% to 36%. This trend was due in particular to the winding-up of major nuclear energy research projects and of funding for civil aircraft development. The increase in basic funding for institutions (35% to 44%) is

---

10   BmbF (1998), p. 15
11   Federal Ministry of Education and Research.
12   Federal Ministry of Defence.
13   Federal Ministry of Economics.
14   BmbF (1998), p. 19

influenced by the foundation of new research institutes and the improvement of research infrastructure in the new Länder. The above-average increase for university-related funding and international co-operation up to 1993 has seen the shares rise from 8% to 13% and from 7% to 9%, respectively. Since 1993, there has been a brief slowing-down in the share of these funding types.[15]

The **structure by group of recipients** reveals a below-average growth rate of the federal expenditure on research in the business enterprise sector. The share of the private industry diminished by sixteen percentage points, from 38% in 1981 to 22% in 1998. In contrast, the share of the private non-profit organisations – particularly due to the above growth rate of the funding of institutions such as the MPG, DFG and FhG – increased from 30% to 40% over the same period. A slightly smaller but also above-average (six percentage points) rise from 25% to 31% was seen in the share for territorial authorities, whereas recipients abroad have a steady share of 9%.[16]

Within the **business enterprise sector** it is the share of the mining and energy sector which is apparently disappearing more and more behind the other sectors. Between 1981 and 1997 we see a continuous decrease of that share, from 10.2% to 1.0%. Diametrically opposed to that, the share of expenditure allocated to the service sector increased from 9.0% to 20.4%. The construction sector has a share of only 0.5%. Due to the funding of renewable raw materials, the share of agriculture and forestry increased from a low level 0.1% to 1.0%. The manufacturing sector is still the strongest sector with a share of 78.1%, with a slight decrease of three percentage points since 1981.[17] The manufacture of electrical and optical equipment as well as the manufacture of air and spacecraft gets the lion's share of 28.0% and 23.7%, respectively. "Other services" (18.5%) and the manufacture of machinery and equipment (9.0%) already show a large gap to the above-mentioned branches. All other branches follow only at a large distance.

---

[15]   BmbF (1998), pp. 33.
[16]   BmbF (1998), pp. 34.
[17]   BmbF (1998), pp. 388.

Figure 12.10 Federal research expenditure in the business enterprise sector by industry in 1995
(source: BMBF, 1998, p. 388)

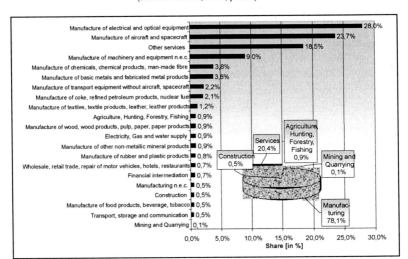

12.3.1.2   Building research

"Departmental research funded by the Federal Ministry for Regional Planning, Building and Urban Development and the Federal Ministry of Transport provides the basis for improving housing and living conditions and for modernising and maintaining the infrastructure in the building and transport sector. The BmbF funds R&D projects aimed at preserving historical monuments to contribute to safeguarding the cultural heritage for future generation"[18] The construction research activities primarily aim at reducing construction costs and rationalising the construction process. Every year the Federal Ministry for Regional Planning, Building and Urban Development invites proposals under a **building research programme** (investigator-initiated research) which defines the various research priorities. Building research currently focuses on "saving construction and housing costs", "avoiding building damage", "accessible dwellings for everybody", "health and environment protection" and "cost-effectiveness and rationalisation". **Road construction** research does not address extensive areas, but rather has to deal with a large number of smaller projects, such as "environmental protection", "road engineering", "road traffic engineering" and "federal trunk road network". There are some further thematic priorities, such as **"building provision and civil defence"**, **"hydraulic engineering** research" and **"regional planning and urban development"**.

---

[18]   BmbF 1998.

Figure 12.11 Expenditure on building research [in Mio. DM] (source: BMBF, 1998, p. 143)

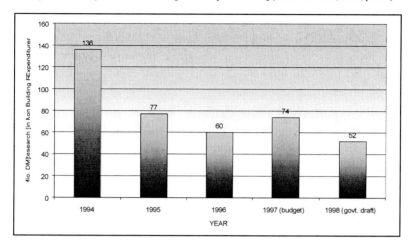

The government expenditures to support building research have continuously decreased over recent years. In 1994 expenditures amounted to 136 million DM, for 1998 this amount decreased by more than half to 52 million DM. In searching for promotion for their innovation plans, in the future the construction companies will have to concentrate more and more on general promotion programmes outside of building promotion. Only time will tell whether this will happen to the expense of innovative ideas in the construction sector.

12.3.1.3 *Expenditure of the Länder Governments on research and joint research funding by Federal and Länder Governments*

The research funding of the Länder focuses on higher education and comprises of all basic funds for research and education as well as the external funds of the Länder. A total of 85.6% of the Länder's funding budget in 1995 flowed into the "higher education sector including university hospitals" (28.3 billion DM). "Non-university science and research", comprising firstly of expenditure on joint research funding by Federal and Länder governments, secondly of funds allocated to financing institutions and thirdly of funding schemes for the business sector, covers 14.4% of the Länder's funding budget. The latter share has risen slightly in recent years.[19]

The joint research funding by Federal and Länder Governments is almost used as basic funding for jointly-financed organisations. The Federal Government's share amounts to 68%.

---

[19]   BmbF (1998), pp. 37.

Figure 12.12 Joint research funding by Federal and Länder Governments in 1998
(source: BMBF, 1998, p. 41)

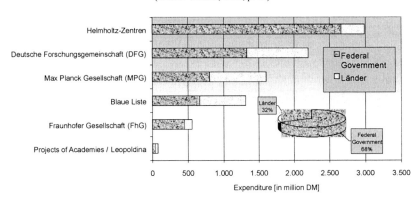

12.3.2   The institutional mapping of the German innovation system

Education and training for increasing human capital, generating and publishing
knowledge from research and contractual co-operation are the main ways in which
the (publicly-funded) sector of research institutions directly improves the
innovation success of firms in different industries.   The direction of the
institutional support can be looked at in four main areas:

- support of Promotion and Supporting Organisations (sections 12.3.2.1
  and 12.3.2.2),
- support of Federal Institutions Performing R&D (section 12.3.2.3),
- support of Universities and University-related programmes (sections
  12.3.2.4 and 12.3.2.5),
- support of Transfer Institutions (section 12.3.2.6).

12.3.2.1   Promotion organisations

The **DFG** (German Research Council, Deutsche Forschungsgemeinschaft) -
promoted jointly by the Federal and Länder governments - is the major promoting
organisation for science and research activities in Germany.   The federal R&D
expenditure on DFG basic funding increased continuously by 24%, from 871
million DM in 1994 to 1,083 million DM in 1998 (govt. Draft).   As a self-
government organisation, its main tasks are the funding of research projects, the
support of research co-operation and the promotion of young scientists.   The DFG
develops and maintains the relationship and co-operation with international
research institutions and it is an important consultant for the policy-maker in terms
of scientific questions.   For that, the DFG dispose of formal instruments such as

postgraduate and postdoctoral programmes; launching special research areas, and supporting teams of researchers from different institutions and fields of interest.[20]

The **DAAD** (German Academic Exchange Service, Deutscher Akademischer Austauschdienst) supports and organises exchange programmes for students, postgraduate and postdoctoral researchers with foreign universities and research institutes. The DAAD is an intermediary for the implementation of foreign cultural and academic policy as well as for educational co-operation with developing countries. Furthermore, the DAAD is the national agency for the EU programmes SOKRATES, LEONARDO and TEMPUS, as well as the IAESTE National Committee for the exchange of student trainees. The organisation is financed by the Office of Foreign affairs, the Ministry of Education, Science, Research and Technology (BMBF), and by some of the foreign co-operation partners. The main instruments are scholarships and financial support for exchange students and researchers as well as the organisation of the administrative formalities. Additionally, it provides information on foreign research programmes, universities, teaching opportunities and so on.

The **Stifterverband für die deutsche Wirtschaft** is an association of firms, various private non-profit organisations and private persons. It supports science and technology projects as well as institutes or other organisations which need additional financial or organisational help for performing R&D. The Stifterverband provides services such as statistics of economic indicators, seminars and infrastructure facilities for scientific activities.

Many other Promotion Organisations exist in Germany, the list above represents just a small selection of the most important ones.

12.3.2.2   Supporting organisations

The Federal Government provides basic funding for the **Max Planck Gesellschaft (MPG)**, the Fraunhofer-Gesellschaft (FhG), the "Blaue Liste" and the Hermann von Helmholtz Gesellschaft. The MPG is a sponsoring organisation with 71 research facilities in Germany. The organisation is mainly engaged in basic research in natural and social sciences as well as arts. New and promising research topics with no adequate focal point at universities are particularly focused upon. It co-operates with universities and provides them with appliances. "Technology transfer seems not to be a priority; rather, it is seen as a by-product or spin-off of the institutes' research activities".[21] The Max Planck Institutes – like the facilities of the FhG - complement the research done at universities and, for that reason, locating near universities and management through university professors is the rule. The MPG consists of 11,500 employees, among them 3,015 scientists.

The **FhG** is also a sponsoring organisation with 47 facilities for applied research and two service facilities, ten of them in the new Länder (1995). Additionally, there are three further offices in the USA. In contrast to the MPG,

[20]   BMBF (1998), p. 89.
[21]   Abrahamson, N. et al. (1997) p. 312.

which receives largely public funding, the share of public funding of the FhG budget amounts to only 40%. That is the reason why the FhG institutes are forced to acquire funding through projects with the business enterprise sector (30%) or through public research programmes (30%). Accordingly, their research activities are mainly focused on the demands of the practice. Basic research should be expanded to the development of prototypes in order to stimulate the transfer of new technologies to the business enterprise sector. Funding given by the Federal Government and the Länder enables the FhG to carry out self-chosen research topics to make sure of their scientific potential.[22] The FhG offers its services to firms and public authorities in the areas of: microelectronics, information technology, production automation, production technology, material and components, process engineering, energy and structural engineering, environment and health technical-economic studies and professional information. The close relationship to universities is institutionalised through the joint appointment of Fraunhofer directors as regular university professors.

The **Hermann von Helmholtz Gesellschaft** was founded in the late 1950s, when the Federal Government tried to establish an active role in technology policy. The Federal Government increased its influence in an area which belongs to the responsibility of the Länder. Originating from civilian nuclear research, other fields have been added, e.g. aeronautics, computer science, biotechnology.[23] The marginal commercial benefit of the public reactor programme and the demise of public support for nuclear energy led the nuclear research to phase out the research programme.[24] Research using large-scale equipment with a focus on specific priority topics, primarily large accelerators, neutron and synchrotron sources, as well as observatories and telescopes, is the special aim of Helmholtz-Centers. There are 16 of them in Germany. The Hermann von Helmholtz Gesellschaft receive about 80% public and 20% self-generated or other external funding. The Federal Government contributes about 90%, the Länder 10% of the institutional funds.[25] The mission of the Hermann von Helmholtz Gesellschaft is mainly engaged in basic research which emphasises a science-push rather than a demand-pull approach. The development of the scientific and technological basis for future high-quality research and the publication results are considered as the most important efficient technology-transfer.[26] In recent times a lot of authors have cast doubts over the economic efficiency of the large-scale research centres in the Helmholtz Gesellschaft. Beise and Stahl (1999) investigate the effects of publicly funded research at universities, polytechnics and federal research laboratories on industrial innovation in 2,300 German companies. They found that big-science laboratories as a source of innovation are almost non-existent.

Besides the large research organisations mentioned above, 84 small institutes with service functions are sponsored by the Länder and the Federal

---

[22]   Licht, G. et al. 1995, pp. 14 and Beise, M., Stahl, H. 1999, pp. 400.
[23]   Meyer-Kramer, F. (1990).
[24]   Keck, O. (1980).
[25]   Abrahamson, N. et al. (1997) p. 314.
[26]   Abrahamson, N. et al. (1997) pp. 312.

Government. In the wake of the German reunification and the resultant set-up of a new overall German research "landscape", the "Blue List" has been expanded to 34 institutes in the new Länder, from 1992 onwards. These semi-public institutes of the so-called **"Blaue Liste** (Blue List)" are engaged in a wide range of heterogeneous activities, e.g. social sciences, economics, natural science and information services. An independent "Science Council" evaluates the Blue-List institutions in certain intervals, as to whether they meet the criteria of scientific work. The budget of the majority of these institutes is met entirely by public funding. Only a few of the institutes carry out some contractual research for the private (business) sector.[27]

12.3.2.3   Federal institutions performing R&D

Most closely linked to the Federal Government are the Departmental Research Institutes ("Ressortforschung"). Here research is aimed at obtaining scientific findings which are directly related to the field of activity of a ministry. The ministries with the most important ties to departmental research institutes are the Ministry of Health, the Ministry of Agriculture, the Ministry of Transport and the Ministry of Defence. A total of 56 Federal institutions performing R&D exist and they are fully funded by the Federal Government. A variety of technical and scientific libraries and information centres also belong to these areas of responsibility and therefore also form part of the knowledge transformation system. "Compared with their counterparts in other countries, German departmental research institutes account for a relatively small share of total publicly funded R&D. Nevertheless, they often play an important role in the R&D landscape; some of them are leaders in special R&D sectors".[28]

12.3.2.4   University-related special programmes

The Länder Governments are exclusively responsible for the education policy in Germany. Therefore higher education belongs to this policy area, too. Nevertheless, the Federal Government supports the building and extension of higher education institutions in order to keep an attractive environment for research and innovation. Additionally, in agreement with the Länder Governments, the Federal Government has the policy of supporting universities in research areas which require rapid and disproportionately high funding for limited periods of time. These special programmes are of use both for teaching and for research and contain postgraduate and postdoctoral programmes as well as special research programmes.

---

[27]   Abrahamson, N. et al. (1997) pp. 320.
[28]   Abrahamson, N. et al. (1997) pp. 248.

12.3.2.5   Universities

In 1996, 335 state or officially-recognised institutions of higher education existed in Germany. Of these there were 90 universities, one comprehensive university, 16 colleges of theology, six colleges of education, 46 colleges of art, 146 general Fachhochschulen and 30 colleges of public administration.   In Germany universities are in general financed at the federal level and by the Länder, but about 19.5% of the higher education institutions are privately funded. About 75% of all 1.9 million students are enrolled at universities, 25% at academies. The budget for teaching and research is estimated at about 34.6 billion DM and expenditure for research and development at about 14.7 billion DM in 1997.[29]

Medicinal sciences - accounting for the highest expenditure on education and research – receive 7.8 billion DM, followed by expenditure for humanities and social sciences (6.0 billion DM), the natural sciences (5.6 billion DM), the engineering sciences (4.9 billion DM) and the agricultural sciences (1.1 billion DM) in 1995. The share of the expenditure on R&D reveals a different "R&D-intensity" of the various fields: the natural sciences hold the lion's share (4.2 billion DM) of R&D expenditure, followed by medicine (3.4 billion DM), natural sciences (3.0 billion DM), engineering sciences (2.9 billion DM) and agriculture (0.7 billion DM). Medicine has the lowest (43.6%) and the social sciences the highest (70.0% share among the different fields of sciences.)[30]

A total of 90.7% (1995) of the R&D expenditure of the higher education is publicly financed. This share covers statutorily for universities as well as DFG (see chapter 0) and Federal and Länder's grants for R&D projects. The business enterprise sector funded 8.2% of the R&D expenditure of the higher education sector in 1995. Since 1991, the funding of this sector has increased markedly from 0.8 to 1.2 billion DM (40.0%).

The organisational structure of German universities is important for an understanding of the research and transfer system. The major bodies in charge of the distribution of institutional funds for teaching and research are the various faculties. These faculties comprise of a number of chairs responsible for different areas of teaching.   Sometimes chairs establish institutes where the professors organise research on their own behalf. The status of these institutes varies widely from completely independent institutes, to closely linked ones. Occasionally the institutes are funded by and work jointly with promoting organisations (see sections 12.3.2.1 and 12.3.2.2).

---

[29]   BMBF (1998), p. 43
[30]   BMBF (1998), pp. 43

12.3.2.6   Transfer institutions

Apart from the two main actors (the business enterprise sector and the sector of promotion organisation and research institutes), a variety of institutions serve as intermediaries of knowledge. These transfer institutions are important "linking-nodes" of the German Innovation System. They can be described in accordance with their proximity to the above-mentioned actors of the NIS.

* Chambers of Industry and Commerce or other professional associations close to the business enterprise sector are present in almost every large city in Germany. They dispose of information on what kinds of knowledge and new technology will be important for the certain industries. Well-organised German industry associations play a similar role in the realm of technology transfer. Apart from their lobbying work, their mediation of technical knowledge through so-called technical committees is an increasing emphasis of their work.
* Transfers institutions closely linked to research institutes or universities. Institutionalised consultancy and independent development centres are also part of the transfer system.
* Last but not least, independent technology intermediaries - like transfer agencies and information centres – are occasionally effective institutions of technology transfer. Supported by the Chambers of Industry and Commerce and in many cases by the city council, they try to support regional enterprises. Some of the most helpful to small and start-up firms are the Centres of Technology and Firm establishment.

In conclusion of this section, the following Figure 12.13 draws a summarising picture of the funding structure of the "Promotion Organisations" and "Research Institutions" in Germany. Additionally, Table 12.13 give information about size, main research areas of the institutions and the financial structure of this institutional system.

Figure 12.13 Structure of the German Research Funding System
(source: ZEW, 1999)

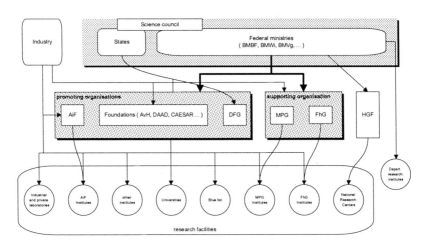

12.3.3   Innovation supporting programmes

The institutions of the German System of Innovation create good conditions for all firms to gain access to the newest technologies. As Winch (1999) puts it, they work to directly promote innovations. Moreover, the companies' innovative efforts are supported by way of additional national and international supporting programmes. The number of such programmes is significant: under the heading "construction" alone, the Ministry of Trade and Commerce's[31] database shows around 120 supporting programmes. Under the expression "innovation", as many as 360 programmes were offered. Altogether, construction firms in Germany are able to enlist the help of almost 400 supporting programmes which directly or indirectly improve the capability to innovate. To represent all of these programmes in the framework of this paper is hardly a feasible task. The differentiation in sponsors reveals that the programmes of the German Federal Government are more widely disseminated than programmes operated by the Länder and the European Union. Therefore, only innovation supporting programmes of the Federal State are described in this paper. The most important programmes are represented once more in Table 12.2 and in more detail.

However, there have only been a small amount of measures up to date which aim at directly supporting innovation projects in the construction industry. The largest part of political innovation subsidies is much more branch-independent and aims at the general innovation capabilities of the economy.

---

[31]   http://www.bmwi.de

The individual programmes show no emphasis as far as supporting a particular phase of the technology life cycle is concerned. A large section of the programmes aim at all phases equally (ALL). Among these programmes – as is the case among programmes supporting research and advanced/best practice - direct and general supporting programmes dominate. Only among the support of standard practice do half of the indirectly-specific programmes come into play. Directly-specific programmes for supporting innovation measures in construction firms hardly exist at all.

The support varies with respect to its type. Among the support given to standard practice, as well as undefined assistance in all phases, instruments of lowering interest rates dominate. In the support of research and advanced/best practice, however, subsidies are clearly predominant.

Figure 12.14 Technology life cycle and the number of innovation supporting programmes (Federal State) (source: BmbF, 1999)

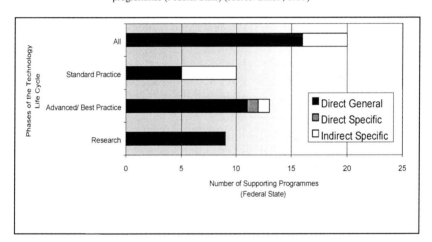

Figure 12.15 Technology life cycle and type of the innovation support
(Federal State) (source : BmbF, 1999)

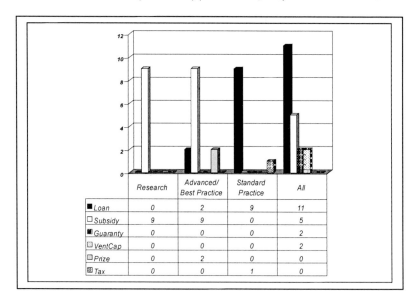

	Research	Advanced/ Best Practice	Standard Practice	All
■ Loan	0	2	9	11
☐ Subsidy	9	9	0	5
▨ Guaranty	0	0	0	2
▤ VentCap	0	0	0	2
☐ Prize	0	2	0	0
▨ Tax	0	0	1	0

12.3.4   Involvement of the construction sector in innovation supporting programmes

In 1996 around 5% of all innovative construction companies claimed public support for their innovation projects.   The share of supported innovators (supporting quota) is therefore much lower than that of the manufacturing industry (about 10%).   Against the background of construction-specific innovation supporting programmes hardly existing, this small supporting quota is easily explained.   Similar to the manufacturing industry, supporting quotas between companies from the new and the old Länder differ from each other.   These differences are very distinctive, however. About 3% of the construction firms in the old Länder receive governmental support, whereas the respective share in the new Länder is about 7%.   In relative terms, large enterprises participate in promotion programmes more frequently than small and medium-sized enterprises.
Subdividing the manufacturing sector into classes of "innovation intensity" produces the following picture: compared to the rest of the manufacturing sector, the number of supported firms grows with their innovation intensity. A large number of high-tech companies take part in the German Federal Government's promotion programmes. Particularly in the new Länder, there is scarcely a firm with an innovation intensity of more than 5%, financing its

innovation without public promotion.[32] At first glance, the structure of innovation promotion provides a positive picture: supported firms are more devoted to innovation processes, record better results, have higher turnover share deriving from new and improved products, have higher turnover share deriving from market novelties, and have higher rates of export.[33]

Figure 12.16 Involvement of the construction sector in innovation supporting programmes (source ZEW, 1998)

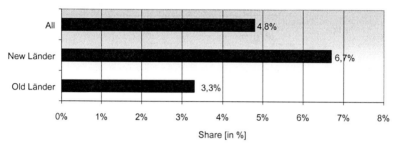

Details weighted according to the number of innovative companies

These results cannot be verified for the construction sector. In contrast to the manufacturing industry, the probability of a promotion does not rise with the innovation intensity of company. In the multivariate maximum-likelihood logit model (see Figure 12.17), no correlation can be drawn between the phase of technology life cycle in which a construction firm is active, and the fact that this company receives support. For a company of standard practice, the probability of getting support is just as high as it is for a company of advanced or research-based practice. In a similar fashion, the company size plays no role in the multivariate analysis. At this point the impression that the promotion programmes aim at the whole spectrum of the construction sector becomes unavoidable. Only the higher probability of support in East Germany, which has already been described, keeps its significance in the multivariate logit model. In East Germany, the probability of support increases in conjunction with decreasing labour productivity. The focus of support in the German construction industry thus clearly lies in the intensification of competitiveness of the unproductive firms in the new Länder.

---

[32]   ZEW, 1998: Mannheim Innovation Panel.
[33]   Beise, M. et al (1999b).

Figure 12.17 Logit model to identify construction firms involved in innovation supporting
programmes (source: ZEW, 1998)

Logit Estimates			Number of jobs	=	47	
			chi2(6)	=	13.09	
			Prob > chi2	=	0.0416	
Log Likelihood =			Prob > chi2	=	0.0416	

		Robust					
Variate	**Correlation**	**Std. Err.**	**Z**	**P>z**	**(95% Confidence Interval)**	
Best Practice	.0819137	1.412021	0.058	0.954	-2.685597	2.849424
Standard Practice	-.7721642	1.457389	-0.530	0.596	-3.628593	2.084265
Company size	.4682195	.4235885	1.105	0.269	-.3619987	1.298438
New Laender	5.586464	2.05852	2.714	0.007	1.55184	9.621089
Labour productivity (LP)	-3.40313	3.977203	-0.856	0.392	-11.19831	4.392045
LP New Laender	-19.84702	10.51575	-1.887	0.059	-40.45751	.7634722
Constant	-4.001126	2.296409	-1.742	0.081	-8.502006	.4997531

12.4   RESULTS

If one lays down the traditional reference numbers for R&D intensity or innovation
intensity as a scale of comparison with the manufacturing or service sector, the
innovation behaviour of the German construction sector is below average.   An
international comparison shows that R&D expenditure is concentrated mostly in
Japan.   In Germany, the automation of construction tasks is a new technology
which is not utilised much yet.   There are first steps to utilise robots on
construction sites, but up to now it showed very little success.   Indeed, the
„industrialisation" of the construction industry is not realised.   It still has much
more the character of craft.   On the other hand the construction industry is an
important customer for most parts of the manufacturing sector.   This concerns
material producers in the stone, clay and glass industry and the construction
machinery industry.   The R&D intensity of these technology suppliers is below

average. Here the construction industry does not seem to give much innovative initiative; there is no demand pull to observe.

But from which firms could such a demand pull arise? In Germany it is mainly the larger companies who are active in research-based or advanced practice. The research activities in the construction sector are still underdeveloped in East Germany when compared to those of West Germany. An expansion of research-based and advanced practice among smaller firms and among firms in East Germany could provide the focus for innovation supporting programmes. In practice, this focus is not established: First, the government expenditures to support building research have been continuously decreasing over the last years. There have only been a small amount of measures up to date, which aim at directly supporting innovation projects in the construction industry. Second, the number of innovation supporting programmes for supporting R&D lies below the number of programmes for supporting standard practice. This could also be a reason for why the probability of gaining support for a company of standard practice is just as high as that of a company of advanced or research-based practice. Even the shares of small and large companies, which are supported, do not differ from each other. Only the aim of the innovation support for the East German construction firms appears to be sufficiently fulfilled. Since the probability of gaining innovation support is increasing as labour productivity decreases there, it is to be doubted whether or not it is employment policy which is being pushed ahead in East Germany as a result of these programmes of innovation support, as opposed to innovation support itself. From these results, however, one cannot come up with the statement that the Federal innovation supporting programmes do not nonetheless lead to an improvement in the innovation capabilities of the supported construction firms.

Recently, the Federal Government opens a new chapter in research policy by funding "Leitprojekte". The purpose of those projects is to help tackle and achieve forward-looking strategic innovation goals more efficiently. In order to attain marketable products and processes business and universities, research institutes and users form consortia and co-operate in networks. All participants are involved in the research process as well as in the implementation of results. The participants set their own tasks in the framework of well-defined subject areas, such as "mobility in agglomerations". The Federal Government stages public proposals for "Leitprojekte" and referees will select up to five projects.[34]

---

[34]   BmbF (1998), p. 134.

Table 12.1 Financial resources and main areas of activity of German research institutions (source: BMBF, 1998)

Institution	Expenditure in Mio. DM [1997]	Public support in Mio. DM	Number of Institutes	Employees	Public support in Mio. DM	Relation of support federal / states	Main area of research
DFG	2,209	2,200			2,200	100/0	Physics 24.4% Biology 35.5%
DAAD	370.5	350			350	90/10	exchange of students & scientists
AvH	120	87.7			87.7	100/0	
Stifterverband	143	Foundation wealth			foundation wealth		
Volkswagen-Foundation	180	Foundation wealth, dividends			foundation wealth, dividends		
DBU	150	Foundation wealth			foundation wealth		
CAESAR	750	685			685		
AiF	170	170	109		170	100/0	

Table 12.1 (Continued) Financial resources and main areas of activity of German research institutions (source: BMBF, 1998)

Institution	Expenditure in Mio. DM [1997]	Public support in Mio. DM	Number of Institutes	Employees	Public support in Mio. DM	Relation of support federal / states	Main area of research
MPG	1,708	1,602	80	10,700	1,602	50/50	Physics 21.3% Biology 31.7%
FhG	1,223	551	50	6,620	551	90/10	Engineering sciences 72%
Helmholtz-Centers	4,159	2,988	16	22,399	2,988	90/10	Natural sciences 67% Engineering sciences 25%
"Blue List"	1,727	1,312	82	11,273	1,312	50/50	
Federal Institutions	2,867	2,867	57	18,682	2,867	100/0	Natural sciences 34% Engineering sciences 29% Agricultural science 14%

Table 12.2 Innovation-supporting programmes of the federal state

Name of the programme	Objectives	Support	Field	Instru-ment	Phase
KfW Programme for modernising living space	To make investments into restoration and modernisation as well as creation of living space in the new Bundesländer incl. Berlin (East) easier, the Government awards credits at favourable rates of interest via the KfW Programme.	Loan	Construc-tion Energy	Indirect-specific	Standard Practice
KfW Programme for modernising living space (Emigrants)	The aim of the programme is to create organisations for the temporary accommodation of emigrants in the Länder of Brandenburg, Mecklenburg-Vorpommern, Sachsen, Sachsen-Anhalt und Thüringen.	Loan	Construc-tion	Indirect-specific	Standard practice
Support for industry-wide training centres	To be able to meet the demands as much as possible, the industry-wide professional training centres are orientated to the regional qualification requirements. The investments and operations of these centres are supported through public funds.	Subsidy	Training	Indirect-specific	All
ERP Construction programme	Loans in the new Länder and Berlin (East) for the setting-up, take-over, expansion, reorganisation or fundamental rationalisation of businesses.	Loan	Investment	Direct-general	All

Table 12.2 (Continued) Innovation-supporting programmes of the federal state

Name of the programme	Objectives	Support	Field	Instru-ment	Phase
Take over of export guarantees (Hermes Bürgschaften)	For safeguarding themselves from economic and political risks of loss of pay linked to export businesses, German exporters can take advantage of the Government's export guarantees for supporting German exports.	Guaranty	Export	Direct-general	All
ERP Credits for capital investment companies	For their participation in working with SMEs in the framework of the ERP-Participation Programme, capital investment companies receive rediscounting credits.	Loan	VentCap	Indirect-specific	All
ERP Export finance programme	As measures of special power from the ERP, loans may be granted for the financing of German exporters' export activities which concern the delivery of investment goods and efforts made in developing countries.	Loan	Export	Indirect-specific	All
Support for own property for young families(KFW)	The programme finances the construction and acquisition of people's own lived-in homes and owner-occupied flats in the old and new Länder by way of loans with favourable rates of interest.	Loan	Construc-tion	Indirect-specific	Standard practice

Table 12.2 (Continued) Innovation-supporting programmes of the federal state

Name of the programme	Objectives	Support	Field	Instru-ment	Phase
KfW Programme for CO2 reduction	The KfW Programme for CO2 reduction serves as a method of long-term financing with favourable interest rates for investments into reducing CO2 levels and energy-saving efforts in residential buildings, as well as the erection of low energy houses.	Loan	Energy	Indirect-specific	All
New mathematical methods in industry and services	In order to forge stronger research co-operation between economics and science, the BMBF supports plans for mathematical pure research oriented towards application.	Subsidy	R&D	Direct-general	Research
Innovation competence of medium-sized companies (PRO INNO)	To stimulate SMEs into expanding their technological and economic competence through leaps in innovation, by way of this programme the Federal Ministry for Economics and Technology (BMWi) supports the collaboration of companies as well as co-operation and research organisations.	Subsidy	R&D	Direct-general	Advanced practice
Middle-class price 1999 for EXPO 2000	To give middle-class manufacturers an incentive for development, market entry and application of product innovations, the Federal Minister for Economics and Technology sets this price.	Subsidy	R&D	Direct-general	Advanced practice

CHAPTER 13

JAPAN
PUBLIC POLICY INSTRUMENTS
AND INNOVATION IN
CONSTRUCTION

Ryoju Tanaka, Shin Okamoto, and Tomoya Kikuoka

13.1   INTRODUCTION

The Japanese construction companies are quite active in research and technological development.   There are more than 20 companies, which possess their own technological research institutes including large-scale research and experimental facilities.

The market share of the major 20 construction companies, however, remains below the level of 18% (1997). In the construction industry as a whole, the same situation as in most other countries, is encountered, which is one dominated by small-scale enterprises. Among a total of 560,000 construction companies (1996), the rate of corporations with a capitalisation of ¥100,000,000 or more is only 1.1% whereas small companies with employees of 20 or less account for more than 90%.

Meanwhile, the ratio of construction investment to GDP (gross domestic product) has tended to shrink since 1992, but still accounts for approximately 15% or 75 trillion yen. Construction is the key industry in Japan supporting a workforce of approximately 10% (about 6.6 million 1998 average) of the total population engaged in industries.

Under these circumstances the public sector has implemented two incompatible policies simultaneously – protection and development of small companies from a viewpoint of promoting the local economy and the fulfillment of a responsibility as an ordering party to procure higher quality at lower cost.

The Japanese economy, which had consistently maintained its growth during the post-war period, has continued to be sluggish for a long time following the bursting of the economic bubble in 1990. Various conventional Japanese socio-economic systems, which supported its economic growth, such as lifetime employment, seniority order wage system, and convoy-style administration management, have been obliged to make fundamental structural reform in accordance with a loss in growth momentum.

The construction industry has also entered into a mega-competition period, never before experienced, due to long-term reduction of public investment caused by restrictions in the national financial situation and re-examination of the role played by the public in fulfilling its responsibility as the ordering party. This situation is also caused by conversion of public policies such as diversification of bidding systems in addition to the environmental changes such as demand structure changes, globalisation of the construction market, low-price orientation, drastic computerisation and ageing as well as serious global environment problems. A reform of the construction industry through the utilisation of technology is inevitable.

In this paper we furnish an overview of the changes in recent socio-economic conditions surrounding the Japanese construction industry and organise and analyse the trend of construction technological development and the role of the public sector to promote technological development in the private sector and these effects. We also highlight the future perspective for the role of the public in technical innovation of the construction industry.

13.2   CONTEXT

13.2.1   General context

13.2.1.1   Geographic characteristics and construction technology

When discussing the technological development of the construction industry in Japan, we would like first of all to point out that it is closely related to the characteristics of the country.

The Japanese Islands are located on the sea at the east of Asia and north-west of the Pacific Ocean: facing east and south to the Pacific Ocean, west to the Japan Sea and East China Sea, and north to the Sea of Okhotsk, running from north-east to south-west in a bow shape. The total area is approximately 377,800 km². The land covers 3,000 km from south to north having weather zones from subtropical to subarctic with complex geological structure. The land is divided into small areas of mountains, basins and plains.

The Japan Islands have constantly been exposed to threats of nature. Disasters such as earthquakes, typhoons, localised torrential downpours, floods, eruptions of volcanoes, high tides and high waves landslides, and fires threatened almost all the areas of the Islands throughout history. Some areas have been bothered by heavy snow, drift sands and high humidity and temperature.

The development of construction technology in Japan in modern times is the history of counter measures and challenges against these natural threats. After the period of high economic growth which shares from 1955, artificial phenomena such as noise, vibration, sunshine, air pollution, water contamination, industrial pollution, and overpopulation in cities have been added and solving these problems poses a fresh challenge.

Under these circumstances, for the development of construction technology in Japan, large-scale public projects to improve social and industrial infrastructures in accordance with the rising economic growth in the post-war era were implemented. These projects were the construction of bullet trains, expressways, and huge amounts of public housing, enhancement of urban infrastructure such as waterworks and sewage, drastic modifications of mountains and rivers for forestry conservancy and flood control, and construction of dams. The private sector also carried out construction activities based on strong capital investment. Through these construction activities, the private firms actively invested in research and technological development and alternates were made to divide situations to the above problems of business, government and academia. As a result, up to now quite a number of problems related to construction technology have actually been resolved. Moreover, recently the following technological development activities were conducted: construction of super-high-rise, building use of underground space, land improvement, construction of energy facilities, enhancement of transportation infrastructure, construction of telecommunication facilities, urban redevelopment, suburb development, and implementation of safety measures. It is said that the Japanese construction technology achieved through this kind of process sits at the top level in the world. Figure 13.1 summarises the change in needs required for socio-economic progress as well as construction technological development.

13.2.1.2 Recent business conditions and outlook for the construction market

The structure of construction investment in Japan is indicated in Figure 13.2. The investors can be divided into the government (47.4%) and private sector (52.6%), and clients can be divided into building (50.4%) and civil engineering (49.6%). The construction investment has mainly been made in the private housing (27.8%) and private non-housing (14.7%) in the building sector as well as governmental civil engineering (39.5%) (1998).

As indicated in Figure 13.3, construction investment stagnated after the peak of about ¥84 trillion (nominal) in 1992, and it is anticipated that FY 2000 will be the level of the latter half of the 1980s or just prior to the bubble economy.

As the public infrastructure is not enough compared with advanced Western countries, the government's capital investment is larger than these countries. But judging from the current national financial situation, it is the general view that although there may be a temporary increase by fiscal stimulation, the governmental investment will continue to decline and the private investment will not recover drastically.

In other words, it is said that the Japanese economy has entered into the maturity phase after passing through the growth phase represented by the post-war Economic Miracle and the bubble economy in the 1980s. For the construction industry, new markets tend to be decreasing, and therefore an increase in the weight of maintenance, repair and modernisation markets is expected.

Ryoju Tanaka, Shin Okamoto, and Tomoya Kikuoka

Figure 13.1: Alteration in major needs for construction technology development

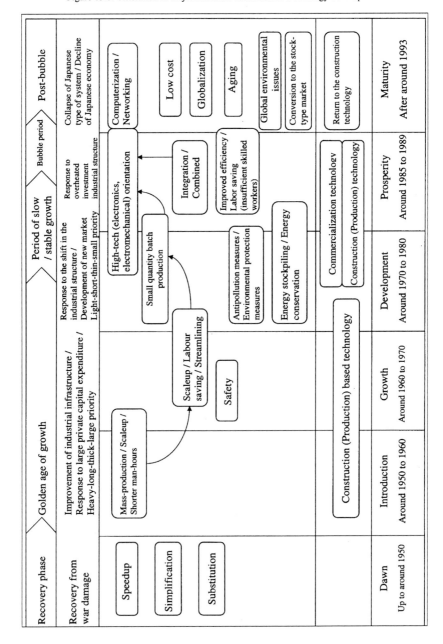

Figure 13.2: Structure of construction investment

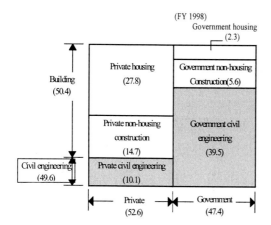

Note: Numbers in parentheses are share (%) of the total investment amount (100%).

Source: Ministry of Construction

Figure 13.3: Changes in Construction Investment

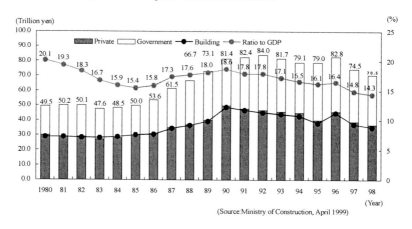

(Source:Ministry of Construction, April 1999)

13.2.2   Construction industries context

13.2.2.1   *Overview of construction companies and R&D trend*

There are approximately 570,000 construction companies in Japan. For the amount of capital, companies with a capitalisation of ¥100 million ($909,091) to ¥1 billion ($9,090,910) occupy only 0.8% and the share of corporations with a capitalisation above ¥1 billion is 0.3%. The share of the small- and medium-sized companies with a capitalisation below ¥100 million ($909,091) is 98.9% (Figure 13.4). The construction companies are roughly divided into building contractors and civil engineering contractors. In Japan major building contractors have established the system of executing design activities internally. This is a feature, which cannot be seen in Western countries (there are also architectural design offices dedicated to design work).

At present the private capital investment is sluggish due to the recession, which is said to be the worst since World War II, and the Japanese construction companies are caught up in the difficult management environment of a shake-out of weaker companies as a result of fiercer competition.

Figure 13.4: Behaviour of licensed builders by scale

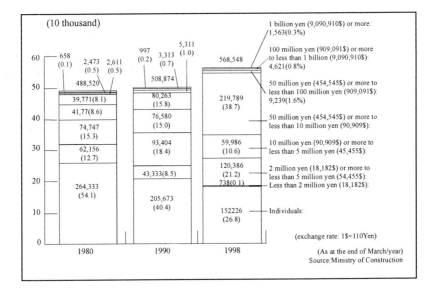

(exchange rate: 1$=110Yen)

(As at the end of March/year)
Source:Ministry of Construction

Under these circumstances we can discern new movement for reform even in the construction companies since around 1999. In other words, in order to survive, Japanese construction companies, mainly the major firms, are starting to engage in completely new business activities: mergers and acquisitions, business affiliations, introduction of an executive director system, and the development of fee business.

Let us explain here the recent situation for the rapid movement toward reform of corporate management.

The Japan Civil Engineering Contractors' Association, Inc., the leading association of this sector, compiled the report entitled *Changes and Responses of the Civil Engineering Construction Market– Scenario of Market and Industry Heading for Maturity* and referred to the necessity of a shake-out of weaker firms suitable to the market size. Furthermore, the Japan Federation of Construction Contractors, the association of major construction corporations, summarised the directions of business organisations and reform of business nature such as spinning of subsidiaries and capital ties as *System of Management of the Construction Industry in the 21ˢᵗ Century*. In this report JFCC also made the proposal for policies in the construction industry as well as public works allocations.

In July, 1999 the Ministry of Construction established the Construction Industry Recovery Programme, in which these movements in the industry were strongly reflected, focusing on approximately 60 major general contractors. This programme showed the direction of strategies of the construction firms proclaiming "reinforcement of innovation and affiliation in the management organisations" to present the specific themes of environmental improvement and support from the administration.

Meanwhile, the major general contractors consecutively announced new management plans by the beginning of fiscal 1999 and initiated a transformation into masculine and slim management nature that included reinforcement of financial position, rapid decision making by the introduction of new executive systems, and restructuring of their organisations.

Autumn 1999 two sound-management second-tier companies agreed on business affiliations in the fields of technological development and use of materials and equipment. This news became a hot topic. There were also mergers of medium-sized firms.

We can see a new movement in the service provided by general contractors. Previously the concept where the service provided by general contractors was free pre-dominated. Recently, however, the idea of fee business has been disseminated and the direction of the idea has been changed where it is adequate and it is possible to receive consideration for high quality service and appropriate response to the customer.

13.2.2.2   Overview of investment in technological development

One of the characteristics of technological development in Japan during the post-war period is that mainly private companies conducted such development actively. Looking at the government statistics on the ratio of R&D investment in Japan as a whole, the share of the private firms is 72 to 75% and the number of researchers in the private sector holds about 65%.

The share of R&D investment by the construction industry in the whole Japanese industries in the aforementioned statistics is relatively small: 2.1% in 1997.

The R&D expenses to sales in the construction industry is 0.39% in 1997, which is smaller than in other industries. Changes of the research investment and R&D expenses to sales in the construction and other industries are summarised in Tables 13.1 and 13.2, respectively.

One of the characteristics of technological development in the Japanese construction industry is its concentration in the 25 major and semi-major firms, which have their own technological development division or research institute.

The concentration of R&D investment by the five major companies is nearly 30% and that of the top 20 firms is almost 60%.

13.2.2.3   Characteristics of construction technological development

When comparing the technological development activities in Western countries to those in Japan, the great difference is as follows: in the case of the Western construction industry, it is universities and public institutes. As the most part who conduct the technological development activities and almost no such activities are undertaken in the corporate sector. On the other hand, in Japan these development activities by general contractors are a frequent occurrence.

Large-scale technical research development activities by the construction firms are concentrated in the major or second-tier firms. These firms possess their own technological development division or research institute. The top five major firms have research institutes that exceed the scale of the government's building research institutes and that invest in excess of ¥10 billion annually in R&D.

13.2.2.4   Overview of post-war technological development in the construction

At present technological development activities by the private sector occupy a major position by the basis of such development. But in construction technological development so far, the private sector has played the major role in the building area and the public sector has played the major role in civil engineering. It was the middle of the 1970s when the private sector made a start in conducting full-scale technological development in civil engineering.

Table 13.1: Changes in corporate R&D expenses for individual
industries in Japan (1980 - 1997) (Unit: million yen)

Fiscal year	Construction	Chemical engineering	Steel	Total Industries
1980	75,690	558,252	147,064	3,142,256
81	72,891	612,354	169,653	3,629,793
82	80,629	687,493	182,772	4,309,018
83	80,629	687,493	182,772	4,039,018
84	101,342	774,532	186,088	4,560,127
85	116,123	852,793	192,091	5,136,634
86	121,103	983,585	255,290	6,120,163
87	128,303	1,095,887	245,176	6,494,268
88	148,462	1,190,226	249,734	7,219,318
89	185,147	1,313,882	268,131	8,233,820
90	212,677	1,416,775	303,805	9,267,166
91	204,604	1,547,707	360,054	9,743,048
92	243,469	1,604,722	311,485	9,560,685
93	248,201	1,561,433	286,114	9,053,608
94	221,999	1,548,794	237,707	8,980,253
95	204,363	1,554,884	213,541	9,395,896
96	224,514	1,593,250	201,476	10,058,409
97	225,162	1,609,252	213,631	10,658,357

(Source: Report on the Survey of Research and Development issued by the Management and
Coordination Agency Government of Japan)

Table 13.2: Changes of corporate R&D expenses to sales in Japan
(1984 - 1997) (Unit: %)

Fiscal year	All industries	Agriculture, forestry and fishery	Mining	Construc -tion	Manufac- turing	Transportation, communication and utilities	Software
1984	1.99	0.24	0.63	0.47	2.34	0.84	
85	2.31	0.24	1.03	0.48	2.69	0.98	
86	2.57	0.24	1.16	0.55	3.03	0.96	
87	2.59	0.31	1.01	0.51	3.14	0.84	
88	2.60	0.38	1.27	0.48	3.15	0.95	
89	2.72	0.21	0.94	0.52	3.29	1.06	
90	2.78	0.50	1.13	0.54	3.36	1.07	
91	2.81	0.25	1.41	0.46	3.47	0.85	
92	2.83	0.28	1.38	0.55	3.52	0.87	
93	2.76	0.43	1.17	0.54	3.47	0.88	
94	2.72	0.39	0.98	0.50	3.39	0.97	
95	2.73	0.43	0.98	0.45	3.43	0.9	
96	2.77	0.39	0.87	0.46	3.43	0.89	9.83
97	2.85	0.53	1.15	0.39	3.67	0.91	7.84

Technological development in post-war Japan put the emphasis on the introduction of technologies developed in Western countries as well as improvements designs to adapt these technologies to the severe Japanese geographical conditions mentioned in (1) and (2) above.

The construction technological development also started from mechanised execution due to the introduction of large machines from the US in the 1950s. From the latter half of the 1950s to 1960, machines and construction methods were imported from foreign countries and improved. In the case of civil engineering, non-Japanese consultants played major roles in introducing these foreign-born technologies.

From the middle of the 1970s, in the Japanese construction industry as well, the technical research institutes have expanded and improved and started to conduct their own technological development. From this time even in the civil engineering field, technical engineering activities led by the private sector have been conducted.

Innovation of post-war construction technology was not carried out by the construction industry itself but was brought about by the technology innovation from other industries such as machinery, steel and cement, which are related to the construction industry.

The characteristics of the technological development in the Japan construction industry, as mentioned above, is that the construction firms have their own technical research institutes.

In the Japanese industrial world, the boom for establishing research institutes occurred (first boom) in and around 1960. During this period even in the construction industry, it was the five major firms mainly who established their technical research institutes. Then from the latter half of the 1960s to the latter half of the 1970s, the second tier firms built their research institutes.

Table 13.3 summarises the background, motives and purposes at the time when the major general contractors established their research institutes. Figure 13.5 shows the changes in the roles and functions of these technical research institutes.

13.2.2.5 Characteristics of technological development in the Japanese civil engineering and building divisions

In this way we find that the situations for technological development in civil engineering and that in building differ slightly. Historically speaking, in the civil engineering sector the technological development was based on public investment, and the technological development activities by the government, public institutes and universities, which are the ordering party, were active and still are so even at present.

The civil engineering works took the directly managed system where the ordering party in the government and public utility corporations performed all the activities from planning, design and execution. The ordering party gave instructions right up to detailed contents of the structure. Therefore, there was no pressing need for subcontractors to perform technological development activities. Active technological development activities of the private firms could mainly be seen in the building sector. The differences of technological development in the building and civil engineering sectors are also covered in Section 13.3.

The technological development in the building sector is closely related to the building production method as well as to the roles of the general contractors.

A researcher in Japanese building production pointed out the following:[1]

The most significant background that the Japanese general contractors are active in technological development and have their own technical research institutes is that the range of technology supported by the Japanese general contractors is very wide. In other words, we can highlight the following situations:

In Western countries, the range of services of the executors incorporated with architects and engineers are wide and the executors have responsibility for solving technical problems. Therefore, it is all right for construction firms to follow the design documents and instructions of the

---

[1]   From the speech of Osamu FURUKAWA entitled "Japanese Style General Contractors" during the 10th Building Production and Management Symposium in 1994 organized by the Architectural Institute of Japan.

executors. The construction firms are not asked to take on more obligations. On the other hand, under the current situation the Japanese general contractors share the responsibilities for preparing a part of the design documents in the implementation design and undertake supplementary activities on a wide scale even in technical issues. The general contractors substitute some actions and services, which are supposed to be performed by the designers, and take various risks after the design. This is the Japanese conventional building production method and the background of services, technological development and design activities of the general contractors. This is the reason why general contractors have their own technical research institutes.

13.2.3   Innovation in the construction industries

13.2.3.1   General trend

For the past few years, due to a sluggish Japanese economy, the construction investment has been slow. There were no future large national projects because the major sales performance had already entered into a maturity phase and the sales performance of the construction industry was poor. Therefore, the R&D investment in the Japanese construction industry has been in decline.

The prevailing situation is that fierce competition to take the lead, which was common in 1980s to the beginning of the 1990s, is weakened.

Now, we would like to discuss the major movements of the organisation and management regarding technological development in the construction industry for the past few years.

Table 13.3: Background to the establishment of central research institutes
in overall industries and technical research institutes in construction industries
(first bom) and situations in which they were established

	Overall industries	**Construction industry**
	1956 - 1961 - 1963	**1956 - 1961 – 1963**
Background	Recognition of importance of technological innovation due to the experience of technology import Improvement of R&D infrastructure in order to reinforce international competitive power to cope with trade liberalisation Potential due to the growth of corporate power Progress of science and technology policies and promotion schemes	Active construction demand Request for recovery to the pre-war technical projects Technical introduction, response to technical innovation from other industries and recognition of necessity of research and technical development Stimulated by the boom to own research institutes of other industries such as manufacturing Changes of industrial environment and nature

Table 13.3: (Continued) Background to the establishment of central research institutes in overall industries and technical research institutes in construction industries (first bom) and situations in which they were established

	Overall industries	**Construction industry**
	1956 - 1961 – 1963	**1956 - 1961 – 1963**
Consciousness of top management	The top management themselves started to emphasise R&D They recognised that prosperity of the company would be brought by expansion and improvement of research.	The top management themselves recognised the importance of technical research.
Other motives	Social recognition as a leading company and improvement of status Me-too consciousness and fear of missing an opportunity	Social recognition as a leading company and improvement of status Me-too consciousness and fear of missing an opportunity
Model	Research institutes in the Western countries. Executive missions to the Western countries and invitations of Western researchers	The organisation was formed and facilities were improved using national research institutes and universities as a model. The second-tier class imitated those of the major class.
Purpose and role of establishment	Development to a new phase of research activities Research of fundamental and new areas: several fields from basic to development research Improvement and adaptation of technologies introduced from overseas The basic research was conducted in full-scale after the latter half of 1980s.)	Support of work-site operations Role as a laboratory to adapt and improve technologies introduced from overseas Improvement of technologies for foundation and body related work
Location	In and around three major cities	Same as left (Tokyo, Osaka)
Personnel	Difficulty in hiring excellent researchers. Recruited from universities and research institutes. After that, the number of researchers has been increasing because universities and graduate schools have produced many scientific-related researchers.	Insufficient number of research engineers Dispatching to and interchange with universities and national institutes and personnel recruitment from these organisations After that, the number of researchers has been increasing because universities and graduate schools have produced many scientific-related researchers.

Figure 13.5 Changes of roles and functions of technical research institutes
of major general contractors

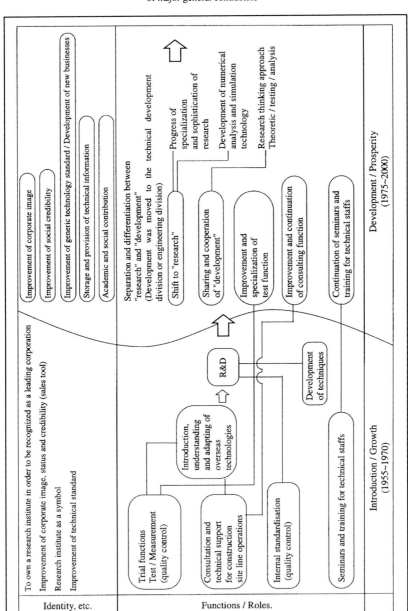

Figure 13.6: Establishment and moving of research institutes of major and second-tier construction companies

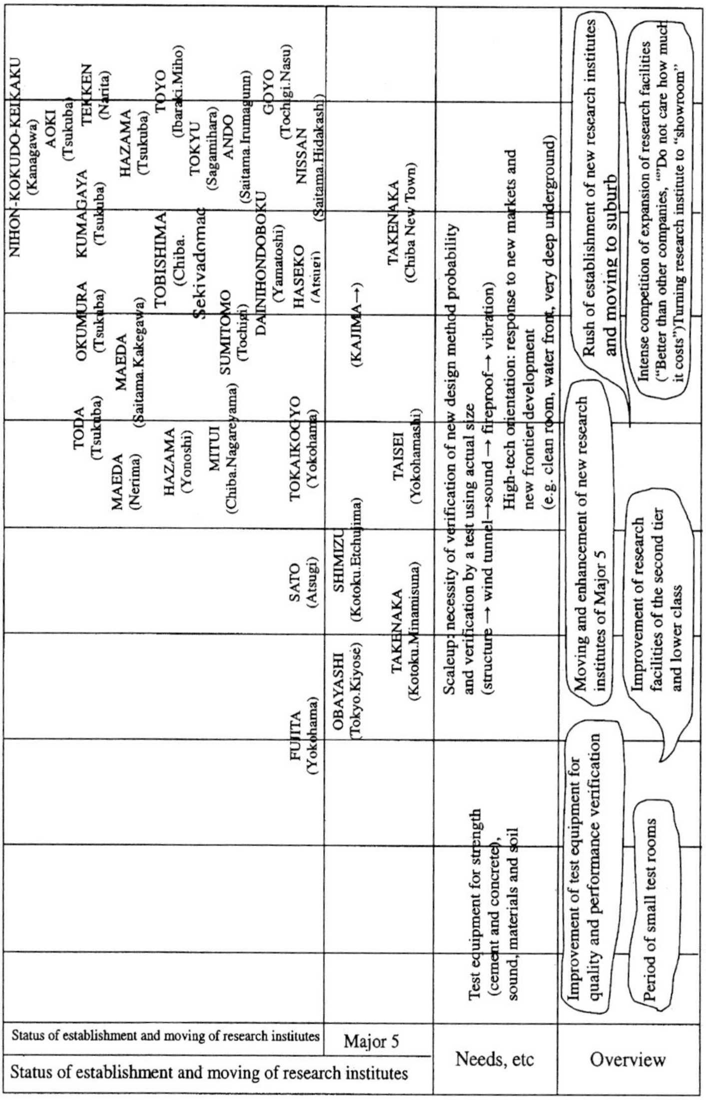

Note: Additional construction of test laboratory at the existing location is not included in principle (e.g. Nishimatsu).
Year for construction of each research institute means when the initial construction was completed. There are several cases that took one to two years to start actual operation.

The biggest factor is that under a tough business environment where the number of orders has been decreasing and profit ratio of the work has been sluggish, individual general contractors place more emphasis on technological development, which supports activities to win orders and connects these support activities to the businesses. Such movement becomes more visible.

Furthermore, we can identify a movement to consider or try joint research and technological development among the industry associations and individual companies. So far individual companies conducted the same research separately at the same time. It was repeatedly pointed out that such practice was inefficient. Due to reductions in the construction market and structural changes, the movement to seek a sharing of management resources and efficient and low-cost technological development has emerged[2].

We would like to introduce several specific trends from the past one to two years.

- Even research and technological development divisions are not sacred areas for corporate restructuring and the research structure has actively been reorganised under the keyword of "improvement of efficiency"
- a major technical research institute reformed its organisation, which had been subdivided by each element of technology, into an organisation classified by the need to respond to market demands
- another major technical research institute removed the framework separating civil engineering and architecture and reformed the organisation in accordance with the market needs
- in order to promote simplification, promptness and efficiency of the organisation, the engineering division was reformed and the name changed to the Technical Research Institute. (The conventional Research Institute was abolished and the Departments of Intellectual Property, Research, Technological Development, and Environment were established)
- the organisation was reformed to promote simplification of the organisation, which has encountered problems caused by segmentation and vertical structures, as well as to demonstrate comprehensive

---

2   Collaboration of Technological Development / Joint Research among Construction Firms Necessity and probability of joint research among general contractors were reviewed by the Technical Research Committee of the Building Contractors Society during the latter half of the 1970s. Then during the latter half of the 1980s, BCS launched the Technical Research Task Force and discussed again the movement to perform joint research for common issues. Furthermore, the Japan Federation of Construction Contractors compiled a report under the same theme. Every opinion insisted that problems should be solved by collecting knowledge, avoiding overlaps and performing operations efficiently without waste, rather than handling them as a single company. For this kind of discussion although the construction firms agree on such purposes in general, in actual practice there is always inconsistency as a result of competition among corporations. The Japan Association of Representative General Contractors, the industrial organization of medium-sized construction firms, established the joint research institute in 1987 as the first movement by the Japanese construction industry.

capabilities in any area, on which the company must concentrate, such as environment and cutting edge technology
- collaborations to proceed with joint fundamental research were announced one after another. Three major general contractors agreed and started to conduct some joint fundamental research. Furthermore, several second tier general contractors affiliated in the field of technological development. Subsequently three second tier and top medium-sized general contractors formed a technical tie-up
- meanwhile, among the top management of the five major general contractors there still remains a strong opinion which affirms, "Technological development and research contain many corporate strategic elements and the source of competitive advantage is technical capability" indicating that joint research with competitors is difficult
- the top management of the major general contractors is oriented towards a slim but strong nature, but emphasises technical capabilities in the management strategy, which concentrate and select resources effectively. They said, for example, "In order to win the competition for securing contracts, technical capability is even more required," "Technical capability reinforces quality and cost competition," and "We will improve and respect the environment of technological development more."

13.2.3.2   Trend of construction engineering innovation

1980s to 1990s

As mentioned above from the 1980s to the 1990s the R&D investment by the general contractors expanded. They made efforts not only for technological development targeted to new markets such as waterfront development, very deep underground space development, biotechnology, seismic, anti-vibration and opening-type dome, but also for rationalisation of production and execution and development of new technology due to the shortage in the labour force. This is because general contractors entered into the development of new businesses using high-tech as the driving force.

At that time differences in technology were essential for general contractors to win the competition for the contract. Comprehensive technical capabilities were required, and consequently technological development became the most important issue for their management strategy.

The construction industry reinforced their attitude to promote technological development in advance of the needs and proposed them actively. At that time a top executive of the industrial association said as follows: "The results of technological development are utilised to increase orders received and to improve profits. The Japanese general contractors have not been defeated by overseas construction firms because they have the basis of research and development (omitted)."

At this time the willingness to invest in R&D became strong and expanded rapidly throughout the Japanese industrial world. The reason why the industries increased the R&D investment was to overcome the tough Japanese economic environment due to high appreciation of the yen and trade friction by being armed with technology to avoid competition with emerging Asian newly industrialised economies (NIES). Under the movement to convert the Japanese industrial structure, each industry emphasised R&D activities for other areas such as biotechnology and electronics. Naturally the construction industry aimed at capital investment markets of these other industries.

In this way from the latter half of the 1980s to the first half of the 1990s, the technical innovations for the demands of quickness, labour-saving, sophistication and scaleup occurred consecutively. The number of staff in the technological development division substantially increased and there was a rush to set up and expand technical research institutes as explained earlier.

From the 1990s to the present post recession

From the latter half of the 1980s technological development such as cost reduction, labour-saving, rationalisation and VE was progressed. The middle of the 1990s was the time when technological development during the bubble economy brought various outcomes at once.

Up to now, following the Japanese economy's entry into recession, and due to the Hanshin Awaji Earthquake Disaster, the construction firms have competed to develop earthquake-proof modifications, seismic and damping technologies. Technologies related to global environment such as CO_2 reduction, energy conservation, and roof greening, cost reduction, computerisation of building production systems, and quality assurance have also been developed.

In the civil engineering, the progress of environmental businesses such as purification of water quality and contaminated soil as well as construction and renewal of final waste disposal facilities, reinforcement and repair technologies for earthquake-proof of civil engineering structures such as bridge piers, and shield technology for tunnels can be pointed out.

A shift to technological development directly connected to obtaining the contracts is also a characteristic of the present day trend. In other words, for the environmental business and newly developed markets, the construction firms reinforce engineering business connected to the enterprising by integrating technical know-how accumulated in the building and civil engineering divisions individually into surrounding technologies.

The following summarises the major movements of technological development from the middle of 1990 to 1999 mainly for the architecture technology:

Building technology

- Major examples of technological development related to rapidity, rationalisation and sophistication:

    - Commercialisation of lifting and material handling system:

        - Vertical and horizontal handling of materials; automation technology for rationalisation; and automated operation system of vertical moving and tower cranes.

    - Commercialisation of new RC:

        - Establishment of super high-rise reinforced concrete jointly developed by the government, businesses and universities under the General Technology Development Projects of the Ministry of Construction. Development of high strength concrete with 600 to 1,000 kgf/cm^2 and super strong reinforcing bars with yield strength of 5,000 to 7,000 kgf/cm^2.

    - Highly fluidised concrete:

        - Development of high performance concrete with excellent fluidity and material separation resistance. No vibration without compaction / labour-saving application.

    - Hybrid method / CFT (Concrete-Filled Steel Tube):

        - Frames such as column RC and beam S require special approval under the Architectural Standards Law and the private firms have engaged in technological development for technical evaluation and accreditation by the Minister.

    - All-weather temporary roofing:

        - These roofs have been developed and commercialised for the purpose of improvement of operations by construction workers, standardisation of term of works, and stabilisation of process.

- Earthquake-proof reinforcement / repair, base isolation and damping technologies:

- After the Hanshin - Awaji Earthquake Disaster, the construction firms have made tremendous efforts to develop earthquake-proof reinforcement, base isolation and damping technologies as well as earthquake movement analysis methods. Furthermore, due to the execution of the Earthquake-proof Improvement Promotion Law, technological development and various proposals were made while anticipating the demand for earthquake-proof improvement. The technical evaluation of the Architectural Disaster Prevention Association is used to provide an incentive for promoting technological development of construction companies.
- Fibre reinforcement method: Various types using carbon fibre.
- Seismic restoration technology: Various seismic restoration technologies while using and/or living in.
- Base isolation technology: Performance improvement, expansion of applications (e.g. multistoried buildings, light-weight buildings, detached houses and poor ground), sliding bearing.
- Vibration control technology / Low-yield-point steel damper: Hybrid-oriented devices, improvement of damping capacity, viscous (oil) fluid dampers, Low - yield - point steel response control wall.

- Environment-related:

  - Recycling and reuse of buildings (seismic retrofit).
  - Energy conservation and reduction of CO_2.
  - Recycling / zero-emission technologies: Aiming at contracts in the market of production facilities to obtain an ISO 14000 Certificate.

- Various design support, business support and diagnostic systems.

Civil engineering technology

- Environment / Environmental engineering:

  - Technologies to construct / renew waste final disposal facilities: seepage control system
  - Recycling of animal waste: compost
  - Technologies to purify water quality and contaminated soil: system to remove secretion disturbing materials such as heavy metals and dioxin
  - Recycling / Zero emission technologies.

- Shield technology:

    - Super large section tunnel construction method: MMST method.
    - Long distance excavation method.
    - The Japanese shield technology can be proud of its place in the world. Before the completion of the last major project in the 20th Century, Japan's first and the world's first new technologies have been established. Some of them were originated by the public (Metropolitan Expressway Public Corporation) and established as methods by the general contractors or adopted for works ordered from the public on a trial basis.

Computerisation / Communication technology

- Response to computerisation, building production information system using intranet and Internet, CALS/EC (introduced by the Ministry of Construction for public works).

13.3   PUBLIC INTERVENTIONS

Based on the above assumptions, we now would like to discuss the major programmes of the Government as they relate to the technological development of the construction industry.

There is the Science and Technology Basic Law (promulgated and enforced in 1995) in Japan. Based on this Law, the Science and Technology Master Plan has been established with the aim that there is the Science and Technology Basic Law (promulgated and enforced in 1995) in Japan. Based on this Law, the Science and Technology Master Plan has been established with the aim that public research and development investment will increase to the level of the major Western countries as a share of GNP at the beginning of the 21st Century.

In accordance with the Science and Technology Basic Law and the Science Technology Master Plan, the related administrative agencies promote technological development under their individual responsibilities. The Ministry of Construction determines the direction, establishes the objectives, and provides support for construction technology. In other words, MOC has established the Technical Policy Task Force under the Construction Technological Development Meeting to examine policies and programmes for promoting the implementation of science and technology research outcomes efficiently and smoothly. The Task Force compiled a report in 1997 to identify the policies of the actions. The directions and objectives are as follows:

1.   Assurance of safety and sense of security
2.   sustaining economic vitality

3.   preservation and recovery of natural ecosystem and global environment.
4.   respect for different values
5.   promotion of computerisation and networking for versatile communication and affiliation, and
6.   contribution to the international society.

The changes in major policies and programmes of construction technological development, which have promoted private technological development, are summarised in Figure 13.7. The structure of construction technological development and utilisation programmes at the current MOC is indicated in Figure 13.8.

The following are the measures for construction technological development in which the Ministry of Construction is currently engaged:

1)   Programmes for conducting research and development with the co-operation of government, business, and universities
2)   Pre-project research on leading construction technology
3)   Programmes for evaluating technical proposals from the private sector
4)   Introduction of methods utilising corporate technical capabilities
5)   Mutual affiliations among business, academia, and related administrative agencies and programmes of affiliations and joint research with foreign countries
6)   Programmes for inspecting, certifying, and evaluating new construction technologies developed by the private sector
7)   Pilot projects utilising the results of technological development in public construction sites
8)   Establishment of test sites to actually implement new technologies
9)   Projects to monitor the conditions of new technological development in the private sector and information on evaluation after the projects have been implemented
10)   Bank financing programmes for new construction technological development
11)   Tax incentive for technological development
12)   Preparation for uniform technical standards and common documented specifications as well as response to the globalisation
13)   Construction of the integrated information system to support public works
14)   Programmes to secure the quality of technology through granting national qualification (national technical control system)
15)   Commitment to technical research and development activities in the affiliated research institutes.

Ministry of Construction to promote these measures is in total about ¥39.5 billion in FY 1998; ¥41.3 billion in FY 1999; and ¥44.2 billion (including transfer from a part of the budget of the Science and Technology Agency and special account budget) in FY 2000. In spite of tough financial situations, the budget shows a great increase due to economic stimulation measures.

For the research institutes of the MOC, there are the Public Works Research Institute and the Building Research Institute, which are affiliated organisations, as well as the Geographical Survey Institute, a national administrative agency. The profiles of these institutes are as follows:

The Public Works Research Institute promotes construction technological development responding to the needs of people and the administration and disseminates the results as the single general research institute regarding civil engineering in Japan. According to the 5th five year program which commenced in Fiscal 1999, the mission of the Public Works Research Institute is positioned to provide technical support in the construction administration as the core research institute related to the land management technology.

The Building Research Institute is the single national test and research institute in the field of building, housing and urban development in Japan. At present, in addition to five goals so far, based on the report of the Building Council in March 1997, the Institute promotes R&D activities focusing on the themes having strong needs in terms of society and administration as well as rationalisation of standard accreditation, international harmonisation, and performance regulation of building standards.

The Geographical Survey Institute plans and develops basic measures for survey, establishes various technical standards related to survey, and conducts related R&D activities as a national administrative agency supervising survey and maps. Recently in accordance with the rapid development of computerisation and space technology, their activities have been expanded into the peripheral fields such as diastrophism observation and environmental research.

An overview of the major programmes contained in Figure 13.7 is introduced below:

- General technology development projects: Among important R&D themes related to construction technology, especially for issues which are urged and which cover numerous fields for the subject of the R&D activities, the research is conducted comprehensively and organisationally with the executive branch operating as the main body in close co-operation with universities and businesses.

- Joint public-private research: Through the joint development between the public and private sectors, application of technologies in other areas developed by the private sector to the construction area should be promoted while avoiding development risk in the private sector with the aim of promoting the construction technological development.

- Joint research programme with the research institutes of the MOC: R&D activities are promoted utilising characteristics of technologies owned by businesses, universities and government individually while maintaining affiliations organically.

- Construction technology evaluation system: In order to promote R&D activities in the private sector and introduce and utilise the results in construction projects, the government presents the targeted level of technological development for the research themes which have been determined on the basis of the administration needs. Then the private sector implements the R&D activities. The Construction Technology Evaluation Committee evaluates and then, based on the results, the Minister of Construction makes an announcement.

- Programmes for inspecting and certifying construction technologies developed by the private sector: In order to promote the utilisation of construction technology developed by the private sector autonomously, the contents of such technologies are subjected to a searching examination and then certified. 15 agencies including Foundations supervised by the Ministry of Construction are certified by the MOC to perform inspection and certification activities.

- Pilot technology application projects: These are the pilot application programmes to utilise technologies developed by the general technology development projects, test and research institutes and private companies on site in actual public works. The individual regional construction bureaus of the MOC conduct such activities.

- Test field programme: For the technologies, which have a strong administrative need for the future and are necessary to improve their completeness through technical appearance on site, test fields are established on the sites directly supervised by the MOC to perform various tests.

- The Japan Development Bank financing programme for new construction technological development: Low interest loans are provided to private enterprises, which conduct joint R&D activities with the MOC, to promote technological development.

- Tax incentives for technology development: There are five programmes including the tax system to promote recycling technological development such as the application of tax credits for testing and research related expenses.

Among the above programmes the most notable is the general technology development projects started in 1972. The themes of these projects, which are progressing as of 1999, and the budget for Fiscal 1999 are summarised in Table 13.4. The themes of the past projects are indicated in Table 13.5. When implementing the construction technological development project, the private sector spends their own budget and allocates their research staff to form the public-private joint R&D structure.

In 1995, for example, the General Guideline for Construction Industry Policy presented the promotion of joint use of research facilities, collaborations between businesses, universities and government, burden sharing, and facilitates technological development of the construction industry together with these actions.

For construction technologies of harbours and dredging, the Ministry of Construction has various programmes including a joint research system, evaluation system of private technologies, and new technological development pilot projects.

Figure 13.7: Changes in private technical development promotion and construction technology R&D programmes-mainly the measures of MOC-

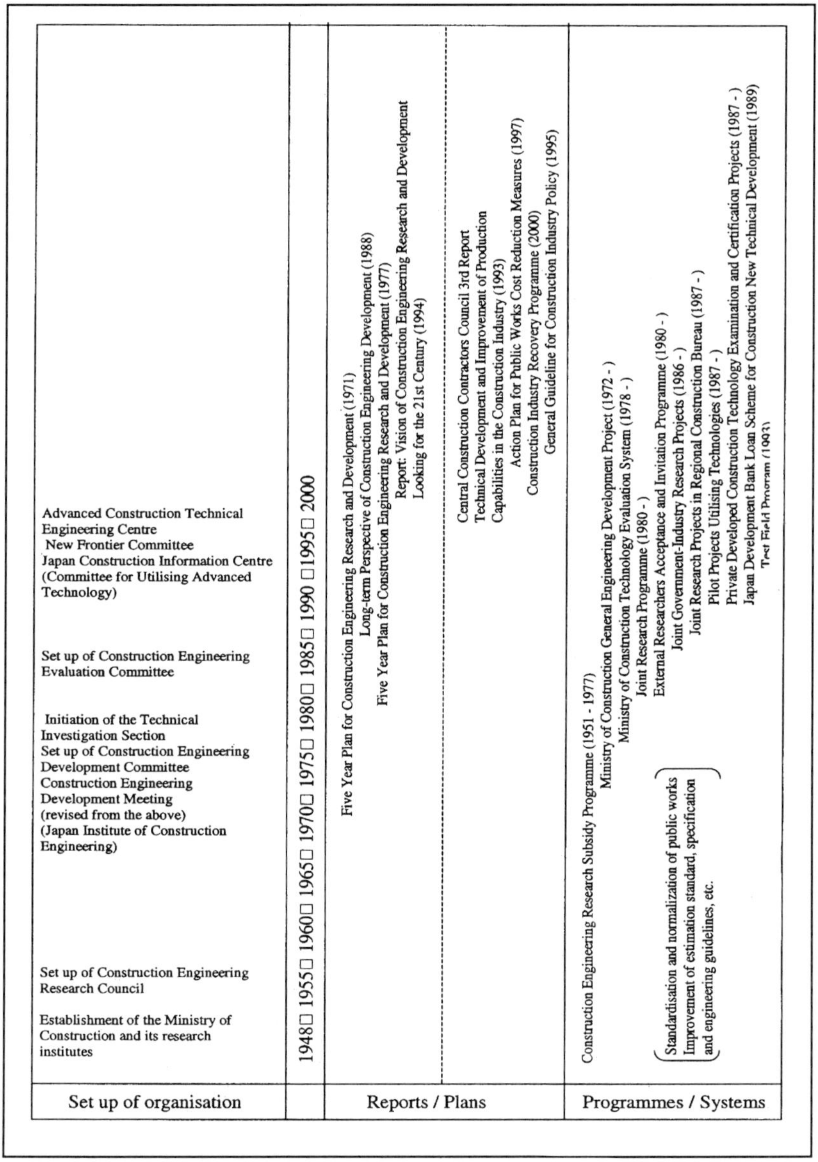

Set up of organisation

Advanced Construction Technical Engineering Centre
New Frontier Committee
Japan Construction Information Centre
(Committee for Utilising Advanced Technology)

Set up of Construction Engineering Evaluation Committee

Initiation of the Technical Investigation Section
Set up of Construction Engineering Development Committee
Construction Engineering Development Meeting
(revised from the above)
(Japan Institute of Construction Engineering)

Set up of Construction Engineering Research Council

Establishment of the Ministry of Construction and its research institutes

Timeline: 1948 1955 1960 1965 1970 1975 1980 1985 1990 1995 2000

Reports / Plans

Five Year Plan for Construction Engineering Research and Development (1971)
Long-term Perspective of Construction Engineering Development (1988)
Five Year Plan for Construction Engineering Research and Development (1977)
Report: Vision of Construction Engineering Research and Development Looking for the 21st Century (1994)

Central Construction Contractors Council 3rd Report
Technical Development and Improvement of Production Capabilities in the Construction Industry (1993)
Action Plan for Public Works Cost Reduction Measures (1997)
Construction Industry Recovery Programme (2000)
General Guideline for Construction Industry Policy (1995)

Programmes / Systems

Construction Engineering Research Subsidy Programme (1951 - 1977)
Ministry of Construction General Engineering Development Project (1972 -)
Ministry of Construction Technology Evaluation System (1978 -)
Joint Research Programme (1980 -)
External Researchers Acceptance and Invitation Programme (1980 -)
Joint Government-Industry Research Projects (1986 -)
Joint Research Projects in Regional Construction Bureau (1987 -)
Pilot Projects Utilising Technologies (1987 -)
Private Developed Construction Technology Examination and Certification Projects (1987 -)
Japan Development Bank Loan Scheme for Construction New Technical Development (1989)
Test Field Program (1993)

Standardisation and normalization of public works
Improvement of estimation standard, specification and engineering guidelines, etc.

Figure 13.8: The Ministry of Construction's Technology Development and Application Program Structure

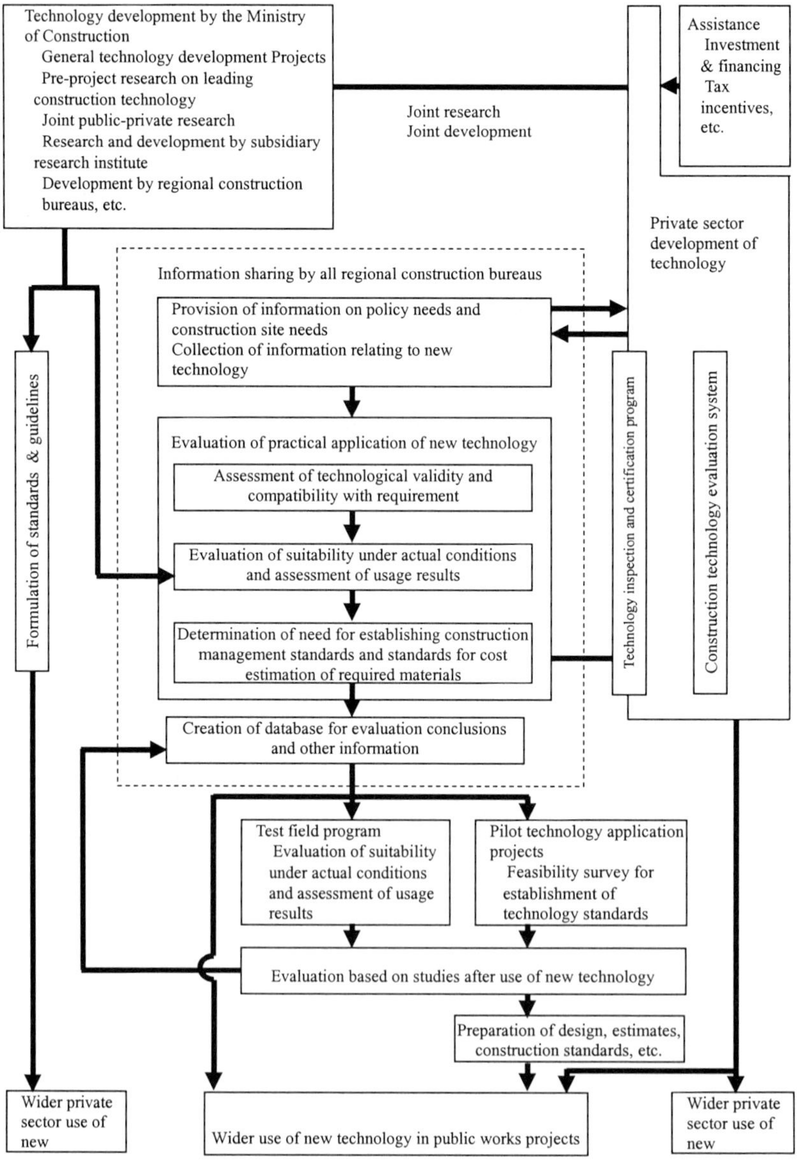

Technology development by the Ministry of Construction
   General technology development Projects
   Pre-project research on leading construction technology
   Joint public-private research
   Research and development by subsidiary research institute
   Development by regional construction bureaus, etc.

Assistance Investment & financing Tax incentives, etc.

Joint research
Joint development

Private sector development of technology

Formulation of standards & guidelines

Information sharing by all regional construction bureaus

Provision of information on policy needs and construction site needs
Collection of information relating to new technology

Evaluation of practical application of new technology

Assessment of technological validity and compatibility with requirement

Evaluation of suitability under actual conditions and assessment of usage results

Determination of need for establishing construction management standards and standards for cost estimation of required materials

Creation of database for evaluation conclusions and other information

Technology inspection and certification program

Construction technology evaluation system

Test field program
   Evaluation of suitability under actual conditions and assessment of usage results

Pilot technology application projects
   Feasibility survey for establishment of technology standards

Evaluation based on studies after use of new technology

Preparation of design, estimates, construction standards, etc.

Wider private sector use of new

Wider use of new technology in public works projects

Wider private sector use of new

As mentioned above, the government promotes technological development with the joint research scheme with the private sector. But due to restrictions on the public sector, compared with joint development in the private sector, including different types of businesses, joint research with public organisations or universities is not easy. Some people even claim that joint research with national universities is especially difficult[3].

13.4 EFFECTIVENESS OF PUBLIC INTERVENTIONS

13.4.1 Effectiveness of public interventions for building technology

As mentioned above, the motivation to promote private technological development is the large-scale public work projects. Especially major projects, in which the construction firms face tough execution conditions and problems that need to be overcome, are a great stimulus.

The concepts for motivation of technological development in the construction firms are summarised in Figure 13.9. In this Section we consider programmes and measures to guide and promote private technological development.

When talking about private technological development and the roles played by the public, there are great differences between the building and civil engineering sectors. So it is necessary to consider them separately. This is because in the building area approximately 90% of the ordering parties, except for some public buildings, are the private sector. On the contrary, in civil engineering, nearly 90% of the ordering parties are the national and municipal governments as well as public organisations, foundations, and corporations.

In that sense, the building technology can be proposed and developed by the construction firms while demonstrating their leadership.

The civil engineering works have long been conducted under direct management of the ordering parties, mainly by the public section. Consequently, for technology, dominance by the public or ordering party is strong. Technological development is also led mainly by the public.

This is a fundamental issue, which can be regarded as a difference in the evaluation of building and civil engineering cultures and characteristics of engineers.

---

[3]   Hironosuke MIYATA, seminar of the Japan society of Civil Engineers – Heading for the 21st Century – Explore the Direction of Technological Development, July 1993 (in Japanese).

Figure 13.9: Map for motivation of engineering development
in recent corporations

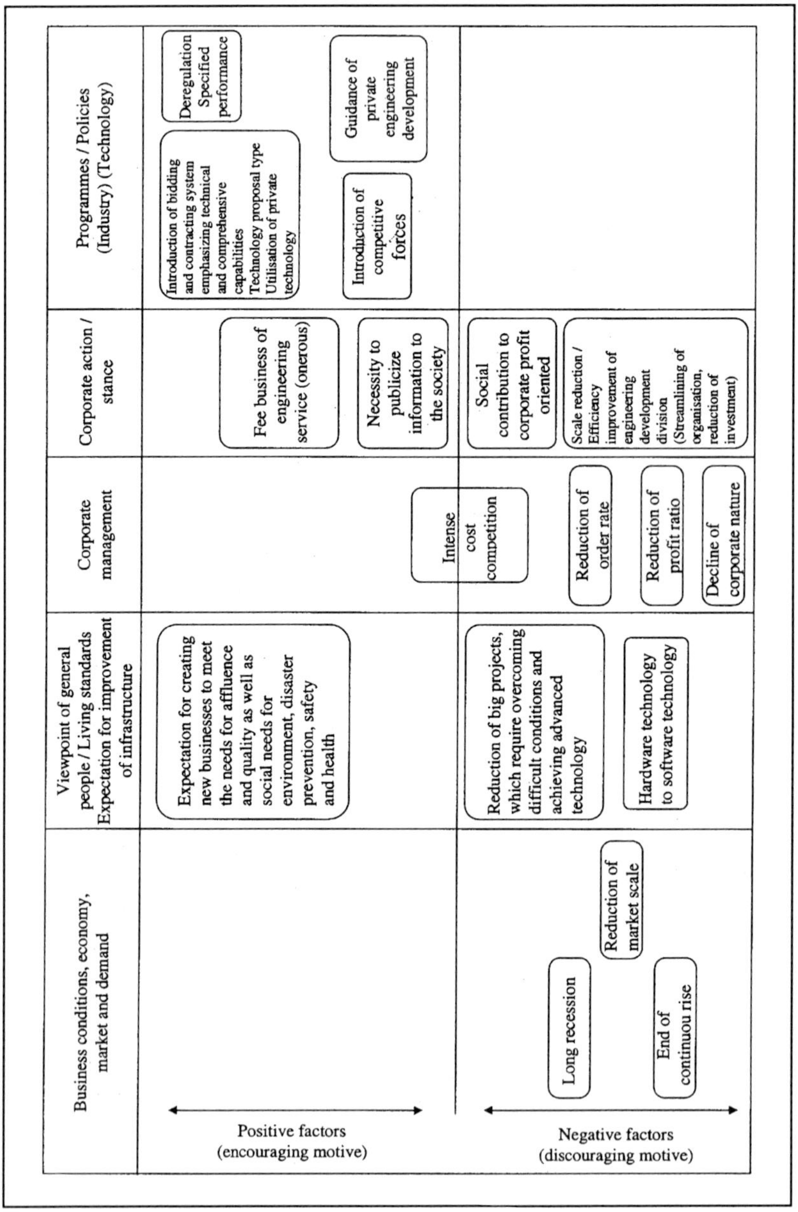

Table 13.4: Summary of construction technology research and development-related budget outlays for FY 1999

Item	(Ministry of Construction)		(Unit: yen)	
	Budget for Fiscal 1998 (A)	Budget for Fiscal 1999 (B)	Increase or decrease (B| A)	Ratio of change (B/A)
1. Construction technology R&D (general technology development projects)	1,147,119,000	1,426,542,000	279,423,000	1.24
(1) Development of technology for preservation of ecosystems and creation of new habitation spaces (Page 16)	100,340,000	104,783,000	4,443,000	1.04
(2) Development of technology related to construction project quality investment-efficient and linger-lasting urban collective housing (Page 18)	214,164,000	216,426,000	2,262,000	1.01
(3) Development of technology related to construction project quality management systems (Page 20)	185,776,000	189,367,000	3,591,000	1.02
(4) Development of technology for reducing construction project costs, including external costs (Page 22)	160,339,000	179,523,000	19,184,000	1.12
(5) Development of assessment and countermeasure technologies for disaster prevention in town planning (Page 24)	125,368,000	167,012,000	41,644,000	1.33
(6) Development of technology for comprehensive analysis of tectonic activity observation (Page 26)	100,283,000	111,621,000	11,338,000	1.11
(7) Development of high national land management technology (Page 28)	0	141,705,000	141,705,000	I
(8) Development of technology for hybrid timber building structures (Page 30)	0	94,188,000	94,188,000	I
(9) Research on technology utilizing artificial satellites for dealing with disasters (Page 32)	0	203,973,000	203,973,000	I
(10) Development of construction method for industrialized infill housing, etc. (Page 33)	0	17,944,000	17,944,000	I
(11) Development of structural safety improvement technology utilizing new-generation steel	102,512,000	0	£ 102,512,000	I
(12) Development of advanced construction technologies using LALS	158,337,000	0	£ 158,337,000	I
2. Per-project research on leading construction technology	71,326,000	74,509,000	3,183,000	1.05
(1) Research on design methods for damage control of buildings (Page 34)	15,372,000	17,032,000	1,660,000	1.11
(2) Research concerning symbiotic river improvement harmonized with natural functions (Page 35)	11,667,000	12,917,000	1,250,000	1.11
(3) Research on methods for creating social infrastructure facilities considering people's sensibilities and regional characteristics (Page 36)	7,757,000	9,320,000	1,563,000	1.20
(4) Development of a method for achieving optimum materials design of concrete (Page 37)	0	14,337,000	14,337,000	I
(5) Development of technology for communicating information on building structure performance pertaining to safety and livability (Page 38)	0	11,355,000	11,355,000	I
(6) Development of technology for rapid restoration of structures after large-scale earthquakes and disasters (Page 39)	0	9,548,000	9,548,000	I
(7) Development of next-generation water-spanning construction technology	15,873,000	0	£ 15,873,000	I
(8) Development of technology to prevent disasters caused by major escarpment collapse	10,722,000	0	£ 10,722,000	I
(9) Development of super durable materials	9,935,000	0	£ 9,935,000	I
3. R&D evaluation	7,484,000	7,484,000	0	1.00
4. Joint public-private research	103,751,000	108,116,000	4,365,000	1.04
(1) Development of soil environment preservation construction technology (Page 40)	21,251,000	21,470,000	219,000	1.01
(2) Development of technology contributing to a wholesome residential environment (Page 41)	16,128,000	18,218,000	2,090,000	1.13
(3) Development of regionally adapted housing technology in conformity with next-generation standards (Page 42)	13,898,000	19,337,000	5,439,000	1.39
(4) Development of technology for increasing the life span and the stock of wooden houses (Page 43)	13,076,000	15,871,000	2,795,000	1.21
(5) Development of high-technology new steel structure building systems (Page 44)	0	14,271,000	14,271,000	I
(6) Development of technological measures to deal with liquefaction of the ground directly under existing structures (Page 45)	0	18,949,000	18,949,000	I
(7) Research on GIS standardization	19,958,000	0	£ 19,958,000	I
(8) Development of new evaluation technology for piles	19,440,000	0	£ 19,440,000	I
Total	1,329,680,000	1,616,651,000	286,971,000	1.22
5. Testing laboratory	16,498,803,000	17,572,115,000	1,073,312,000	1.07
(1) Pubic Works Research Institute	3,593,320,000	3,834,150,000	240,920,000	1.07
(2) Building Research Institute	2,460,543,000	2,584,269,000	123,726,000	1.05
(3) Geographical Survey Institute	10,445,030,000	11,153,696,000	708,666,000	1.07
Total	17,828,483,000	19,188,766,000	1,360,283,000	1.08

Note: Fiscal 1999 figures for the Minister's Secretariat, Engineering Affairs Management Division, Public Works Research Institute, Building Research Institute, and Geographical Survey Institute include only technology development-related spending.

Table 13.5: General technology development projects: end results

(Unit: yen)

No.	Project Name	Period	Budget	No.	Project Name	Period	Budget
1	Development of new seismic design methods	1972-1976	513,882,000	23	Development of new wooden housing construction technology	1986-1990	342,989,000
2	Development of technology for the construction of offshore structures	1972-1976	449,799,000	24	Development of underground space utilization technology	1987-1991	501,109,000
3	Development of new road transport system	1973-1976	151,147,000	25	Development of disaster information systems	1987-1991	344,100,000
4	Development of comprehensive housing performance evaluation systems	1973-1977	294,703,000	26	Development of technology to improve the urban/housing environment in an aging society	1987-1991	271,358,000
5	Development of new construction methods for small-scale housing	1974-1975	57,300,000	27	Development of new materials utilization technology in the construction industry	1988-1992	516,499,000
6	Development of new soil improvement technology	1975-1979	412,539,000	28	Development of super-lightweight, super-high-rise technology for structures made of reinforced	1988-1992	348,661,000
7	Development of new freight transport system	1976-1980	361,093,000				
8	Development of groundwater utilization technology	1976-1980	410,127,000	29	Development of new construction process technology	1990-1994	825,490,000
9	Development of urban fire prevention methods	1977-1981	320,932,000	30	Development of technology for maintaining, renovating and upgrading infrastructure	1991-1995	662,199,000
10	Development of energy-saving housing	1977-1981	355,315,000				
11	Development of technology for improving construction work environments	1977-1981	298,526,000	31	Development of landslide and pyroclastic flow disaster prevention systems	1992-1995	278,750,000
12	Development of various technologies related to improvement of the roadside living environment	1978-1982	324,718,000	32	Development of resource- and energy-saving construction technology	1991-1995	591,261,000
13	Development of technology for improving buildings' durability	1980-1984	373,523,000	33	Development of methods for controlling construction by-product generation and methods for utilizing recycled materials	1992-1996	592,193,000
14	Development of waste utilization technology in the construction industry	1981-1985	339,907,000				
15	Development of repair technology for buildings and infrastructure damaged by earthquakes	1981-1985	344,533,000	34	Development of landscaping technology	1993-1996	401,931,000
16	Development of general lake and marsh water management technology	1982-1986	234,139,000	35	Development of technology for earthquake disaster prevention in large metropolitan areas	1992-1997	649,264,000
17	Development of fire safety design methods for building	1982-1986	268,178,000	36	Development of assessment methods for fire safety performance	1993-1997	376,906,000
18	Development of methods and technologies for city planning in snowy regions	1982-1986	203,554,000	37	Development of new engineering framework for building structures	1995-1997	491,281,000
19	Development of systems for upgrading building technology through the use of electronics	1983-1987	295,087,000	38	Development of structural safety improvement technology utilizing new-generation steel	1996-1998	291,461,000
20	Development of technology to improve durability of concrete structures	1985-1987	543,846,000	39	Development of advanced construction technologies using CALS	1996-1998	436,695,000
21	Development of new wastewater treatment systems employing biotechnology	1985-1989	650,974,000		(Source: Ministry of Construction)		
22	Development of technology for creating and preserving marine space	1986-1990	341,212,000				

In the case of building technology, the major cases to provide incentive to technological development of the private sector by the public can be pointed out as follows:

- Although there are certain restrictions on materials, structures and methods under the Urban Planning Law and the Building Standard Law, new methods and materials developed by the private sector have been adopted under the special measures such as Ministerial certification.

- The major and second-tier general contractors tried the outcome of their technological development under the approval of the government. Based on the trial result, the public sector further improved the standard as the theme of the General Technology Project. Then the guidelines and manuals were prepared with the co-operation of the government, businesses and universities to promote dissemination and generalisation of such technology. The sound development of technology was achieved in this case.

- In accordance with the guidance of the technical result developed by the public sector, the private sector also conducted technological development and progress in construction has accomplished mainly by the public housing supplier.

In the Ministry of Construction the executive branches other than the construction technological measures such as the housing administration executive branch take measures to guide and promote construction technological development. To quote a case by way of example, there is a competition proposals for production and supply technology of new housing. Various policies to promote industrialisation of housing production in the 1970s became active as a part of this and a technological development proposal competition was conducted. Meanwhile, when these industrialisation and standardisation had been established uniformly, the Supervisor of the Housing and Urban Development Corporation (former name) considered that a certain standardisation of public housing methods might cause enthusiasm for technological development to stagnate. Therefore, they changed to adapt the order system based on performance to respect the proposals from private enterprises.

There is a medium and high-rise housing proposal project implemented by the housing administration executive branch in the first half of 1990s. One of the standards applied to select contractors was their strong willingness to invest aiming at promoting and guiding technological development in the private sector.

At present the housing administration executive branch develops the policies for energy conservation, high durability (long life), environmental symbiosis, barrier-free measures for ageing society, and reduction of housing construction costs. For technological development in this field, they guide and promote joint technological development between the public and private or private enterprises alone and provide various tax incentives for dissemination.

The Ministry of International Trade and Industry also implements projects to promote R&D activities for industrialised and next generation housing. As a recent case, there is a housing project implemented from the middle of the 1990s. Under the guidance of MITI, general contractors, house makers, suppliers of energy such as electricity and gas, steel, housing parts and materials have established a technical research union and engaged in technological development while receiving advice from researchers of universities.

The revision of the Building Standard Law might be an incentive to stimulate future technological development in the private sector. It means the conversion from specification provisions to performance provisions. The public works in the civil engineering section are also moving from specification to performance.

13.4.2   Effectiveness of public interventions for civil engineering technology

As mentioned above, in civil engineering it is conventionally the public sector, which means the ordering party, who is the main body for technological development. Historically the private sector has merely executed the construction works based on the results obtained by the public sector. Even at present this basic image is not changing.

During the post-war period, in the development of construction technology, which supported the social capital improvement, the proprietors of the projects mainly introduced technology from foreign countries and then improved it so that it was suitable for geographical and natural conditions in Japan. There were not many cases where the executor was the main purposes of the construction method. In other words, in general public projects, the ordering party has prepared the design, specifications and various drawings. This type of situation obstructs the construction firms from improving their competitiveness for execution technology. In design technology such a tendency is even more pronounced, and restrictions such as standards and structural orders are prime factors in preventing progress of technology development in the private sector.

However, from the 1970s, the major general contractors started to have their own technological research institutes or divisions to improve their technical capabilities.

In accordance with the improvement of private technical capabilities up to now, the technology proposed by the ordering party was accomplished by the private sector as a specific construction method in many cases. Such technology was adopted in the public work on a trial basis to promote generalisation.

The shield construction is a typical example. The number of cases where the private sector proceeds independently with technological development and the public sector follows are on the increase. In this case what has frequently occurred is that a government and business joint committee has been created to improve the technical contents and to incorporate them in the public works. In the civil engineering field, several firms get together to organise the association of construction methods for developing such methods jointly. This association of construction methods which is generally called "*koho kyokai*" in the construction industry has been formed to disseminate newly developed technologies and methods. By licensing patents owned by the member firms of the association, the other members can use applicable technologies and methods.

In many cases at the time when *koho kyokai* was established, there had been technologies completed to the stage when it was feasible to make a patent application. One of the reasons for the existence of this association is to cope with the bidding system in the public works[4].

So far specialised technology available by a single company or for which there is no substitute is hard to adopt in public works. So after generalisation and common sharing, such technology is adopted as a standard order specification of the public works for the first time.

Because of that, the competition due to technical advantage is not fierce. Previously it was pointed out that incentives did not work for the technological development in the private companies – it rather discouraged the willingness for such development.

Regarding the order format, there are similar problems. Because design and execution are separated, it is hard for the system to stimulate technological development as a whole.

Furthermore, under a society aiming at order and equality using the "convoy system," there is an aspect where competition and technological development are difficult to advance[5].

However, recently even in contracts for public works, technical proposal systems and various bidding subcontracting systems have been adopted and tried based on the assumption to compete while promoting differentiation of technologies.

In this way in order to promote public technological development, the structures and schemes of the administration and public sector are necessary. Thus, the Ministry of Construction introduces and examines various programmes from both the sides of technological development and industrial measures.

One of the examples of such programmes is the contract system to utilise private technical capability.

Specifically, the introduction of various methods such as construction management, value engineering, design build, design competition, and proposal methods has been examined.

Furthermore, under the MOC New Action Plan for Public Works Cost Reduction Measures, technological development for cost reduction including labour saving has been promoted.

On the other hand, an external organisation of the Ministry of Construction (one of the organisations indicated in Figure 13.7) established the Construction Technological Development Award in 1999 under the support of the

---

[4]   Hironosuke MIYATA, seminar of the Japan society of Civil Engineers – Heading for the 21st Century – Explore the Direction of Technological Development, July 1993 (in Japanese).

[5]   Masahiko KUNISHIMA, interview article *NIKEEI CONSTRUCTION*, June 11, 1993

MOC for the purpose of stimulating R&D willingness and improving technical standards in the private sector. While introducing the technology developed into the public works and improving the bidding contract system, there is a tendency to grant mild incentives such as manifestation and awarding. There is a system to grant "official authorisation" where the developed technologies have been certified and/or evaluated by an official organisation under the administrative agency in both the sectors of civil engineering and building. In addition to this, measures to provide mild manifestation and encouragement are also implemented.

13.5   TRENDS AND CONCLUSIONS

In order to respond to the drastic changes of our social economic situations and promote the realisation of an affluent life for the general public and society, the assignments and roles, which the construction technology should handle in the 21st Century, are of notable significance.

In accordance with the reorganisation of the ministries and agencies under the administrative reform in Japan, the national research institutes such as BRI and PWRI will be transformed into independent administrative corporations in 2001 and the construction industry themselves will be required to alter its culture. It is necessary for both the public and private sectors to commit to technological development, which means more than simply extending previous achievements.

Some of the issues, which should be handled under the affiliations and co-operation between the government, business, and universities, are as follows:

- New technological development of constructions in response to the changes in the Japanese industrial structure
- Technological development of construction aiming at reorganisation for cussing the technologies of the construction industry
- Development of technology which is aimed at sustainable economic development that achieves a balance between maintaining economic viability and environmental protection
- Promotion of technological development of construction that meets the people's new and diversified needs.

The anticipated roles of the public sector to handle the above measures are as follows:

- Strategic direction for construction technological development of construction as well as the formulation and enforcement of technological development policies
- Introduction and dissemination of various management techniques (ISO 9000, ISO14000, PM, CM) in accordance with the globalisation of construction markets

- Review of various official technical standards, guidelines and regulations in accordance with deregulation and with performance based regulation
- Provision of incentives and support to private technological development
- Institutions reform and further promotion of technical proposal systems for the active application of the results of private technological development in the public works
- Development of electronic procurement technology in accordance with rapid progress of computerisation
- Review of burden sharing between the public and private sectors in basic, leading and applied research in R&D activities to conform with the trend for affiliated research institutes to become independent administrative institutions.

CHAPTER 14

R&D IN CONSTRUCTION: THE PORTUGUESE SITUATION

Fernando Branco and Adriana Garcia

14.1   INTRODUCTION

The construction industry in Portugal has been suffering rapid changes in the last years. In fact from the 50´s till the 80´s, Portugal lived in a closed economy where the main construction companies worked for national public works, with important activities in the African colonies.

With the adhesion of Portugal to the European Community, important funding was received for infrastructures, that led to a large development of the construction sector in the 90's. Since then several challenges occurred, mainly arising from the necessity to improve the productivity of the companies, due to the international competition that now exists in the Portuguese market.

This report begins with a characterisation of the construction industry in Portugal in the beginning of the XXI century. The data presented is based on a recent analysis of the situation, performed for a prospective study of the industry in Portugal (E&T 2000 Project). The information was obtained from national statistical indicators and from an enquiry performed among the 50 greatest construction companies.

The study analysis the construction cluster in terms of the market situation, internal organisation of the companies and changing of the production capacities, regarding the problems associated with the innovation implementation in the industry.

The new challenges associated with sustainable construction, quality, productivity and new materials and construction technologies are then analysed and the necessary correlation with R&D activities and education are presented.

The national financing policies for R&D and their effectiveness are particularly discussed, through the presentation of the position of the main actors, namely the government, the construction companies and the research centres.

14.2   THE CONSTRUCTION INDUSTRY

14.2.1   Country general context

Within the European Community, Portugal with a population of 10 million, has still a low economy level, but with a high increasing rate, associated with a low unemployment level, a low inflation rate and a controlled public debt.

Basically the country is changing from a classical agricultural situation to a country with developed specialised services and several clusters of industries. This led already to the existence of important regions with life levels above the European average, in parallel with interior regions with low development.

14.2.2   The construction industry context

14.2.2.1   The role in the national economy

The Construction Industry in Portugal has a relevant importance as it represents almost 13.5% of the GDP and employees 8.8% of the active population Figure 14.1 and Figure 14.2, respectively.

Figure 14.1 Variation rate of the construction production in relation with the GDP

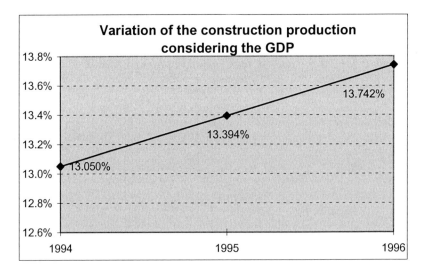

Figure 14.2 Employment in the construction sector

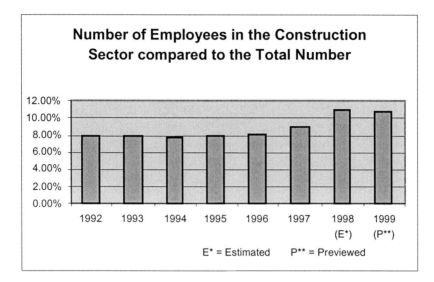

Besides these direct indicators, the construction industry has a large "Drag Effect" as compared to other sectors in the national economy (the "Drag Effect" indicator shows how other sectors of the economy are affected for each unit produced by the construction industry). In Portugal, this indicator is around 1.75 (one unit produced in the construction industry affects 0.75 unit in other sectors). The most related sectors are Commerce, Glass and Mineral No-Metallic Articles, Machinery and Material Transportation and Petroleum Sub-Products.

14.2.2.2 Recent evolution

During the last few years, the construction sector in Portugal had a major boom due to the realisation of several public works namely the EXPO'98, the 12 km long Vasco da Gama Bridge and several kilometres of expressways, among others.

Besides the public intervention, the residential sector has been having an important role in the construction industry. In the last few years the rate of licenses to new residential constructions has been increasing constantly (Figure 14.3).

This new construction is mainly in the suburbs and can be explained by the overcrowded downtown area and its high rental costs, by the bad conditions of rehabilitation of the existing constructions and by the governmental credits to the youth.

Figure 14.3 Licenses for new residential constructions

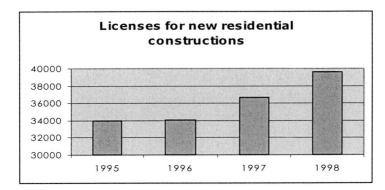

14.2.2.3   Global structure

The construction structure in Portugal is very diffused. According to the National Institute statistics, in 1996 there were around 50,000 companies. This number includes all kind of construction companies from multinationals to small subcontractors.

Basically small companies (less than 100 employees) make the universe of construction in Portugal (Figure 14.4).

Figure 14.4 Comparative number of construction companies

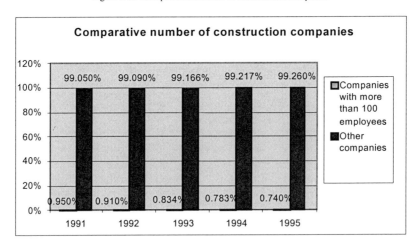

Figure 14.5 Comparative employment

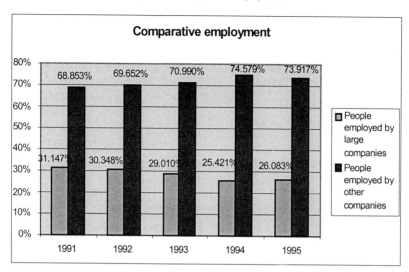

Figure 14.6 Comparative production

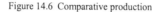

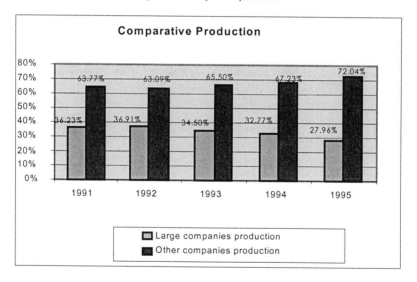

Although large companies represent less than 1% in number, they hire more than 25% of the employees in the sector and are responsible for 33% of the production (Figure 14.5 and Figure 14.6, respectively).

During the last few years, the number of small companies (less than 10 employees) has been increasing while the number of people employed by those companies has been decreasing (Figure 14.7 and Figure 14.8, respectively). This shows a new trend, related to the development of small companies with few employees of their own that hire temporary labour workers whenever necessary.

Figure 14.7 Number of companies by its size

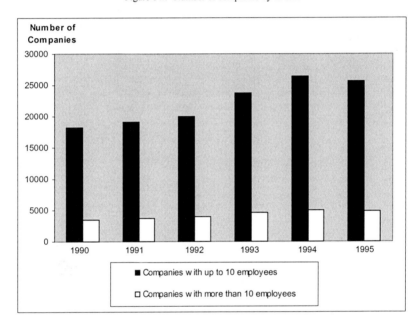

Figure 14.8 Comparative number of companies by the number of employees

This new trend can lead us to wrong conclusions when analysing the productivity of the sector (the productivity is obtained by the construction production divided by the number of employees in the sector). The analysis of the evolution of the productivity (Figure 14.9), in fact, does not necessarily represent an increase as the companies are dismissing their own employees and subcontracting temporary labour.

Figure 14.9 Variation of productivity in construction

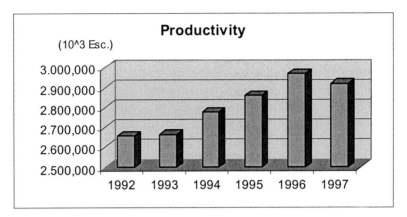

14.2.2.4 Main characteristics

a)   Regional Characterisation

Continental Portugal is divided in 5 regions: North, Centre, Lisbonne and Tejo´s Valley, Alentejo and Algarve. Besides the continental area, there are 2 islands that are also Portuguese territory: Madeira and Azores.

The greatest concentration of construction companies in Portugal is in the North region (37%), followed by Lisbonne (30%) and, at last, the Centre region (20%) Figure 14.10. This corresponds to the greater development of the coastal zones as compared with the interior.

Figure 14.10 Number of companies by region

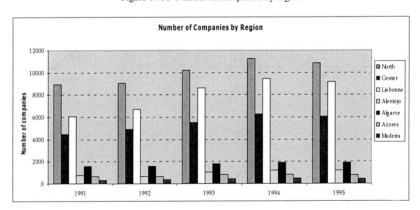

b)   Production Characterisation

The sector of new constructions is the one that increases at highest rate in Portugal differently from other countries in Europe. The rehabilitation sector is only beginning to develop but yet very slowly.

During the last few years there were major investments in transportation infrastructures in Portugal to maintain and upgrade the road systems, to cope with increasing trade and movement of people, to reduce the cost of transport and to promote development and cohesion in the less developed regions.

The share of the construction sector in Portugal is very different from the European Community (EC) Figure 14.11 and Figure 14.12), respectively. The total national volume of construction is shared 40% by the public sector, 30% by residential buildings, 23% by non-residential buildings and renovation, maintenance and repairs represent only 7%. In EC those numbers are 23, 23, 21, and 33%, respectively.

Figure 14.11 Construction production in CE

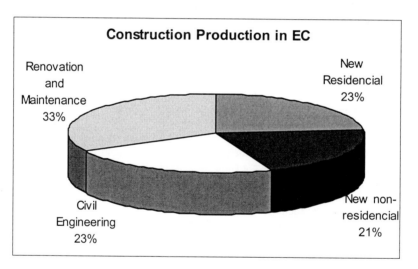

c)   Main Actors

The main actors that compose the construction cluster are presented in Figure 14.13, showing their interconnections and their participation during the constructions lifetime.

The main promoters are the government and housing private investors. Recently government is changing from direct investments to a policy of concessions, leading to important private investments in public works.

Figure 14.12 Construction production in Portugal

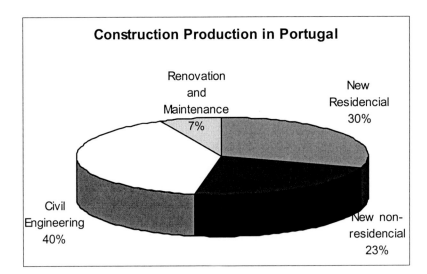

In terms of services, the civil engineering design sector in Portugal is quite developed, with high international level, that can be seen by the design of very important constructions (some national bridges were world record) and by the international work performed in several countries.

The materials production companies have also reasonable standards as they produce materials and elements in quality and quantity for the national construction and they also have a significant level of exportation. Besides, some national companies bought international ones and have presently international groups (cement production, precast industry, etc.).

The construction equipment (machines in general) is essentially imported, as very little national production exists.

The construction companies, themselves, have the characteristics presented in the previous points. The know-how of the greater companies is perfectly updated and allow them to participate, not only in the national construction but also in international works all over the world (namely in Asia, Africa and South America).

14.2.2.5   The technological role – results of a national enquiry

A university research regarding the main technological aspects of the construction sector was recently implemented. This research was part of a prospective work with the main objective of identification of the priority areas and strategies for Engineering and Technology development. The results are expected to contribute to the increase in the competitiveness of the Portuguese construction companies.

Figure 14.13 Construction cluster – main actors

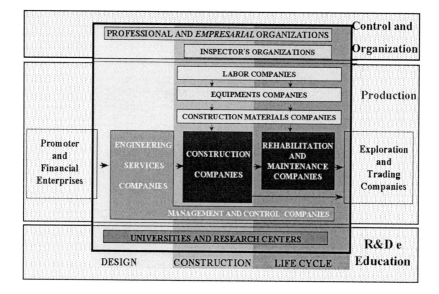

It is interesting to analyse some of the results obtained for this research from a survey performed with the 50 greatest national construction companies. The questions and the associated answers are now presented.

a)   Considering the construction processes, do you consider your firm as antiquated, equivalent or innovative when compared to other firms in the sector.

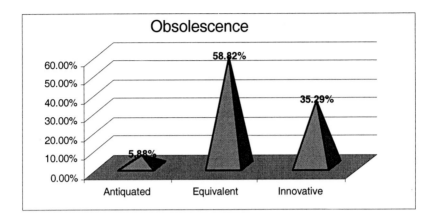

b.1)   There was in your company any important technological change during
          the past three years?

                    Yes   76.47%
                    No   23.53%

b.2)   What kind of changes were they?

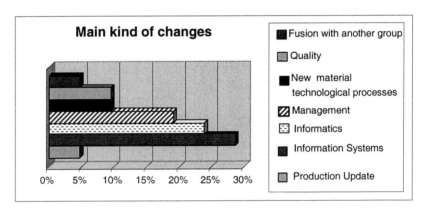

c.1)   Which areas incorporate more technology?

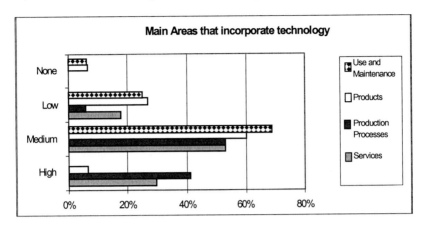

c.2)   Where does the technology come from?

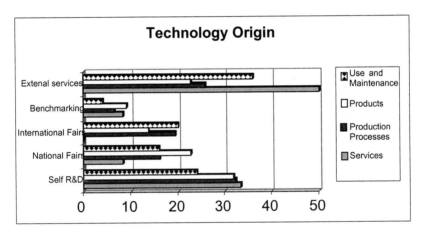

d)   Which are the main reasons for the technological improvement?

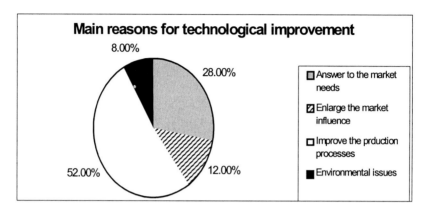

e)   Which are the main institutions used for the improvement of the technological know-how?

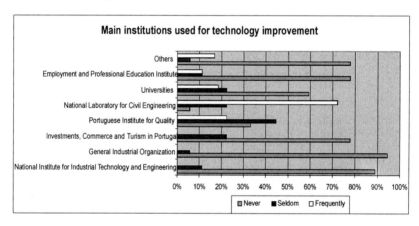

f)   Which areas do you consider as the most important regarding the
        technological matters in the future?

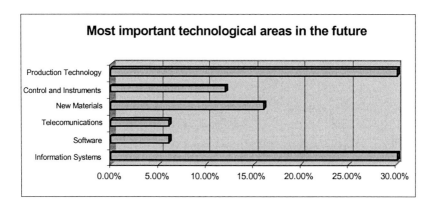

g)   Where should the Portuguese construction sector seek for technological
        support?

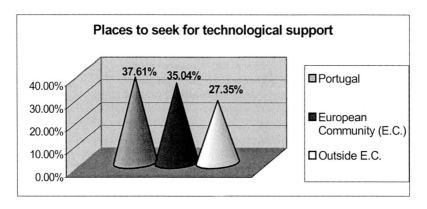

14.3   POLICIES IN CONSTRUCTION

14.3.1   R&D in Portugal – general context

The national innovation investments in R&D represent only 0.65% of GDP (1999), which is considerably less than in other EC countries (Figure 14.7).

Figure 14.14 Comparative investments in R&D

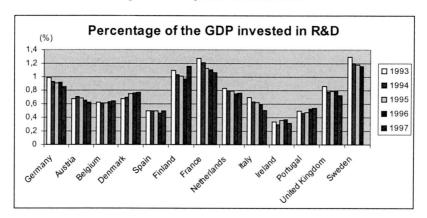

But the scenario is changing. The Portugal Government has shown an increasing interest in R&D. In 1988 the public investments represented 0.43% of the GDP and in 1999 it was 0.65%, what means a growth of 60%.

This scenario change is due to the scientific and technological system modification which promotes a quality R&D held by increasing public investments, estimulates the international co-operation and promotes the exchange of information between the research centres and industry.

Within the R&D investments, the construction sector is one of those that less invests in R&D. From the total amount invested, only 0.10% refers to the construction (Table 14.1 and Figure 14.15).

These public investments in construction R&D lead to research mainly performed by the National Laboratory for Civil Engineering (LNEC) and by the universities. These institutions perform research, give technical education and develop also consulting and quality control work related to the industry.

Table 14.1 R&D Investment Evolution by Economic Activity
(source: Statistic Summary – Inquire to the National Potential Technological and Scientific – 1995
Observatory for the Science and Technology and Ministry for Science and Technology, Lisbon, 1995)

		1992	1995
1	Agriculture, Silviculture, Hunting & Fishing	0.04%	0.05%
2	Extractives Industries	0.24%	0.37%
3	Manufacturing Industries	43.47%	41.15%
31	Food, Beverage and Tobacco	2.26%	2.38%
32	Textile, Clothing and Leather	1.30%	1.73%
33	Wood and Cork	0.19%	0.72%
34	Paper, Graphic Arts and Publications	3.29%	2.31%
35	Chemistry, Coal and Petroleum Sub-Prod., Plastic and Rubber Prod.	7.71%	9.64%
36	Mineral No Metallic Products	0.72%	0.33%
37	Metallurgy	1.61%	0.09%
38	Metallic Products and Machinery, Transportation Equip. and Material	26.41%	23.78%
39	Other Manufacturing Industries	0.00%	0.17%
4	Electricity, Gas & Water	0.72%	3.72%
5	**Construction**	**0.05%**	**0.10%**
6	Commerce, Restaurants & Hotels	0.11%	0.14%
7	Transportation, Storage & Communication	7.14%	8.64%
8	Banks and others Finance Institutions and Insurance	2.06%	2.94%
9	Social Services, Personal Services and Services in General	2.70%	1.73%

Figure 14.15 R&D Financial Evolution by economic activity (source:
Statistic Summary – Inquire to the National Potential Technological and Scientific – 1995
Observatory for the Science and Technology and Ministry for Science and Technology, Lisbon, 19)

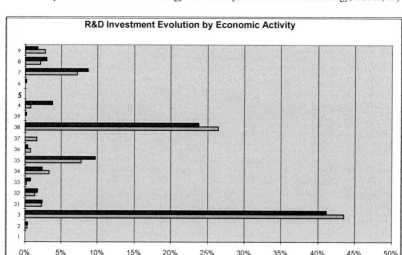

14.3.2   Public financing of R&D in construction

14.3.2.1   The governmental research policy

Some of the main goals of the new public policy are: to promote a quality and
relevant scientific investigation, increase significantly the public investment,
stimulate the international collaboration and create a close net between researchers
and the industry sector.

The public financed research, related to the construction cluster in
Portugal, is dependent on the following Ministries:

a)   Ministry of Research:
         It promotes and finances research activities, in all scientific areas, being
the Construction included in the Civil and Mining Engineering Area. The main
actions of this Ministry are related to:

•   Annual financing of university research centres based on periodical
         scientific evaluation. There are presently around ten university centres
         with activities related with civil engineering.
•   Financing of scholarships for Master and PhD degrees. Annually there
         are around ten scholarships for the Civil and Mining engineering areas.

- Promoting tenders for national research projects for all the scientific areas.

Within this Ministry it was implemented an *Innovation Agency (AI)* that was created to gather national and international projects with the main objective of being used by researchers and companies as a link to international research programs and to make research closer to industry. It acts at two levels: before and after the project.

Before the project the objectives of the AI are to inform, to help choosing the research program and help preparing the documentation. After the project the objectives of the AI are to promote the commercialisation of the products and to make contacts with industrial potential users.

b)   Ministry of Public Works and Building:
It finances partially the activities of the National Laboratory for Civil Engineering. Besides, some of the ministry departments finance also small development projects and technical courses related to their activities, which are usually implemented by LNEC and the University research centres.

c)   Ministry of Environment:
Some of the ministry departments finance small development projects and technical courses related to their activities, which are usually implemented by LNEC and the University research centres.

d)   Ministry of Labour:
Some of the ministry departments finance mainly technical courses related to their activities, which are usually, implemented by the University research centres.

14.3.2.2   *Characteristics of the national R&D projects*

The national research projects have typically a duration of 2-3 years and finance basically the consumable, travel and scholarships for young researchers. They do not finance the university researcher hours.

Typically, the projects involve a few PhD researchers with links among several research centres and have an average financing of 20.000 Euro/year, half of which is spent with the scholarships. Typically a national research project gets a financing of 10% of a corresponding European project

In each tender usually around twenty are given to the civil and mining engineering area, and here most of them are related to design of structures and not really to construction problems.

The proposals rarely involve industrial partners, and when they enter, most of the times, they just present a letter showing their interest in the results. Very rarely they co-finance these research projects.

Besides research, the diffusion of standards to the industry works well as this activity is performed by research centres and universities within their professional education programs.

14.3.3   Private financing / self R&D

R&D performed in Portugal within construction companies is very small and very few have their own research department or even someone in charge of research/innovative activities.   Figure 14.16 presents the budget percentage invested in R&D and whether this amount was for internal development or for acquiring R&D in the market.

Figure 14.16 Budget's percentage invested in R&D

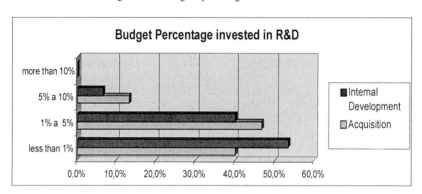

A small exception for this situation can be found in the construction materials sub-sector, namely cement industry, architectural materials and precast industry were some innovative research can be really found, usually performed with the company own means and eventually with the support of some research centre.

Besides the increasing investments in R&D by the Government, there is not any public special policy to support R&D activities performed by the construction companies.

The R&D research projects frequently appear when industry wants to introduce a new product in the construction market and so it looks for some development studies, performed by research centres, to support the technological performance of the new product.

The use of research centres by the construction industry to develop new technologies is not a current situation in Portugal.   In fact this type of contacts occurs more frequently to solve problems that, appeared in the construction process, than to do innovative research.

These projects paid by the industry are typically development projects and consist in using the research centres to develop a new product or technology, or more frequently just to check their properties.

14.3.4   International financing

14.3.4.1   European financing

There are several European financing projects in the construction sector, namely BRITE, CRAFT, EUREKA; INNOVATION, among others.

The Portuguese construction sector participates in European projects mainly in the "New Products and Technology", "Future Cities" and "Environmental Technologies" areas.

European R&D projects are quite different from the national ones. The main characteristics associated with these projects are:

- They must have an international representation. This means they must involve partners from several European countries, interested in doing research in the same problem.
- Usually they must lead to a final product close to commercialisation. This means that must exist a strong component from the industry and the industry controls the development of the project. Besides, as the financing is only partial, it is necessary to find industrial partners interested in putting some money in the risk of the project.
- The competition for the projects, within Europe is very high and small countries, with small companies, have difficulties to get good projects.
- Finally the construction scientific area is not directly one of the main research areas, so projects need to find special topics to enter within the research domains of the Commission.

The advantages of these projects are associated with a good financing, paying 50% of all the types of research costs and usually lead to products that are close to commercialisation, what motivates the researchers.

In Portugal the Government policy related to these international projects was only to create the Innovation Agency and to have some contact persons to help in the preparation of the proposals.

14.3.4.2   Other international financing

Besides the European Community funded research projects, there are other entities that also finance research for national centres, within the Construction Research Area, but with a much less significant expression.

Here can be referred the NATO Program for Peace and several bilateral agreements between countries, usually associated with the Ministry of External Affairs. These programs finance mainly the researcher's exchange of know-how.

The characteristics associated with these projects are similar to those referred for the European projects.

14.3.5   Research areas

The research areas performed within research centres and when possible in association with the industry are related to the main issues that lead presently the development of the construction sector. Some of these are here discussed.

14.3.5.1   Sustainable construction

According to the definition of the Brundtland Report 1987 sustainable development is *"development that meets the needs of the present without compromising the ability of future generations to meet their own needs"*.

Sustainable construction, in fact, is more than the presently main research area, it is a new way of thinking the construction industry, including design, production, education, etc. In fact it is something that must be associated to all activities of the construction cluster.

The construction industry and the built environment must be counted as two key areas if we are to seeking the sustainable development. As an example, in the European Union, buildings are responsible for more than 40% of the total energy consumption and the construction sector is estimate to generate approximately 40% of all man-made wastes.

Sustainable construction can be defined as a way for the building industry to respond towards achieving sustainable development on the various environmental, socio-economic and cultural facets.

This concept is very wide and can be adopted during several stages of the construction as for example the management, product issues, resources consumption, etc. The main challenges are introduced below.

Design process - The most important decisions are taken during this phase and small changes at this point can result in major benefits. It is very important that the design process includes a multidisciplinary team so suggestions can come from all parties.

The information flow should be done in an electronic way so there is no time waste or information loss. The main idea is *"do it right the first time"*.

A recent analysis has found that 60% of the construction anomalies arise during the design phase due to:

Inadequate specification:	20%
Late specification:	20%
Incorrect specification:	20%
Construction processes:	30%
Faulty materials:	10%

Re-engineering of the building process - There are a lot of innovation actions that can contribute to the sustainable construction during this stage. Product and building issues are concerned with how to optimise the characteristics of building and products in order to improve the sustainability performance. This also includes the recycling of construction debris.

Education - The first step towards the sustainable construction is the awareness of all the involved parties about its importance. There must be a sensitising campaign for all the stakeholders in order to create a sustainable consciousness as a factor of competitiveness.

Figure 14.17 shows a summarised diagram of Sustainable Construction and its challenges.

Figure 14.17 Diagram of sustainable construction

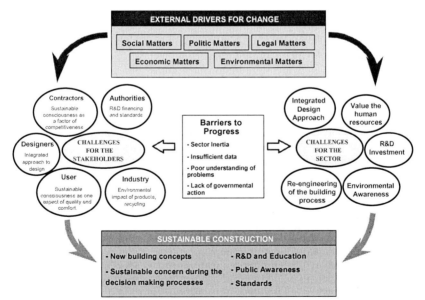

14.3.5.2 Technical research clusters

The R&D projects financed by the European Union are gathered by Target Research Actions (TRA) that summarise the five priority objectives, on which R&D should focus:

- *The Product of the Future* - Solve user problems and research customer satisfaction through higher added value product.
- *New Generation of Machines* - Create machines that has as characteristics be more efficient, easier to maintain and recover, be safe, user-friendly and reliable with low costs.
- *The Extended Enterprise* - Ensure effectiveness of supply chains and production network with multi-skilled and knowledge intensive enterprises.
- *The Modern Factory* - Create an efficient, agile and clean factory towards modernisation and adaptation to change in response of customer needs by the development and integration of advanced technologies and processes.
- *Infrastructures* - Create a condition for an EU economic success and a means for creating wealth and security though a life-cycle approach.

Based on these areas, the construction group defined several clusters for R&D projects, showing the following main areas of research:

Cluster 1	Environmental Technologies for Construction Materials and Components
Cluster 2	Testing and Quality Assurance for Construction;
Cluster 3	Construction Process and Management of the different Life Stages of Construction
Cluster 4	Seismic and Vibration Isolation
Cluster 5	Improved Performance of Concrete in Structures
Cluster 6	Wood Properties and Technologies for Construction
Cluster 7	New Technologies in Geo-technical Engineering
Cluster 8	Steel Research
Cluster 9	Road Research
Cluster 10	Recycling in Construction

14.3.6   Employment and technical education

14.3.6.1   The construction workers

Theoretically, the most important long-term aims for the sector are associated with improving working conditions and job satisfaction and hence raising recruitment. Nevertheless parts of the sector are known to be users of cheap labour, often temporary immigrants in a semi-legal manner.   To improve quality and productivity the industry must attract and retain competent people.

Recruitment of young qualified people into construction sector will be an increasing problem because it is now seen as a poorly paid, danger, uncertain and unhealthy sector.   Work in construction has become less attractive to well educated school students, than occupation in manufacturing and services, and so the general level of education has declined.

Health and Safety - Many, both construction firms and material producers see health and safety measures, as costly.   The industry must take an active role in promoting health and safety measures to reduce the direct costs of time lost through sickness, disability and accident as well as to improve the image of the sector.

There is a problem of improper competition when some contractor will evade health and safety measures to cut costs or hire illegal labour.   Rigorous control of such measures is essential and requires adequate inspection.

Illegal Work - The other point that needs special attention is the illegal work as a way to minimise tax, social security payments and other work taxes.   In the past few years, Portugal and other European countries had a major immigration movement mainly from African ex-colonies and East Europe countries.   This extra labour force is heading for the construction sector.

Education - The training is also a problem, which raises some challenges mainly because the final customer, which is the main driver for change, does not demand yet a quality construction and buys in a "cost-based" view. Furthermore, the use of new technologies, products and regulation create an ongoing need for training which is not yet met.

The main priority for training is seen as site training.   Because of the structure of the industry, with mainly very small firms and mobile workers, and the characteristics of the markets which have a volatile demand and aggressive competition there are strong disincentives to training by individual firms.   A supportive public policy is needed backed by professional institutions.

The prime responsibility for training should rest with the industry because the companies are in the best position to identify the skills they need. Nevertheless, there is a strong disincentive for firms to do training when the labour force is mobile.   The pioneer firms are penalised because workers leave to join firms who spend less on training but pays better.

A possible solution to these problems could be public tax benefits to companies that performance work training.

Figure 14.18 presents some results of the enquiry to the companies where was asked if there was any training investment during the three last years and the areas affected by this training. The great majority (94.44%) did some kind of training program during the last years but, considering the areas affected, it was not really site training for construction.

Figure 14.18 Priority training areas

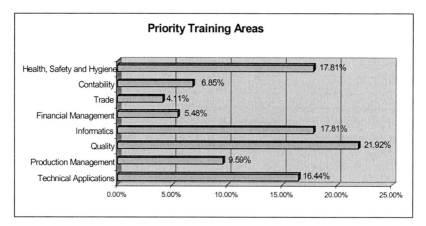

14.3.6.2 *The researchers*

The universe of researchers in the area of civil engineering is mainly composed by the research staff of LNEC and the University staff of the Civil Engineering Departments leading to a global number of around 200 of PhD and senior researchers. From these, those that work in construction related problems are around 30.

Despite these small numbers, related to the importance of the sector, it is positive to see that the number of researchers in Civil and Mining Engineering that get the PhD's is increasing significantly in the last years as shown in Figure 14.19

Figure 14.19 Number of PhD in civil and mining engineering

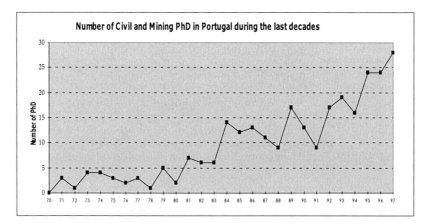

Nevertheless the construction area has some difficulties in recruiting research people as this profession leads to a 5-8 years of studies before getting the PhD, with salaries around 50% of what they can get working in the industry, and to get, at the end, a title that the industry still does not specially evaluate. To overcome this economical problem, typically, research is performed by the researchers in association with consulting work performed outside the research centres.

14.4   EFFECTIVENESS OF R&D POLICIES

The causes for the problems with innovation in the construction industry, arise from all the partners: Government, Research Centres and Industry.

14.4.1   Government policies

a)   Ministry of Research
- The government presents small investments in R&D and the investments are mainly oriented to theoretical research projects from universities and research centres.
- It is necessary to promote projects chaired by the industry, partially financed by the industry (to feel some risk and so to have a real involvement) and clear oriented to obtain a final product to be used by the industry.

- Innovation Agency is practically unknown by the industry, what means that its goals are not reached.

b)   *Ministry of Public Works and Building*
- Should promote the imposition of quality designers;
- Should promote the imposition of quality levels in the public works;
- Should be open and support innovation solutions.

c)   *Ministry of Environment*
- Should promote the imposition of quality designers;
- Should promote the imposition of quality levels in the public works;
- Should be open and support innovation solutions.

d)   *Ministry of Economy*
- Should implement taxation reductions for activities related to R&D industry

e)   *Ministry of Labour*
- Should impose rules for the use of non-qualified labour, and define qualifications for all the labour levels;
- Should Implement and control the Safety and Health rules

f)   *Ministry of Internal Affairs*
- Must implement a solution for the large number of illegal workers, without stopping the industry. Solutions of temporary work permits, or acceptance of skilled labour led to reasonable results elsewhere.

g)   *Ministry of Education*
- The career of the Researchers, with high payments in the industry, against a degree of 5-8 years not leading to a specially well paid job, may lead to a significant decrease of researchers. This situation should be analysed considering the particularity of this industry.

In fact all theses points related to Government policies show that should exist an integrated policy from the Government to the construction industry, covering all these problems. This global policy should be implemented and co-ordinated by the Ministry of Public Works and the Ministry of Research.

14.4.2   Research centres policies

The main problems related to the research performed in the research centres are related with the following points:

- The research centres and universities usually perform research with low practical application and are considered by the industry as "theoreticians".
- in this area it is important that researchers go to the field (put some mud in the boots) to see what are the real problems to be solved
- it is also important that they understand that companies need to apply the research as soon as possible, because R&D costs money, so the projects have to be performed with tight scheduled programs
- it is important that researchers bring to industry the news of research results from other zones of the world, promoting seminars and publishing papers.

14.4.3   Construction industry policies

14.4.3.1   The innovation problem

The position of the industry related to innovation is mainly related to the following points:

- Most of the industry still thinks that profit can be obtained from costs reduction and investments in R&D is to loose time and money.
- When innovation is felt necessary, the best way is to buy it ready to be used from other countries.
- When approached by a research centre to develop a research project, usually the companies have difficulties in identifying a problem to be investigated to make the company more competitive. There is practically no collaboration between the national construction companies and the research centres what frequently leads to the development of projects with low practical application.
- The concept of permanent innovation to be competitive is only slightly felt in some sub-sectors of the industry (precast and materials). This traditional way of seeing the industry, must clearly be changed, namely to promote the new challenges that are arriving with the XXI century.
- Companies need to have Innovation departments, asking questions to the workers. Very few companies have a research department and this kind of research can only be noticed in the main construction companies and in large companies through the precast and materials (cement) sub-sectors. So, there is a real problem of dissemination of technology and product knowledge to small firms and to older workers.

The relation between companies and research centres, to promote innovation, is still poor, and companies typically refer some difficulties with the use of innovation, namely (Figure 14.20):

- Difficulty to accept new solutions by the owners (even public departments)
- low profit margins in the industry (special for small companies) which does not allow large investments in R&D
- difficulty to obtain technical support to innovate.

Figure 14.20 Main problems to introduce new technologies

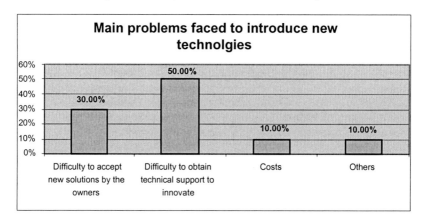

14.4.3.2 The education problem

The training is also a problem in Portugal as there is not a professional training investment culture in the companies. The main reasons are:

- The final client does not demand yet a high quality construction and buys considering just the price. This kind of behaviour does not stimulates the contractor to invest in his employees.
- The sector is mainly composed by small companies that cannot support this kind of investment and does not have a long-term management view.
- The workers are not truly engaged in the companies they work for, so there is high rate of professional volatility and it does not stimulates training investments.

Therefore, the training problem means that many new innovations are very slow to be adopted and national traditions of construction change very slowly. Trying to introduce new technology more quickly than the training of the labour force can adapts leads to several problems of productivity lack.

14.5   TRENDS AND CONCLUSION

The evolution of the construction industry in Portugal in the next years will be characterised by three main vectors: The markets, the company's internal organisation and the production technologies.

The markets will still be essentially national, associated at the public level (public and concession funding) with transportation infrastructures (roads, railways, metro and a new airport), environment infrastructures (sanitary, and coast protection) and special investments (stadiums for EURO2004, Public buildings for PORTO 2001).   At the private level investments will appear mainly in housing (namely social housing and rehabilitation) and tourism infrastructures.   At the housing new concepts begin to appear related to the users (old aged people, tele-work, intelligent buildings, etc.), which is a sector needing important innovation.

Mainly larger companies feel the necessity for improving internal organisation and adoption of new management strategies.   The main vectors are associated with a policy of fusion to became larger and with more power for international competition as well as with diversification of markets (construction companies begin to have many types of business).   Besides, investments are also being performed in management related to the implementation of new methodologies (just in time, lean construction, etc.).   This is also a sector for investments in development.

The vectors for innovation in production are related with sustainability, quality, new materials and technologies.   For this, important investments are needed in research and development as well as in education.   Both of these supports need a significant increase in the co-operation between construction companies and research centres plus universities supported in better policies from the Government.   Innovation in the production sector will be one the most important factors in the competitiveness of the construction companies in the XXI century.

CHAPTER 15

INNOVATION POLICY AND THE CONSTRUCTION SECTOR IN THE NETHERLANDS: BROKERING AND PROCUREMENT TO PROMOTE INNOVATIVE CLUSTERS[1]

Joris Meijaard

15.1 INTRODUCTION

Among Dutch construction firms the idea is rather widespread that being innovative does not pay off. Research institute TNO identified so-called negative dynamics in the balance between competition and co-operation in the Dutch construction sector (Jacobs et al, 1992). Since then, the situation has improved, but not dramatically: most construction firms still focus on cost-based price competition and their local markets. Essentially they are selling construction capacity, putting all efforts in maintaining their 'network' and simply securing future projects. Few firms focus on product improvement and innovation. Customer needs are important but seldom considered critical. Continuing excess demand in the late 1990s has kept fuelling this, mostly due to the scarce building space and strict spatial planning, especially in the Randstad[2]. Only very few, mostly smaller firms try to invent and transfer best practice from other sectors (mostly project development and engineering firms). Yet even to them, the question remains how to get real pay-offs?

Innovation is a key part of the dynamic efficiency of economies, industries and production processes (Metcalfe (1995)). Innovation opens up 'higher' production frontiers. It shifts the scale and scope for the allocation of inputs and the production of outputs. However, incentives to engage in innovative activities are often systematically weak and sub-optimal especially in industries

---

[1] The author would like to acknowledge the generous support of SEOB, foundation for economic research in construction at the Erasmus University Rotterdam.

[2] The Randstad is the Western middle part of The Netherlands. It encompasses the largest cities (Rotterdam, the Hague, Utrecht, Amsterdam) and the smaller cities in between. More than 7 million people live in a very small area.

where the returns to innovation are hard to appropriate. Innovators act in the anticipation of future profits, and, if it is hard to protect knowledge and its pay-offs, firms tend to limit the active pursuit of innovation. Patents and 'natural' mechanisms like secrecy and knowledge and competence platforms may protect innovations, but imitation is often easy and cheap.

A distinction between process and product innovations is helpful to understand some of the complexities involved in construction innovation. In many industries secrecy protects pure process innovations by hiding them behind factory walls. Construction takes place 'on site' and in co-operation with subcontractors, which limits the opportunities to hide innovations from straightforward copying. Product innovations (unless patentable) are difficult to protect in any industry. Lead-time and a required knowledge base may 'protect' product innovations, but in construction, the 'limited edition' of most products makes this difficult as well. The construction of buildings, infrastructure and other products involves an organisational and competitive environment that is often thought to *require* an active role of the government. Dynamically efficient levels of *innovation*, *production* and *growth* are only achievable if information exchange and innovation are actively stimulated. The construction process itself and the 'normal' conditions for the sale and marketing of construction products are what economists call imperfect and monopolistic competition. This should not have to be a bad thing, but it *does* demand some active intervention policies to improve industry efficiency.

Governments can use a number of tools to lend market processes a hand. The appropriateness of the interventions depends on the present (and future) processes of production, the (desired) degree of co-operation and the features of competition: how often do individual customers buy the products, how clear is the quality at the time of sale, how durable is the good and so forth. In construction, significant generic difficulties to 'protect' innovative know-how exist that may induce firms to underinvest in innovation. Since outcomes of research and development efforts are typically uncertain anyway, firms may refrain even more from investing in innovation opportunities that are attractive in principle. Finally, some innovation opportunities may require the combination of a range of complementary competencies. It may be problematic in terms of trust and power to bring these competencies together, which again may obstruct investment in these attractive innovation opportunities. In summary, since most positive externalities (creation of general knowledge, spin-offs and spill-over effects) are not relevant to the individual firm and since time- and context-dependent uncertainty *is* very relevant to the individual firm, governments can choose to play an active role in innovation in order to improve dynamic performance. Each of the potential policy measures may have negative side-effects, but the benefits often outweigh the disadvantages (discussed at length in e.g. Metcalfe (1995) and OECD (1999a)).

When evaluating government policies with respect to innovation, one should bear in mind that many of the institutional conditions have evolved gradually[3]. They exist within the system of other regulatory and economic conditions and their removal may have considerable side effects. Obviously standards to evaluate the adequacy of policy measures are subjective on specific effects (negligibility). Various rationales for innovation policy must be made explicit in order to understand what policy makers can (and perhaps should) do.

In this chapter, we attempt to provide an objective overview and evaluation of Dutch innovation policy as it impacts the construction industry. We start with a short overview of Dutch construction. Section 15.3 then discusses the most important programs of relevance to innovation in the Dutch construction sector, section 15.4 evaluates the policy measures, identifying key trends and prospects and section 15.5 presents a summary and conclusions.

15.2   DUTCH CONSTRUCTION IN BRIEF

15.2.1   Conditions[4]

In understanding the Dutch national construction sector, historical, geographical and climatological conditions must be kept in mind. We list the most important ones.

- Dutch weather is wet and windy, stimulating technologies for solid and windproof construction (especially housing).
- Dutch soil is often unstable with highly fluctuating water levels especially in the metropolitan west of the country, stimulating specific foundation technologies (traditionally straw and wooden piles, more recently mostly concrete piles).
- Sand, gravel and river clay are traditionally cheap and abundant in The Netherlands, wood and natural stone are relatively scarce and expensive. This has held back skeleton-based construction.
- Dutch natural gas is cheap for local use, presenting additional advantages for production of brick, roofing tile and cement.
- The sea and rivers have been central to Dutch economic development. Trade and transhipment have promoted key technologies in logistics, dredging and other areas.
- Large waterworks have been an important part of Dutch construction (especially between 1953 and 1993), stimulating the related core technologies.
- High population density combined with very active Dutch policies to limit the space destined for housing, offices and industrial buildings has

---

[3]   Utterback (1994) and Mowery and Rosenberg (1998) explain the path-dependence of technological development in depth. The Dutch construction is a good example of this.

[4]   In 1992, Jacobs, Kuijper and Roes presented an excellent overview of the economics of the Dutch construction industry. Much of this chapter relies on their work (eight years is relatively short).

stimulated technologies to build 'compactly', and, to effectively maintain and renovate existing buildings.

- High population density and rather high environmental taxes have benefited developments in sectors of the construction industry that deal with waste and water treatment.

15.2.2 Sales and value added

The Dutch construction sector (NACE code 45) produced total sales of $52.7 billion (117 billion guilders) in 1998, almost 10% more than in 1997. These total sales account for almost $1/7^{th}$ of the Dutch GDP. In the national accounts, the construction industry counted for $36.4 billion, a rise by 7% (firms for which construction is not the main activity are excluded here). The first half-year of 1999 showed increasing sales by almost 11% compared the first half of 1998. In 1998, the value added generated in the construction sector was estimated to be $15.9 billion, i.e. about $1/3^{rd}$ of total construction sales[5].

Over the past ten years the real value added produced in the Dutch construction sector has been rather stable, with a slight rise over the last three (perhaps five) years (see Figure 15.1). Contrarily, total sales have been rising with accelerating growth in the late 1990s (more than 5% per year). This seems to indicate that especially companies providing inputs into the construction process have been the sources of the growth in construction sales (and the ones earning the extra money). The last few years do seem to show a shift towards increasing value added for the firms the construction industry (construction costs are now rising more slowly than construction sales). Of course, this closely relates to the dramatic rise in Dutch real estate prices over the last few years.

Figure 15.1 Value added for the Dutch construction sector (million $ ("1999 dollars")) (source: Statistics Netherlands)

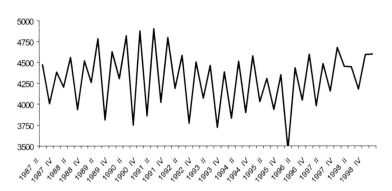

---

[5]   All data are from Statistics Netherlands (1998, 1999), $ values calculated as $1.00 = ƒ2.22.

15.2.3   Industry structure

In 1998, almost 55,000 companies were active in the construction sector. 47,000 of those are small firms, of which 25,000 self-employed entrepreneurs. There were 7,400 medium-sized construction companies (10-100 employees) and 400 large construction companies (> 100 employees).

As in almost any other country, the government is a major customer of the Dutch construction industry. In 1997, it invested $6.7 billion (gross) in 'built' fixed assets, of which $4.2 billion civil works and $2.5 billion government buildings[6]. Corporate investments in fixed assets amounted to $27.0 billion, of which $16.8 billion in housing, $5.6 billion in office buildings, and $3.7 billion in corporate civil works.

Figure 15.2 Proportions of total construction production (1997)
(source: Statistics Netherlands (1998))

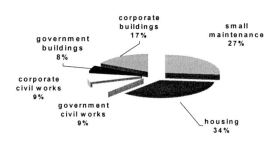

Like most of Dutch industry, the construction sector has a fine-meshed structure of industry associations. A great number of organisations exist for almost every imaginable subsector of the construction industry, including separate associations for specific size-classes and special combinations of activities. The associations promote information sharing between their members. Some of them even organise limited R&D efforts. The associations often compete with each other for members, budgets and status, and firms are often active members of several of them.

15.2.4   R&D statistics

According to Statistics Netherlands, the equivalent of $294 million was spent on *innovation* in the construction industry in 1996. This accounts for 2.7% of all innovation expenditures in The Netherlands. Most of these expenditures concern

---

6   Some of the government investments do not account for construction sector production in the statistics, since employees of the 'lower' governments execute them.

the adoption of innovative equipment and processes. Only 20% are actually on internal corporate *R&D* (in manufacturing this is almost 50%). Private R&D expenses in construction seem to be rising though. This relates to increased attention to activities in CAD, ERP, e-commerce and environmental regulations.

The $294 million innovation expenditure accounted for almost 2% of the gross value added of the construction sector activities (0.6% of total construction sales). Another $94 million was spent on construction-related R&D by the government (including university research). It should be noted that (in addition to this) substantial upstream R&D activity actually concerns inputs into the construction process. Changes in construction industry sales and profits (i.e. opposed to value added) are partly determined by these upstream technological developments, e.g. in fireproof materials, insulation materials and electromechanical elements. Environmental services are also an R&D intensive industry for which the construction sector performs the building activities. It should also be noted that engineering firms and architects execute many of the preparatory efforts in construction innovation that strictly speaking could be accounted as R&D (although these firms may not see their work as such).

Figure 15.3: Innovation expense, % share of sector value added
(Netherlands in 1996) (source: Statistics Netherlands (1998))

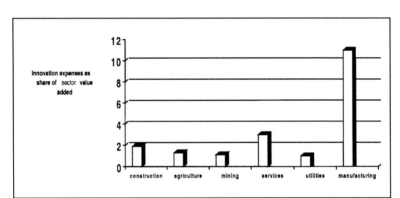

In 1996, R&D expenditures in the construction sector rose by 18%. In the same year, manufacturing R&D remained about constant, agriculture and mining R&D decreased, while R&D in services and in utilities grew even more strongly (R&D in services by almost 50% (!)). Benchmarking the composition of construction R&D against R&D in the rest of the Dutch economy, 24% of Dutch corporate innovations take place in inter-firm alliances and 25% with the aid of government subsidies. In the construction sector these percentages are only 16% and 11%. Of all sectors only publishing has a lower frequency of partnerships. Publishing and only a few service sectors (retail, consultancy and banking) have lower proportions of *subsidy* 'consumption'. The construction companies that work on innovation act relatively often in co-operation with competitors (43% in construction vs. an average 35% in other industries). Relatively few innovations are developed together with the customer (17% vs. 44% for other industries). Innovations in the construction sector usually do take place rather fluently. Only 21% encounter a serious bottleneck (vs. an average 37% in all industries). This may mean that the innovations are just quite straightforward, but -more likely- it indicates that too few innovations are taking place. Only the very clear opportunities and threats are acted upon (mostly me-too innovation).

15.3   PUBLIC PROGRAMS AFFECTING INNOVATIVENESS

Given the sketched circumstances of Dutch construction, how are the actual public programs impacting construction innovation? We will list the most important programs and policies directly or indirectly influencing corporate innovation in construction: how does the Dutch government try to stimulate innovation and what are the effects for construction firms?

Over the last years, Dutch innovation policy has increasingly focused on technological co-operation. Numerous (generic) industry-oriented instruments were launched to stimulate innovativeness through technological co-operation. Furthermore, the structure of government funding for public research organisations has been changing. So-called 'centres of excellence' were established to stimulate and support co-operation between the large industrial firms and the academic research groups. Specific to the construction sector a programme to stimulate the creation of innovative clusters and consortia was initiated recently. A clear change of governmental procurement procedures is an important element in this.

15.3.1   General programs with direct impact

At first sight, the Dutch innovation policies that directly stimulate corporate innovation seem heterogeneous and unstructured. Very few of the subsidies and programs are directly aimed at the construction sector. Many of them *should* still be relevant to construction firms. Table 15. in Appendix A presents a list of policies that are active now (October 1999).

A closer look at the direct subsidy programs shows several dominant types of innovation and investment support. Targeted innovation policies exist at four levels: local, regional, national and European. On each of these levels, small and medium-sized enterprises (SMEs) are specifically subsidised in their innovation efforts. Regionally this takes shape in the so-called "Stimuleringsregelingen" (Stimulation Funds). These funds provide financial support for investment in risky capital, innovation and R&D (usually up to 50% of total costs). Sometimes this may even include the costs of consultants to study strategic change and/or international orientation. Broadly speaking, firms based *outside* the metropolitan area de Randstad can apply for a wider array of funds. The programs in specific support of innovation efforts by small and medium sized enterprises include the direct support of feasibility studies, R&D and (international) marketing of the innovative products. Various investment support programs directly benefit innovative SMEs as well. Specifically, the adoption of several target technologies is being stimulated directly (see below). Last but not least, SME co-operation in innovation is being stimulated intensively.

For larger firms, the government also has targeted programs for R&D co-operation. This concerns key clusters in emerging fields of the economy. Pre-competitive co-operation is being facilitated, both actively and by providing public consultancy. Next to the above mentioned initiatives to stimulate corporate R&D by more or less direct support, the government also directly finances university R&D. Recently, it has been trying to better match the public research infrastructure and the private needs. Tendering programmes and co-financing at Centres of Excellence are the key policy measures.

On top, a whole system of institutions facilitates innovation development and innovation adoption across the economy. The system is meant: (1) to enhance the rate of experimentation, (2) to improve the effective transfer and sharing of knowledge, and, (3) to support and integrate knowledge- and skill-generating activities (usually NOT by actually doing the research). Under the umbrella of "Syntens", SMEs are being helped in their innovation efforts, to find the right partners, to find the right customers, and also, to find the right innovations.

In addition to national innovation support measures, the EU policies are very relevant to innovation in the construction sector. Funds under the Fifth Framework program are intended to stimulate international co-operation by firms in five target areas (of which several are very relevant to construction, especially the program "The City of Tomorrow"). Explicitly for international innovation, substantial funding is reserved for international co-operation by SMEs.

15.3.2   General programs with indirect impact

To begin with, environmental legislation defines some of the clear restrictions that construction companies have to work with in innovation. Some types of buildings need specific environmental licenses (sound, smell etc.). There is a soil protection act. A tight legislation with respect to labour conditions also exists. These laws and restrictions may be sources for innovative activity, but as a rule they put restrictions on the type of ideas and projects that are pursued. The time involved in getting licenses of any sort impedes swift and dynamically optimal levels of innovation. It is clear that many of the existing regulations are justified, but the procedures to get approval are often complex, slow and tedious.

The "Stichting Bouwkwaliteit" co-ordinates certification in construction. This organisation also advises the Ministry of housing, spatial planning and the environment (VROM). Voluntarily, many construction companies choose to certify (parts of) their processes and products to signal compliance. This opens up opportunities for innovation adoption while avoiding dealing with the local codes (Ministry of Economic Affairs, 1998).

National and European economic policies on competition and sustainable development are clearly part of the conditions for innovation in the Dutch construction sector. Regional development stimulation was already mentioned in section 15.3.1. Among other things, competition policy has made large Dutch construction companies a bit reserved towards co-operation in innovation, perhaps even towards innovation in general. The new Dutch competition law is expected to enhance the economic dynamism in the construction sector (Roelandt et al, 1999).

15.3.3   Construction-specific programs with direct impact

University research and funded research at semi, or ex-governmental institutions (especially TNO) are responsible for a large part of Dutch construction related R&D (see section 15.2). Half of the public R&D expenditures are directly related to government-funded construction projects, the other half are more generic. With respect to the latter research, it is often hard to separate construction related R&D from other fields like material science, environmental studies and even some medical studies. Overheads are even harder to separate (actual figures may be substantially larger than $95 million).

As mentioned, "Syntens" is the umbrella organisation for innovation brokering. Advertising itself as the national innovation network for entrepreneurs, it advises SMEs on product-, process-, organisational and ICT-innovation. In short it wants to be the "launching pad" for innovation (15 locations). The main pillar of activities is linking entrepreneurial demand for innovation with the corporate (and academic) supply of know-how. Next to Syntens, several institutes execute regular and project based research directly concerning the construction industry. In Table 15.1 below, a number of the relevant knowledge transfer institutes are listed.

There are further initiatives targeting construction firms, especially in relation to sustainable construction (*"Duurzaam bouwen (DuBo)"*). Subsidies are still available for the adoption of and the consultancy on new sustainable technologies although the relevant subsidy programs have changed recently. Subsidiable efforts should save energy and/or the environment. Some initiatives are industry-wide, sharing information through the technology centres mentioned above.

Within the energy program there is a subprogram aimed at long-term research regarding the built environment: research efforts contributing to the integration of ways to save energy and to the promotion of a sustainable energy supply. Design of energy efficient offices and housing blocks is one of the major areas. A specific program exists for the building materials and glass industries concerning the reduction of energy consumption according to industry-wide covenant agreements. Finally, solar energy is a program that is sometimes of direct relevance to construction. There are subsidies up to 60% of total costs for 'basic' R&D, up to 40% for demonstration projects, and up to 25% for market introductions.

Table 15.1 Knowledge centres for construction industry innovation

Name	Type	Mission	Approx. Size
SBR	Institute for research and knowledge transfer (residential and non-residential buildings)	Develop, spread and promote know-how on buildings and building processes (nine target areas of research)	25 ft.
ISSO	Institute for research and knowledge transfer (building installations)	Satisfy the need for technical development and innovation in building installations.	10 ft.
CUR	Knowledge Network Centre for research and legislation in civil engineering technology	Develop, spread and promote know-how on civil engineering technology (includes a number of subcenters)	25 ft.
CROW	Centre for research and legislation in traffic, transport and infrastructure	Facilitate and transfer knowledge and innovation in the 'GWW' sector	15 ft.
Intrabouw	Platform for information, communication and interaction in the construction sector	Distribute and transfer news, knowledge and knowledge management services in the new networked construction sector.	12 ft.
NOVEM	Energy & environmental research and subsidy coordination	Execute, transfer and support research on energy and the environment (a large branch focuses on sustainable construction)	400 ft.
Dubo-Centrum	National centre for sustainable construction (a joint initiative of several of the above centres)	Inform companies, organisations, institutions and governments about sustainable construction	5 ft.

Recently, a specific partial program has been initiated: "Innovation in Construction". The program aims to improve the level of innovation in the construction industry by stimulating the market-induced incentives to the creation of innovative clusters and consortia. The ministry does not intend to have a strong orientation towards subsidising clusters or limiting rivalry in the market. It intends to focus on the facilitation of synergies in the market. Primarily, this means correcting imperfections in the Dutch innovation system through stimulation of dialogue and interaction between researchers, corporate managers, policy makers and scientists. The program also includes substantial attention to local procurement policies, such that innovation and high quality are promoted.

15.3.4   Construction-specific programs with indirect impact

Housing legislation is defined in "De Woningwet". The law states that municipalities are obliged to define minimum technical standards with respect to safety, health, utility and energy consumption of housing and other buildings. Each municipality is obliged to have a building code, specifying the building regulations applicable to housing. In 1992, the so-called Bouwbesluit ("construction resolution") came into effect, defining national guidelines to simplify and revise the local building codes (for both housing and other buildings). The basic technical standards are now defined on a national level. In this context, there are also European guidelines for the use of building products (mostly safety and energy). Nevertheless, municipalities still have substantial leeway to introduce their own amendments and to interpret the national standards according to the local circumstances (Ministry of Economic Affairs (1996)).

The (local) building codes determine restrictions (and some stimuli) on the willingness to develop and adopt innovations. Particularly the fact that the code still differs substantially per municipality blocks incentives to experiment with innovations. With the 1992 resolution, uniformity and clarity have improved, but the systems are still rigid and over-specified. On targeted issues the local governments are pushed by national policy to be a responsible and innovative 'client'. Stimulate 'better' procurement is a key part of the program 'Innovation in Construction' mentioned above. Recently, the Ministry of Economic Affairs (1998) published guidelines 'Innovatief Aanbesteden' (innovative procurement). More targeted programs are in the making (primarily facilitating and not financing).

Legislation regarding contractor liability, labour conditions and competition are important in the system of institutions determining the innovativeness of the construction sector. However, as mentioned, the 'right to build' and the decentralised procurement of construction projects create a very specific and peculiar context for the behaviour of Dutch construction companies (especially those that also develop projects). Many opportunities to innovate are simply not rewarded (and therefore hardly relevant to the individual firm).

15.4   EVALUATION OF THE CURRENT INNOVATION PROGRAMS

In this section we discuss the impact of Dutch innovation policy on the performance of the construction sector. Present and future conditions are clearly path-dependent. Institutional settings and natural conditions are therefore taken into account to evaluate the way in which the current direct and indirect policies affect innovativeness. In the case of The Netherlands, it is important to remember that building codes are strict, building space is limited and linkages are complex, long-term and rigid.

Both the previous section and the overview in Table 15.2 of the Annex show that the Dutch government puts large priority on *lowering the barriers to innovation*. Barriers to risky investment, barriers to co-operation in R&D and barriers to entry by innovators are lowered through subsidies, tax-credits and active facilitation. On top of this, improvement of the synergy between public R&D and private sector innovation receives large priority.

More downstream, programs exist for the adoption of better practices, particularly with respect to environmentally conscious and energy-saving technologies. Furthermore, for small firms in targeted regions more general support to adopt best practices is available. The small firms are also especially stimulated to develop innovations and to hire educated personnel.

In general, the stimulation of inter-firm alliances in the pre-commercial stage of innovation is becoming the most important element of Dutch innovation policy. The basic idea behind this is that individual firms are lacking the knowledge and the incentive to develop many innovations that could be worthwhile as such. Co-operation and exchange of knowledge can make these innovations feasible and profitable, but firms are often naturally reserved to share information and knowledge on innovations. Also, they often simply do not know the right potential partners. The government wants to assist lifting these barriers.

Up to now, the impact of innovation policy on the technological progress of the Dutch construction industry has been limited. Some environmental innovations are widely adopted (partly due to the policies) but in general other institutional and market conditions block incentives to much construction innovation. Most of the well-meant policy measures are only drops on a glowing plate.

To improve growth and technological progress of the construction sector, it is more important to try to change the institutional framework than it is to apply targeted innovation-promotion measures. The innovativeness of the construction industry is blocked by a lack of incentives to innovate from a profit perspective. Primarily, construction firms acquire projects and safeguard future profits by building good relationships with local governments. Of course these governments evaluate the quality and cost-effectiveness of construction projects, but only as one of many criteria. Political considerations and issues like local employment weigh heavily as well and are likely to reduce the efficiency overall.

New procurement is meant to correct some of the effects of the institutional imperfections, but not nearly enough (in my opinion). Of course it is good to convince municipalities to be demanding clients, but short-sighted political considerations should be made less important in the decision-making process on development, quality and specific contractors. If the authority of individual municipalities were to be reduced, this would strongly benefit the efficiency of the market mechanisms in the construction industry, especially if a more long-term view on construction could be introduced (transformability).

On top of this, it is important to note that the government determines the specific use of land in The Netherlands. Farmland cannot be changed into housing land freely. Landowners have a 'right to build' when the municipality (or government) were to change the destination of the land. This creates two dramatic advantages for construction firms holding land positions. Firstly, the firms can earn good money by selling the land to the municipality (CPB, 1999). Secondly they have a legal right to actually build the things the municipality wants to build (if they are able to). These advantages clearly reduce the incentives to innovate substantially. There is no competition on these projects, leaving few incentives to be efficient. It is obvious that this is an institutional condition that must be changed if optimal technological progress and performance of the construction industry are desired. The extreme scarcity of housing land in The Netherlands probably holds the most important asset in solving this imperfect situation. More than half of both The Netherlands and the Randstad is still farmland. A change in legislation combined with a change in destination for part of the farmlands can make competition more based on cost and quality effectiveness. Furthermore it could improve the innovativeness of firms and the technological progress of the construction sector as a whole, by removing the large tension on the housing market. Everybody would benefit by this. Regrettably, such drastic changes require a very bold and convincing government.

15.5   SUMMARY AND CONCLUSIONS

Over the last decade, the Dutch government has preferred generic technology policy measures to industry-specific ones. Policies focus increasingly on technological co-operation. Furthermore, the structure of public research funding is gradually changing with the establishment of centres of excellence for university-industry co-operation.

Table 15.2 presents an overview of generic programs and their impact on construction. Roughly, Dutch innovation policies are equally distributed over three categories: (1) policies aimed at promoting and triggering R&D, (2) policies aimed at promoting early adoption and experimentation ('unproven' innovations, developed by others), and (3) policies aimed at promoting adoption of best practices ('proven' innovations, not widely used yet). Many instruments are specifically aimed at SMEs and regions. European policy is also relevant, primarily aimed at international cooperation.

The impact of the system of generic innovation policies on the Dutch construction industry is very limited. The nature of the construction process and the nature of construction products seem to be the most important reasons. Innovation is mostly located upstream at the material producers, service suppliers and prefab producers. In the sector itself, innovation is difficult. Construction innovations typically do not fit the generic requirements of subsidies. Also, awareness of the subsidy programs is still a big problem in construction. Additionally, learning processes are seldom seen as important competitive advantages. Efficiency and networking skills still appear to pay off much more directly than innovation, but the government is trying to change this by way of new procurement directives (Ministry of Economic Affairs, 1998).

The Dutch government (Ministry of Economic Affairs, 2000b) knows that it should better understand the interactive process that generates innovation. It has launched a research program on innovative clusters in order to tailor technology policies in the future. The program will produce generic instruments, targeted at specific types and areas of innovation partnering and co-operation, for instance the type of linkages *that are most relevant to innovation in the construction industry*. As mentioned construction is intended to be one of the pilot sectors... The existing innovation transfer and brokering centres (recently reorganised under the umbrella Syntens) will continue to help the process of innovation adoption and diffusion, hopefully increasingly well.

APPENDIXES

Table 15.2 Positioning public policy instruments: The Netherlands
(developed by CIB TG35, May 1999, Data: Senter, The Hague, The Netherlands)

	Name	Resources in Construction 1999 Estimate	Objectives	Means	Contribution to Innovation in the Construction Sector
Programs to support and trigger R&D	Top-instituten	$0.7 million	Improve the social rate of return on public R&D expenditures by facilitating cooperation between industry and knowledge centres	Academic research in close cooperation with corporate initiatives. Still needs to grow.	Large companies are able to outsource basic R&D otherwise not executed.
	Innovatie in de bouw	N/A within the government & through other programs	Stimulate formation of innovative clusters of firms and consortia	Facilitation of interaction between industrial partners, governmental bodies and research organisations (also technology brokering and improved procurement policies)	More cooperative innovation of strategic importance to the construction industry
	B.T.S.	$350,000 out of $15 million	Promote systematic (strategic) technological cooperation between firms and between firms and knowledge centres	Max. 37.5% of relevant costs. Conditions: at least 50% of project costs basic or industrial research. Inter-firm alliance, exemplary, new to The Netherlands and sufficiently promising (economically)	More cooperative innovation
	E.E.T.	$750,000 out of $50 million	Promote 'larger' scale basic R&D, industrial research, pre-commercial development projects linking Economics, Ecology and Technology	Min. +/- $500,000. Conditions: *cooperative project*, 62.5% for basic R&D, 40% for industrial R&D, 25% for development work	Large scale cooperative innovation of strategic importance to the construction industry

Table 15.2 (Continued) Positioning public policy instruments: The Netherlands
(developed by CIB TG35, May 1999, Data: Senter, The Hague, The Netherlands)

	Name	Resources in Construction 1999 Estimate	Objectives	Means	Contribution to Innovation in the Construction Sector
Programs to support and trigger R&D (Continued)	W.V.A. (WBSO)	$3 million out of $300 million. (2nd half 1999	Promote training, R&D (also hiring of long-term unemployed. Childcare facilities, etc.)	Tax credit for R&D personnel: 40% of wage bill, up to $74,000 per company, above that 17.5%. Also $2,200 per employee for training	R&D is stimulated in general
Programs supporting advanced practices & experimentation	Industrie-faciliteit	few	Promote innovative, risky investments by healthy firms	Loans at market rates for large, innovative, risky phases in corporate development ($5-25 million)	support innovative construction firms
	T.O.K.	$500,000 out of $36 million	Remove liquidity problems for risky innovation at companies with less than 25,000 employees. The (potential) financial consequences should be large relative to the company size	Risk-sharing loan: max 40% of costs at market interest rates. Remission after 10 years of no new sales. Report on technical and economic feasibility can be co-financed 50% for <100 employees.	Difficult and risky projects can be executed. Construction firms hardly use the arrangements
	MKB-Initiatief NL (ESF and EFRO)	$250,000	Promote advanced practices and experimentation in SMEs by promoting 1. the use of test and demonstration centres, 2. hiring consulting firms, 3. cooperation and 4. training use of technology	Co-financing by EU, ministries and SMEs. EU financing up to 50%. Only for subsectors of construction that (may) push overall economic growth (*regional!*).	Improved incentives for SMEs to execute R&D and innovative projects.

Table 15.2 (Continued) Positioning public policy instruments: The Netherlands
(developed by CIB TG35, May 1999, Data: Senter, The Hague, The Netherlands)

	Name	Resources in Construction 1999 Estimate	Objectives	Means	Contribution to Innovation in the Construction Sector
Programs supporting advanced practices & experimentation (Continued)	M.P.O.	$350,000 out of $2 million	Promote the development of environmentally conscious products at SMEs	40% of total project costs up to $240,000: direct project costs	Quality improvement of construction products
	K.I.M.	few in construction	Increase the innovative capacity of SMEs	Subsidy for SMEs to hire a well-educated 'starter' to execute research plan for innovation (max. $10,000 in wage costs, max 2 higher educated personnel), and firm older than 2	Improve innovative power of construction firms
Programs to support adoption of best practice	Regionale stimulerin gsregeling	$0.8 million construction	Promoting SME activity, investment and initiative in targeted regions	Co-financing investments, greater use of benchmarking and diagnostic tools, consultants to upgrade firms' organisational abilities	Improve adoption of best practice by construction firms in targeted regions
	Innovatie in de bouw	N/A Government	Aimed at govern-mental departments: change in procurement policy and practice to stimulate innovative contractors	Procurement increasingly restricted to innovative, high quality project development and construction. Local governments as 'responsible' clients	Innovative and 'best practice' companies are stimulated and rewarded
	CO_2 (construction specific)	$420,000	Promote the adoption of CO_2 emission reduction in housing and office projects. Especially in the exploitation phases of construction.	Direct subsidy, relative to cost efficiency and innovativeness: new ways using less wood, energy, etc.	The best new energy efficient construction projects are promoted

Table 15.2 (Continued) Positioning public policy instruments: The Netherlands
(developed by CIB TG35, May 1999, Data: Senter, The Hague, The Netherlands)

	Name	Resources in Construction 1999 Estimate	Objectives	Means	Contribution to Innovation in the Construction Sector
Programs to support adoption of best practice (Continued)	Subsidie-regeling Haalbaar-heids-projecten MKB	$300,000 out of $6 million	Promote the adoption of new technologies of production and organisation in SMEs	Co-financing technology uptake and feasibility studies: 50% of relevant costs at third parties, maximum $12,000	Improved rate of adoption of new technologies and systems by SMEs
	SBT, SP, Innova-tion relay centre	$100,000, rise in 2000	Awareness-building activities and mediating activities for adoption of innovations (including export/import of innovations)	Subsidy for technology centres (SBT), awareness-building (SP) and non-pecuniary support for innovation diffusion/adoption	Improved information and awareness regarding best practices and systems
	EMA, EINP AZS, EIA, SRM, ZON, NETTO, etc.	at least $2.5 million cumul.	Promote the adoption of energy-saving and sustainable technologies (primarily but often not exclusively aimed at SMEs)	Manifold: tax credits, direct subsidies, partly regional, SME-specific	Improved practices regarding energy saving and environmental techno
	EMA, EINP AZS, EIA, SRM, ZON, NETTO, etc.	at least $2.5 million cumul.	Promote the adoption of energy-saving and sustainable technologies (primarily but often not exclusively aimed at SMEs)	Manifold: tax credits, direct subsidies, partly regional, SME-specific	Improved practices regarding energy saving and environmental techno

Table 15.3 List of subsidy programs for corporate initiatives (October 1999)
(source: Ministry of Economic Affairs, copyright Elsevier bedrijfsinformatie BV 1999
(For details of each subsidy program: http://www.minez.nl/subs/))

NCM RHI Regeling Herverzekering Investeringen
AA Krediet
IBTA Subsidieregeling Investeringsbevordering en Technische Assistentie Oost Europa: technische assistentie
IBTA Subsidieregeling Investeringsbevordering en Technische Assistentie Oost Europa: investeringsbevordering
T&S Subsidieregeling Programma Technologie en Samenleving
EINP Subsidieregeling Energievoorzieningen in de Nonprofitsector en bijzondere sectoren
EMA Subsidieregeling Energie Efficiency en Milieuadviezen Schoner Produceren
Subsidieregeling Voorlichting en Doorlichting Schoner Produceren
BIT Bedrijfsgerichte Internationale Technologieprogramma's
SEM Subsidieregeling Exportmedewerkers MKB 1998
Stimuleringsregeling Energiebesparing Energiebedrijf STIMECK Onderdeel Stimulering
Energie Efficiente Commerciele Koelinstallaties
Stimuleringsregeling Energiebesparing Energiebedrijf STIZON Onderdeel Voorzieningen met betrekking tot Zonneboilers
TOK Besluit Technische Ontwikkelingskredieten
EIA Energie Investeringsaftrek
SRM Subsidieregeling Referentieprojecten Milieutechnologie
BSE Besluit Subsidies Energieprogramma's Algemeen
BTS Besluit Subsidies Bedrijfsgerichte Technologische Samenwerkingsprojecten
MKB Initiatief Nederland
AZS Subsidieregeling Actieve Zonthermische Systemen 1998 (ZON)
EFI Regeling Exportfinanciering Indonesie
KREDO Besluit Kredieten Elektronische Diensten Ontwikkeling
MPO Kredietregeling Milieugerichte Productontwikkeling
WBSO Afdrachtvermindering Speur en Ontwikkelingswerk (S&O)
ROF Rente Overbruggingsfaciliteit
Subsidieregeling Haalbaarheidsprojecten MKB
PSB Subsidieregeling Programma Starters Buitenlandse Markten 1999

Table 15.3 (Continued) List of subsidy programs for corporate initiatives (October 1999)
(source: Ministry of Economic Affairs, copyright Elsevier bedrijfsinformatie BV 1999
(For details of each subsidy program: htt://www.minez.nl/subs/))

EET Besluit subsidies Economie Ecologie en Technologie
Stimuleringsregeling Energiebesparing Energiebedrijf STIMAD Onderdeel Stimulering
EnergieEfficiente Aandrijftechniek: Frequentieregelaars
SMO Besluit Subsidies Maritiem Onderzoek
PESP Programma Economische Samenwerking Projecten 1999
MF Matchingfonds Exportkredieten Lichte Matching en Zware Matching
BBMKB Besluit Borgstelling MKB Kredieten Bodemsanering
KIM Subsidieregeling Kennisdragers in het MKB
BBMKB Besluit Borgstelling MKB Kredieten
Stimuleringsregeling Energiebesparing Energiebedrijf ISO HR Onderdeel
Isolatiemaatregelen en HR Verwarmingstoestellen
Industriefaciliteit
IFOM Investeringsfaciliteit Opkomende Markten/OM Krediet
SBT Subsidieregeling Branchecentra voor Technologie 1998
Stimuleringsregeling Energiebesparing Energiebedrijf STIMEV Onderdeel
EnergieEfficiente Binnenverlichting en Buitenverlichting
IOP Subsidieregeling Innovatiegerichte Onderzoekprogramma's

CHAPTER 16

SOUTH AFRICAN PUBLIC POLICY INSTRUMENTS AFFECTING INNOVATION IN CONSTRUCTION

Rodney Milford, Chris Rust, and Mabela Qhobela

16.1 BACKGROUND AND CONTEXT

16.1.1 General context

Although South Africa is classified as a developing country, it is atypical in many respects. South Africa has one of the highest per capita incomes in Africa, but globally only Brazil has a higher income inequality coefficient. The infrastructure needs in South Africa are large, and the majority of South Africans have limited or non-existent access to basic infrastructure.

A major revamp of the policy environment has taken place in South Africa since the democratic elections in 1994 that is required for the necessary transformation in South Africa – resulting in a plethora of national Green and White Papers. Most of these policies build directly on and support the government's Reconstruction and Development Programme (RDP) (South African Government, 1994) and the Growth and Development Strategy (GEAR) (South African Government, 1996), in which key elements of the policy include:

- promoting competitiveness and employment creation;
- enhancing quality of life;
- developing human resources;
- working towards environmental sustainability; and
- promoting an information society.

This South African report concentrates primarily on public policy instruments affecting innovation in construction subsequent to the 1994 elections in South Africa, which have overshadowed previous policy interventions.

16.1.2 Construction industry context

As illustrated in Figure 16.1[1], the period 1970 to 1980 in particular saw strong economic growth within South Africa – in which the construction industry

---

[1] $1 US approximately equal to R6

played a significant role. South Africa was developing rapidly, and significant
investments were made in infrastructure and industrial projects. However, much of
the development during this period (and especially in the 1960s and the 1970s) was
a consequence of South Africa's inward industrialisation strategy, in a period of
increasing isolation from the rest of the world – a strategy could not be sustained.

Figure 16.1 Contribution of construction related GDFI to GDP

Figure 16.2 Employment in civil engineering

        After the heyday of growth in civil engineering in the 1970s, the 1980s
and 1990s saw a recession in the industry as South Africa's economy stalled –
resulting in the shedding of jobs, a loss of skills, and a downsizing of the civil
engineering industry (see Figure 16.2).

Notwithstanding that South Africa is a developing country with huge infrastructure needs (and backlog), the civil and building sectors are today undergoing their worst recession in about 20 years.

As indicated in Figure 16.1, construction related GDFI presently accounts for about 8% of total GDP. To place this into perspective, the GDP by economic activity is given in Figure 16.3 (South African Reserve Bank, 1998), in which it is seen that the manufacturing sector accounts for the highest economic activity (23% of GDP), followed by finance, insurance, real estate and business services (17%) and wholesale and retail trade, catering and accommodation (15%). However, it should be noted that GDP by economic activity includes expenditure on construction related GDFI by economic activity, and for this reason construction related GDFI has been included as an additional item in Figure 16.3. The breakdown of construction spending is about 50% public sector and public corporations and 50% private sector.

Figure 16.3 GDP by economic activity

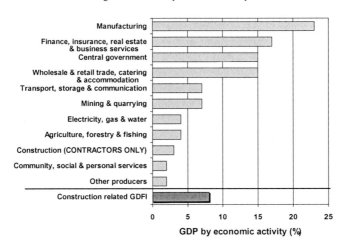

GDP by economic activity (%)

The formal sector of the construction industry in South Africa is well developed, made up of a relatively small number of highly qualified professional and technical people. The capabilities of the construction industry are comparable to that of many of the more developed countries, and over the last 10 years or so, many South African civil contracting companies have aggressively entered the African and international market. Within the context of South Africa, an important component of the formal sector is what is referred to as the "emerging sector" – which is typically characterised by previously excluded enterprises that are registered or are in the process of acquiring registration, but typically lack experience and access to the market. In addition, a large "informal sector" exists, where the enterprises are usually not registered and often weak in the quality of work.

The formal sector of the construction industry (i.e. civil and building) is currently estimated to employ about 350,000 people (which is significantly down from the peak in employment levels of the 1970s), while it is estimated that the informal sector provides employment for a further 600,000 to 800,000.

In addition to the national, provincial and local government structures and the formal financial sectors which impact directly on the delivery of construction products, South Africa also has a well established technical infrastructure, including industry and professional associations, standards generating and regulatory authorities, the tertiary education sector (universities and technikons), and public sector R&D organisations.

16.1.3   Innovation in the construction industry

For the purpose of this paper it is necessary to place a boundary around "innovation in construction" and to differentiate it from the broader issue of "innovation in the built environment". Construction (or more correctly the construction industry) is defined here as:

- the design (architectural and engineering), construction, maintenance and demolition of physical infrastructure; as well as
- the manufacture of materials and products involved in construction.

This definition therefore excludes:

- the upstream activities of spatial planning (including urban planning, town and regional planning, transportation planning, water resource management, etc.); and
- the downstream activities of operation of physical infrastructure (including storage and transportation, traffic safety, etc.).

No specific studies have been undertaken to date on the nature and characteristics of innovation in the South African construction industry. However, it is reasonable to assume that innovation in the construction industry in South Africa displays similar characteristics to many overseas countries, and in particular to that displayed by western countries. In this regard, the recent Canada studies (IRC, 1997a, 1997b, and Milford, 1998) are particularly relevant, and are discussed briefly below.

Most innovation in the construction industry tends to be informal and co-incidental and follows a recognition of the need to learn, explore and apply new knowledge through which to ensure incremental improvements in business practices. Notwithstanding this, the role of the local and international formal technical infrastructure (including government research organisations, tertiary education institutions, standards organisations and industry associations) is particularly important for enhancing innovation in the construction industry.

Throughout this paper, investment in formal R&D is used as an indicator of innovation. As will be illustrated in this paper, and as demonstrated by numerous international studies, private sector investment in formal R&D in the construction industry is very small compared to other industries. One of the main reasons put forward for the low private sector investment in formal R&D is that the construction industry is characterised by a social rate of return that is significantly higher than the private rate of return (Seaden,1995). (The private of return refers to the gains realised by the firm of organisation undertaking R&D, whereas social returns are based on total benefits, including those flowing to consumers and other producers.)

Where there is a significant gap between private and social rates, firms will invest less in formal R&D than is socially optimal as is particularly characteristic of fragmented industries - such as the construction industry. Traditionally, it is often seen as the government to then invested in formal R&D in such cases of market failure and high rates of social return - provided that the returns to government are commensurate with the investment.

In-house private sector investment in R&D in South Africa is very difficult to quantify, and also depends very much on how R&D is defined. However, as will be shown in Section 16.2.7, private sector investment in construction related R&D at the science councils and universities, as well as in co-financing public sector incentive schemes is very small - amounting to less than R1 million per annum (less than about US $200,000 pa). (Note that this should be compared to the total turnover of the construction industry as measured by construction related GDFI of about R40 billion per year.)

A more significant form of investment in R&D in the private sector is possibly the investment in R&D through trade and industry associations, such as (in South Africa):

- the Cement and Concrete Institute
- the South African Institute for Steel Construction
- the Southern Africa Bitumen Association
- the Concrete Manufacturers Association, and
- the Clay Brick Manufactures Association.

Of the above associations, only the Cement and Concrete Association has any significant in-house R&D capacity, although this has been scaled down significantly in recent years with the break-up of the cement cartels in South Africa.

Notwithstanding this, the industry associations have made significant contributions supporting innovation in the construction industry in South Africa over the years, primarily through the development of codes of practice and industry or materials specific guidelines and specifications.

Because of the low investment in formal R&D in the construction industry, public sector policy instruments in support of the construction industry, and in particular public sector investment in formal R&D, is therefore a meaningful indicator of innovation in the construction industry.

Several statutory science councils exist in South Africa, of which the Council for Scientific and Industrial Research (CSIR) has a strong focus on the construction industry. It is currently comprised of ten business units including CSIR Building and Construction Technology (Boutek) and CSIR Roads and Transport Technology (Transportek).

Over the years, and in particular prior to the restructuring of the CSIR in 1987 to be a more market orientated organisation, the CSIR's Boutek and Transportek made numerous contributions to innovation in the construction industry (NBRI, 1987, and Boutek, 2000). Most of these innovations introduced by Boutek and Transportek have been placed in the public domain, resulting in improvements in the performance of the industry, lower construction costs, improved quality of construction products, etc. However, in some cases, the innovations have been appropriated by one or more companies, resulting in the growth of sub-industries and employment creation - such as companies specialising in the design of stiffened raft foundations, the production of flyash, etc. Most of these innovations referred to above were funded, effectively, out of the CSIR's Parliamentary Grant, with little or no financial input from the public or private sector.

A breakdown of the CSIR's current external contract income is given below (see Section 16.2.3 to follow), of which the small investment by the private sector in R&D (i.e. other than specialised services) should be noted:

- 51% public and private sector specialised services (i.e. non-R&D)
- 34% public sector R&D
- 13% international R&D, and
- 2% private sector R&D.

Over the years the universities and technikons have also made substantive contributions to R&D within the construction industry. Much of this R&D has been undertaken by means of post-graduate research, in which National Research Foundation (NRF) grant financing and the Technology and Human Resources for Industry Programme (THRIP) co-financing play an important role (see Section 16.2.4). In the THRIP Programme, contributions are provided by industry and government to fund the research efforts of the academic partners, provided that these projects involve the training of students.

However, as discussed in Section 16.2.4, the construction sector receives very little support from these financing mechanisms in comparison with other industry sectors – which again is an indication of low levels of innovation. In particular, post-graduate research in the construction sector receives less that 1% of the total NRF grant financing and THRIP disbursements to construction related activities amounted to less than 1% of the total THRIP allocations. The low level of THRIP funding in the construction sector is a further indicator of the low levels of private sector co-investment in R&D – which is a requirement for THRIP funds.

In addition to the direct funding of innovation discussed above, as has been illustrated by several studies (IRC, 1997a), innovation in the construction industry and the adoption of innovative solutions in South Africa (as elsewhere) is also strongly influenced by codes of practice, standards, specifications and regulations, etc. – which has had a significant impact on innovation. The primary standards generating and certification bodies in South Africa are the South African Bureau of Standards (SABS) and Agrément South Africa.

The South African Bureau of Standards[2] (SABS) is a statutory organisation, and is South Africa's official body for the preparation and publication of standards for products and services – including those impacting on construction. South Africa's engineering codes of practice are updated or revised on, typically, a 10 to 15 year cycle, and have resulted in ongoing improvements in construction process techniques.

Agrément South Africa[3] is an independent agency which evaluates the fitness-for-purpose of non-standardised building and construction products and systems by applying performance criteria in its assessment procedure. During the period 1970 to 1999, more than 287 certificates have been issued by Agrément South Africa, which has resulted in several new and innovative products being successfully introduced into the market.

In addition to the South African Bureau of Standards (SABS) and Agrément South Africa, several national government departments (including the Departments of Housing, Transport and Water Affairs) also play a strong role in the development of guidelines and standards which influence innovation in construction.

16.2   PUBLIC INTERVENTIONS

It is beyond the scope of this paper to present an exhaustive overview of public policy instruments, and the focus here is on those public policy instruments within the manufacturing sector – of which the construction industry is effectively a sub-sector.

16.2.1   A national system of innovation

The White Paper on Science and Technology (DACST 1996) produced by the Department of Arts, Culture, Science and Technology sets out the government's vision and framework for a national system of innovation (NSI), which seeks to harness the diverse aspects of S&T through the various institutions where they are developed, practised or utilised.

This vision embodies *"a co-ordinated effort to achieve excellence in serving the national goals. It is a broad vision in that it focuses simultaneously on*

2
3 South African Bureau of Standards, http://www.sabs.co.za
    Agrément South Africa, http://buildnet.csir.co.za/agrement

*maintaining cutting edge global competitiveness and on addressing the urgent
needs of those of our citizens who are less able to assert themselves in the market."*
    The White Paper identifies the following five broad interrelated themes
as the focus for the government's S&T policy (which are consistent with other
government priorities and objectives – see Section 16.1.1):

- promoting competitiveness and employment creation
- enhancing quality of life
- developing human resources
- working towards environmental sustainability, and
- promoting an information society.

    A key focus within the White Paper is that public investment in R&D
(which is currently primarily through the Parliamentary Grant through the Science
Council Budget - see Section 16.2.3) is overtime to be redistributed away from the
support of activities within the government's own facilities (i.e. the Science
Councils) and towards *"more comprehensive support of R&D executed in the
private sector"*. Nevertheless, the White Paper acknowledges that this long-term
need must be seen in the light of the government's current responsibilities, namely
to take a lead:

- in pre-competitive research, until a culture develops in the private sector
  where such research is seen as a business imperative
- where entry barriers related to equipment and human resources are high
- in areas where the activity is considered to be a service which the
  government has a duty to provide, and
- in areas of public good in which, to achieve the greatest benefit, the
  research results and technology transfer need to be placed in the public
  domain.

    A prime objective of the NSI is to enhance the rate and quality of
technology transfer and diffusion from the science, engineering and technology
(SET) sector by the provision of quality human resources, effective hard
technology transfer mechanisms and the creation of more effective and efficient
users of technology in the business and governmental sectors.
    The impact of the White Paper on policy affecting the construction
industry is discussed in the following sections.

16.2.2   National foresight studies

The National Research and Technology Foresight Project: The National Research
and Technology Foresight project (DACST 1998) is one of a number of initiatives
being undertaken by the Department of Arts, Culture, Science and Technology as
part of its mission to review and reform South Africa's science and technology
system. The outcomes of the Foresight project, along with other policy initiatives,

is expected to contribute to new directions for science and technology in South Africa.

The aim of the Foresight project is to *"help identify those sector specific technologies and technology trends that will best improve the quality of life of all South Africans over the next 10 to 20 years"*.   The project encompasses technologies that impact on social issues and wealth creation through product or process development. In particular, the project seeks to:

- identify those technologies and latent market opportunities that are most likely to generate benefits for South Africa
- develop consensus on future priorities amongst the different stakeholders in selected sectors (industrial, socio-economic or service)
- co-ordinate the research effort between different players within selected sectors
- reach agreement on those actions that are needed in different sectors to take full advantage existing and future technologies.

The following sectors were selected for foresight studies:

- agriculture and agroprocessing
- biodiversity
- business and financial services
- information and communication technology
- health
- environment
- manufacturing and materials
- mining and metallurgy
- safety of citizen and society, and
- youth.

As can be seen from the above, notwithstanding the importance of the construction industry to socio-economic development in South Africa, the construction sector was not identified as a key sector for a national Foresight study. A contributing factor to the decision not to include the construction sector in the Foresight project was that, at the time of identifying the various sectors, a foresight study was in effect being undertaken by the Department of Transport (see Section 16.2.2.ii to follow). The transport study, however, is very specific to the transport sector, and does not include the broader construction sector.

Moving South Africa: The national Department of Transport began the Moving South Africa project (DOT, 1999) in June 1997. The project encompassed a 14 month process to take the goals of the 1996 White Paper on National Transport Policy and develop a twenty-year strategy to achieve them.   The project's mandate was "to develop a strategy to ensure that the transportation system of South Africa meets the needs of South Africa in the 21st Century and therefore contributes to the country's growth and economic development."

A key focus of the report was on passenger and freight transport, and the role and activities of government as well as the financial sustainability of the transport system. The Moving South Africa report does not however deal with issues of construction of roads (other than policy and planning issues).

16.2.3   The science councils and the parliamentary grant

Several statutory science councils exist in South Africa, of which the Council for Scientific and Industrial Research (CSIR) is the largest and which is of direct relevance to the manufacturing sector and the construction industry. The CSIR was restructured in 1987 to be more market orientated and to be more responsive to the needs of industry, and in which a greater emphasis has been placed on contract income as a measure of technology transfer. In line with international trends, the CSIR has also experienced increasing pressure to reduce its dependency on direct grant funding from government (see Section 16.2.1).

The CSIR is Africa's largest scientific and technological research, development and implementation organisation, and is currently comprised of ten Business Units, namely:

- bio/chemical technologies
- building and construction technology
- defence technology
- food technology
- manufacturing and materials technology
- microelectronics and communications technology
- mining technology
- roads and transportation technology
- textile technology, and
- water, environment and forest technology.

In addition to the contract income generated by the CSIR, government funding in the form of a Parliamentary Grant is made available to the CSIR as an investment into R&D for innovation and for technology support programmes in line with national socio-economic objectives. The Parliamentary Grant is allocated to the CSIR in terms of the business plan put forward by the CSIR. In a separate process, the CSIR allocates the Parliamentary Grant to its business units (and to other corporate programmes and activities), in line with the business units business plans (which themselves are in line with national and sectoral socio-economic objectives).

The total Parliamentary Grant to the CSIR and the CSIR's contract income for the period 1991 to 1999 is shown in Figure 16.4 (in constant 1999 Rands), in which the decrease in the Parliamentary Grant over time is clearly evident. It must, however, be noted that with effect from 1998/99 financial year, the CSR had access to the Government Innovation Fund (see Section 16.2.5.i), which was effectively top-sliced from the Parliamentary Grant. The CSIR's successful bids into the Innovation Fund largely compensated for the decrease in the Parliamentary Grant in 1998/99 and 1999/00 – although the strategic direction of the CSIR's R&D was therefore determined by the priorities of the Innovation Fund and not necessarily by the CSIR's own strategic priorities.

The allocation of the Parliamentary Grant to the CSIR's business units (which broadly align with market or industry sectors) is shown in Figure 16.5, in terms of the total allocation of the Parliamentary Grant to the CSIR line Divisions. Of particular relevance to innovation in the construction industry is the allocation to CSIR Building and Construction Technology (Boutek) and to CSIR Roads and Transportation Technology (Transportek).

Although the full allocation of the Parliamentary Grant to the CSIR's Boutek and Transportek has been given in Figure 16.5, it should be noted that the scope of these Divisions is far broader that the construction industry as defined in Section 1. In particular, Boutek focuses on urban and rural development and facilities planning and management, as well as construction. Transportek focuses on transportation planning, road engineering and traffic safety. The percentage allocation of the CSIR's Parliamentary Grant to construction activities is included in Figure 16.5, for which the total amounts to about 7% of the total Parliamentary Grant allocated to the line Divisions (and which compares favourably with the ratio of construction related GDFI to GDP – see Figures 16.1 and 16.3).

Figure 16.4 CSIR Parliamentary grant and contract income

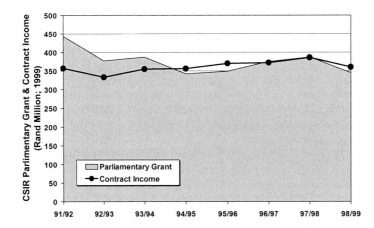

Figure 16.5 Allocation of parliamentary grant to CSIR business units

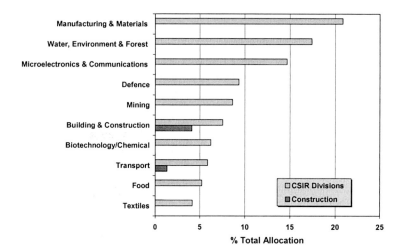

The restructuring of the CSIR in 1987, the increasing dependence on contract income, and the government's strategy of shifting the Parliamentary Grant away from the CSIR to competitive funding has however seen a shift in the use of the CSIR's Parliamentary Grant in Boutek and Transportek away from investments in public and private good, and towards investments that support the CSIR's own competitive advantage through specialised services and contract research. The CSIR's ability to do work in the public good and to influence innovation in the construction industry has therefore increasingly been influenced by client's needs and objectives, which for the most part have been dominated by short-term objectives. In fact, while Boutek and Transportek have achieved strong growth in contract income since 1987, very little of this contract income has in fact been of a research and development and nature, but rather from specialised services for the public and private sector.

The nature of Boutek's and Transportek's contract income is shown in Figure 16.6 in which the external contract income for 1998/99 has been subdivided into specialised services for the public and private sectors, public sector contract R&D, private sector R&D and international R&D. From Figure 6 it can be seen that private sector funding of R&D in the construction industry within the CSIR is less that 1% of the construction industry related contract income. In fact, the only significant and sustainable construction related contract R&D being funded by the private sector within the CSIR at present is that which is being funded by an industry association - the Southern Africa Bitumen Association (SABITA). Public sector funding of R&D however significantly exceeds the private sector funding. (The international R&D shown in Figure 6 arises from contracts associated with Heavy Vehicle Simulator (HVS) that is marketed internationally by the CSIR's Division of Roads and Transport Technology's.)

Figure 16.6 Construction related contract investment in R&D at the CSIR

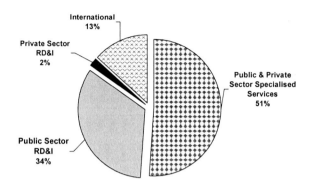

16.2.4   R&D at tertiary education institutions and the National Research Foundation

The major income of tertiary education institutions is that of government subsidies for students (which amounted to about.6.3 billion in 1998) as well as student registration fees. In addition to the subsidy from the national Department of Education, the National Research Foundation[4] (NRF) administers a co-ordinated system of grant financing of higher education institutions (which currently amounts to about R115 million per year), as well as the Technology and Human Resources for Industry Programme (THRIP) on behalf of the Department of Trade and Industry (which amounts to about R72 million per year).

   National Research Foundation Grant Funding: Over the years the Universities and Technikons have made substantive contributions to R&D within the construction industry. Much of this R&D has been undertaken by means of post-graduate research, in which the NRF grant funding is an important component. The NRF grant funding is based on the following principles:

*   all areas of research are eligible for support, as will basic and applied research and activities of technological development
*   a primary objective of the grant system is to promote individual and institutional capacity for research within tertiary education
*   particular attention is given to the introduction of processes to facilitate the financing of problem-oriented research involving participants from many disciplines, and

---

[4]   National Research Foundation, Pretoria. http://www.nrf.ac.za/

- the primary criteria for support is the quality of the research proposed, the relevance of that research to the goals and objectives of South Africa's vision for the future, and the contribution the activity will make to redressing the human and institutional imbalances of the past.

Grant financing by the NRF to higher education intuitions is made within a number of open and directed themes, of which the themes supporting R&D and innovation are highlighted below:

a)   *Open Research Programme:* The Open Research Programme offers its grant-holders complete freedom to undertake research that the grant-holders themselves believe to be important in the context of their chosen areas of endeavour - i.e. the research supported via the Programme is largely curiosity driven, rather than needs driven, and thereby complementary to the more focused programmes in the directed themes of the NRF.

b)   *Competitive Industry Programme:*   The Competitive Industry Programme aims explicitly at developing expertise required for wealth creation and sound industrial entrepreneurship.

c)   *Sustainable Environment Programme:* The Sustainable Environment Theme is aimed at processes, principles, policies, environmental quality indicators and standards, harvesting/production systems, chemical analysis of plants, biodiversity inventories/ classifications, change monitoring, and decision support for integrated environmental management.

d)   *Improved Quality of Life Programme:* The Improved Quality of Life Programme focuses on a wide range of topics, including:

  - the more effective use of indigenous plants, animals and foods; innovative new uses for established crops, animal products and agricultural waste; and the development of new technologies as well as new agricultural and food products, and

  - the provision of energy (with a focus on renewable energy), housing, water and sanitation, as well as effective transport systems, optimal land-use and rehabilitation methods, and waste and pollution management.

The total grant administer by the NRF in support of the above Programmes in 1999 was about R62 million, of which the allocation to construction related activities was less than about 1.5% of the total allocation! While support for R&D through human research development in the construction industry is possible through all these themes, in practice very little support is being given to construction related activities through the NRF.

The Open Research Programme is the largest of the NRF themes, accounting for about 35% of the total NRF grant. A more detailed breakdown of the distribution of grants within the Open Research Programme is given in

Figure 16.7, in terms of the percentage of total grants (i.e. not monetary value) and using 1998 data. Included in Figure 16.7 is the percentage of total grants which can be interpreted as being aligned with the construction industry. While it is noted that the objective of the Programme is largely "curiosity driven" rather than needs driven, of interest to note is that the engineering theme (of which civil engineering is a sub-set) is the least supported – taking a lowly place after themes such as mathematical sciences, biochemistry and microbiology, chemistry, the earth sciences, physics, biological sciences and statistics.

Figure 16.7 Allocation of NRF grant within the open theme (1998)

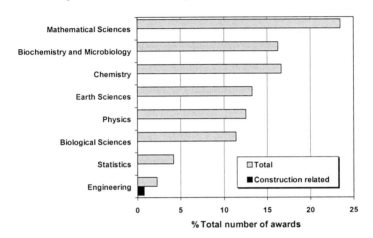

The reasons for the lack of support for construction industry related activities (and in particular within the NRF Competitiveness Scheme) are complex, and relate to:

- a lack of a strategic approach within the NRF in terms of allocating funds with regard to industry sectors and national priorities, as well as
- a lack of quality construction industry related research proposals from tertiary education institutions.

Technology and Human Resources for Industry Programme: The NRF also administers the Technology and Human Resources for Industry Programme[5] (THRIP) on behalf of the Department of Trade and Industry. THRIP is aimed at enhancing the competitiveness of South African industries through the development of appropriate skilled people and the development of technology, and encouraging long-term strategic partnerships between industry, research and

---

[5] *Technology and Human Resources for Industry Programme.* National Research Foundation, Pretoria. http://www.frd.ac.za/thrip/

education institutions and Government. In terms of this Programme, contributions
are provided by industry and government to fund the research efforts of the
academic partners, provided that these projects involve the training of students.
The Government will contribute R1 for every R2 from industry, and if certain
criteria are met, R1 for every R1 from industry could be granted.

    The total funds disbursed through the THRIP Programme in 1998 was
R71.7 million, which was matched by industry contributions of R46.7 million.
Detailed information is only available for 1997, for which a broad breakdown of
the total funds disbursed through THRIP per market sector in Figure 16.8.

Figure 16.8 THRIP disbursements per market sector (1997)

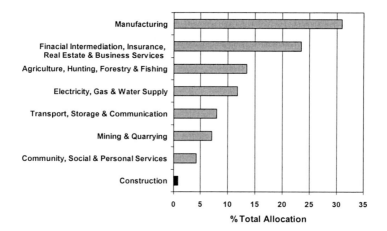

    From Figure 16.8 it can be seen that about 30% of the total allocation
has been in support of the manufacturing sector and about 23% in support of
financial services (which includes computers and software). However, THRIP
disbursements to construction related activities amounted to less than 1% of the
total allocated, and industry contributions to THRIP in support of construction
related activities also amounted to less than 1% of total industry contributions.

    As THRIP is based on a matching principle between government and
the private sector, the low investment in construction related R&D activities is a
reflection of the low investment in R&D by the private sector.

    Further significant sources of funding of post-graduate research is
through the Water Research Commission contracts and the Department of
Transport's research and human resource development schemes, which are
discussed in Sections 16.2.5.ii and 16.2.5.iii, respectively.

16.2.5   Public sector competitive R&D funding

Various government Departments and statutory organisations support industry development, competitiveness improvement, performance improvement through competitive R&D funding within their largely sectoral mandates. Some of the more initiatives are discussed below.

The Innovation Fund: Arising from the White Paper on Science and Technology, an Innovation Fund[6] was established and administered by the Department of Arts, Culture, Science and Technology in 1997. It is a program of support that addresses problems *"serious enough to impede socio-economic development or affect our ability to compete in products and services"*. The aim of the Innovation Fund is to encourage and enable long-term innovation projects with the following objectives:

- collaborative research and technology development programmes (i.e. academia, business, and R&D and technology development institutions)
- a multi-disciplinary approach to problem solving, and
- application based research programmes.

Awards are made within the following core themes of the Innovation Fund:

- National Crime Prevention Strategy (1997 to 1999)
- value-added, focusing on development of new materials, process technology and environmental management (1998 to 2002)
- Information Society including advanced software development and decision support for government (1999 to 2002), and
- biotechnology (1999 to 2002).

A breakdown of the allocation of Innovation Funds per theme for 1999 is given in Figure 16.9, in which a total of R75 million was disbursed. From a construction industry perspective, while the Innovation Fund does not have a theme which directly supports innovation in the construction industry, the "value-added" theme clearly could support applications supporting the construction industry. However, to date, not a single project has been approved supporting R&D in the construction industry.

Water Research Commission: The Water Research Commission[7] (WRC) was established in terms of the Water Research Act in 1971 (Act 34 of 1971). The terms of reference of the WRC are basically to promote coordination, communication and cooperation in the field of water research, to establish water

---

6   *Innovation Fund.* Department of Arts, Culture, Science and Technology, Pretoria, South Africa. http://www.dacst.gov.za/default _science_technology.htm

7   Water Research Commission, Pretoria. http://www.wrc.org.za/

research needs and priorities, to fund research on a priority basis, and to promote the effective transfer of information and technology.

The Water Research Fund derives income from levies on water consumption (which are collected for the WRC by the Department of Water Affairs and Forestry). In 1998 the WRC financially supported to 275 projects to a total value of R45 million in a wide range of water related fields.

The total value of research projects that can be considered to be in support of R&D in the construction industry is difficult quantify, because of the subject nature of the definition the "construction industry". However, a conservative approach is that less than about 3% of the total WRC funds are allocated to construction related projects. (A more rigorous approach suggests less that 1% of the funds are allocated to construction related activities.)

While the WRC can conduct its own research, in practice the WRC funds research under contract with other agencies, including universities, technikons, statutory research agencies (including the CSIR), government departments, local authorities, NGO's, water boards, consultants, and industry. In 1998, the universities were involved in 52% of the total number of contracts, private sector consultants in 23% and the CSIR in 13%.

Figure 16.9 Allocation of Innovation Fund to key themes

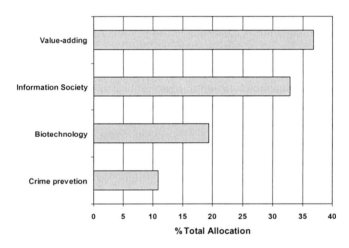

Department of Transport: Research in support of the needs of the transport industry are funded by the Department of Transport's Research and Development Directorate[8] The main functions of the Directorate are to:

• manage the Department's research programmes so as to extend the available knowledge on transport

---

[8]   Department of Transport, Pretoria. http://www.transport.gov.za/docs/transpt7.html

- encourage human development in transport, in the form of post graduate studies, to ensure that appropriate skills are available to government and industry
- provide an effective information service both to the Department of Transport itself, to other levels of government and to the transport industry, and
- contribute towards the development of longer term strategies for achieving the transport vision outlined in the Department's White Paper.

In support of its R&D strategy, the Department has established five main Centres of Development, one at the CSIR Division of Roads and Transportation Technology and four at leading transportation departments at South African Universities. A total of R3 million was allocated to research projects and human resource development in 1998 - which was significantly less than the budgeted allocation. Of the R3 million allocated from the Department of Transport's Research and Development Directorate, less than R1 million was allocated to construction related activities – with the remainder being allocated to traffic safety, transportation planning, etc.

An analysis of the Department's spending on construction related R&D (which follows similar trends to the Department's total R&D spending) is shown in Figure 16.10, from which significant learning can be obtained in terms of the impact of public policy instruments on R&D.

Figure 16.10 Department of Transport investment in construction related R&D

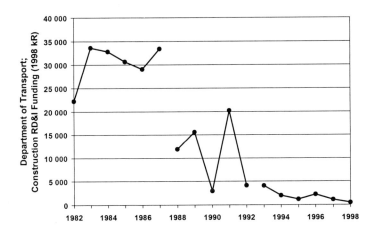

Three distinct phases are identified in Figure 16.10, namely:

i) *1982 to 1987*, in which the construction related R&D funding was largely held constant at about R30 million per year, and in which the

	sole contractor was the CSIR's Division of Roads and Transport Technology.
ii)	*1988 to 1992*, in which the Department reduced its level of funding, and more importantly opened up its funding to the public and private sector on a tender basis. The number of suppliers increased from 1 in 1987 to 29 in 1992, and the average size of the research contracts awarded decreased substantially. These changes in the funding model had several implications:

- the change in the leel of funding for the CSIR from the period 1982 to 1987 to the period 1988 to 1992 had disastrous consequences for the research capacity that the CSIR could retain, and a significant R&D capacity was lost to the industry
- the increasing number of suppliers (and in particular the increasing number of private sector consultants) resulted in a dispersion in the development of research competence
- the decreasing contract size resulted in a loss of focus and a reduction in the value obtained
- by 1992 it was recognised by the Department that the system had in effect collapsed, and a new funding model was introduced.

iii)	*1993 to date*, in which the Department introduced the Centres of Development approach discussed above – in which the funding was again concentrated in selected areas. The level of funding, however, is significantly less than during earlier periods.

16.2.6   Public sector support incentive schemes

The Department of Trade and Industry (DTI) has initiated a number of schemes to provide incentives for economic growth, performance improvement and enhanced competitiveness of South African industry. Some of the more important schemes, or schemes which are relevant to innovation in the manufacturing and construction industry include (DTI, 1998):

- *The Support Programme for Industrial Innovation (SPII)*, which is designed to support technology development of South Africa's manufacturing industry through support for innovation of competitive products and/or processes. Assistance takes the form of a grant of 50% of the actual direct cost incurred in the pre-competitive development activity, reaching a maximum grant of R1 million per project.   The support programme is accessible to all private sector enterprises in the manufacturing sector with the ability to develop and commercialise a product.
- *The Partnerships in Industrial Innovation scheme (PII)*, which is designed to promote large scale technology development in

manufacturing industries through support for innovation of competitive products and/or processes. Assistance takes the form of a grant of 50% of the actual direct cost incurred in pre-competitive development activity, reaching a maximum grant of R1 million per year.

• *The Competitiveness Fund*, which is designed to encourage South African firms to be competitive, both as exporters and in defence of their local marketplace. The fund will support the introduction of technical and marketing know-how and expertise to firms. The scheme will insist on a 50% contribution by the firm itself.

• *The Sector Partnership Fund*, which is designed to support sub-sector partnerships in preparation of technical and marketing programmes with the aim of improving competitiveness and productivity of firms. The scheme is available to any partnerships of five or more organisations within South African manufacturing. Up to 65% of the costs of projects up to R1.5 million are covered.

Only the Support Programme for Industrial Innovation[9] (SPII) is currently fully operational, and is being administered by the Industrial Development Corporation (IDC) on behalf of the Department of Trade and Industry.

A breakdown of the awards made from April 1993 to March 1998, amounting to a total value of R98 million per sector is given in Figure16.11, from which it can be see that the scheme is dominated by the electronics, software, chemical/ pharmaceutical, mechanical machinery, plastics and metals sectors. To date, there has been no support for innovation in construction through SPII.

---

[9]   Support Programme for Industrial Innovation; http://www.spii.co.za

Figure 16.11 Allocation of SPII funds per sector

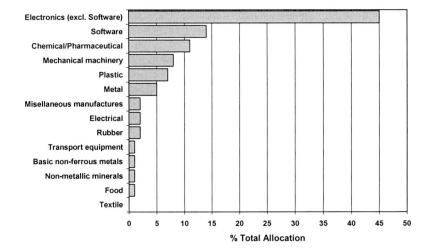

16.2.7   Private Sector participation in public sector policy instruments

A range of public sector policy instruments affecting innovation in the construction industry has been discussed in the previous sections, and an attempt to quantify the private sector investment in the formal public sector programmes and at the CSIR is given in Table 16.1. Note that the private sector investment in R&D at the CSIR excludes problem solving services and product development, which would be regarded as specialised consulting.

As already noted (Section 16.1.3), it is seen from Table 16.1 that private sector investment in R&D at the CSIR, universities and the DTI SPII incentive schemes is very small - amounting to less than R1 million per annum! (Note that this should be compared to the total turnover of the construction industry as measured by construction related GDFI of about R40 billion per year.)

Table 16.1 Private sector investment in R&D

Programme	Private Sector	R&D Amount (Rands)
Thrip (Section 2.4.ii)	Industry Associations	R300k
	Individual organisations	R100k
CSIR (Section 2.3)	Industry Associations	R300k
	Individual organisations	R50k
SPII (Section 2.6)	Industry Associations	R0k
	Individual organisations	R0k
Total	Industry Associations	R600k
	Individual organisations	R150k

The low investment by the private sector in the public sector competitive funding initiatives which are in fact specifically designed to encourage collaboration between the private sector and the public sector R&D institutions points to either:

- a failure of these public instruments in the construction industry (either because such an approach is not appropriate to the construction industry or because the schemes have not been designed to be attractive to the construction industry; or
- a failure of industry to take advantage of such schemes.

16.3 THE CONSTRUCTION INDUSTRY DEVELOPMENT BOARD (CIDB)

The public sector accounts for about 50% of all construction expenditure in South Africa. and is therefore a significant role-player in the construction industry. Furthermore, public sector procurement is therefore a significant potential instrument in achieving the aims and objectives of the government. However, it is immediately apparent from the previous sections of this paper that the government policy instruments outlined here have to date largely not included a specific focus on the broad construction industry – other than standards and certification.

However, the lack of a coherent policy focus on the construction industry in South Africa is changing, and the Department of Public Works in 1999 released the White Paper "*Creating and Enabling Environment for* Reconstruction, Growth and Development in the Construction Industry" (DPW, 1999). In terms of this White Paper, a statutory Construction Industry Development Board (CIDB) will be established as a permanent vehicle to drive industry development and transformation in a spirit of partnership between all stakeholders.

To give effect to this process, a Construction Industry Development Task Team was established in 1998, and it is envisaged that the CIDB will be

established in 2000.   In support of the objectives of the CIDB, it is envisaged that
the CIDB will consist of the following functional Programmes (see Figure 16.12):

Figure 16.12 Proposed functional structure of CIDB

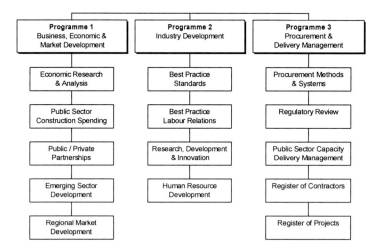

a)   *Business, Economic and Market Development:* The objective of the
          Business, Economic and Market Development Programme is to promote
          a stable delivery environment, increasing expenditure in fixed
          investment and sustained targeted real growth for construction products
          and businesses.   Furthermore, the Business, Economic and Market
          Development Programme will promote the targeting of public sector
          delivery to maximise employment opportunities.

b)   *Industry Development:* The objective of the Industry Development
          Programme is to promote programmes aimed at the enhancement of
          industry performance, yielding value for money to clients and protection
          to the public.   To achieve these objectives, the Industry Development
          Programme will:

          • develop and promote the implementation of best practice standards
            aimed at quality, productivity, health and safety and environmental
            management, as well as design best practice and buildability
          • promote appropriate best practice in labour relations
          • promote research, development and innovation; and
          • establish close collaboration with the skills training organisations in
            order to ensure that human resource development and skills training
            strategies are aligned with the overall industry development
            strategy.

c)   *Procurement and Delivery Management:* The objective of the Procurement and Delivery Management Programme of the CIDB is to improve the construction procurement and delivery environment and the capacity of public sector delivery agencies. In support of this programme, the CIDB will introduce:

- a Contractor's Register and a Best Practice Contractor Recognition Scheme, and
- a Project's Register and a Best Practice Project Assessment Scheme.

It is the writers' views that the CIDB has significant potential to impact on innovation in construction. In particular, to date, the construction industry does not have a single identifiable platform from which it can engage government on appropriate funding mechanisms to promote innovation within the industry, and the CIDB will afford the industry with this opportunity.

16.4   DISCUSSION AND CONCLUSIONS

The objective of this paper has been to investigate the impact of public policy instruments on innovation in construction in South Africa. In doing so, it is useful to first revisit the South African government's strategy with regard to fostering innovation as outlined in the White Paper on Science and Technology, namely:

> *The White Paper is based on a view of the future where all South Africans will:*
>
> - *enjoy an improved and sustainable quality of life*
> - *participate in a competitive economy by means of satisfying employment, and*
> - *share in a democratic culture.*

In order to attain this vision, the following three goals pertinent to the creative use and efficient management of innovation will have to be achieved:

i)   the establishment of an efficient, well co-ordinated and integrated system of technological and social innovation within which:

- Stakeholders can forge collaborative partnerships and interact creatively in order to benefit themselves and the nation at large
- resources from engineering, the natural sciences, the health sciences, the environmental sciences and the human and social

> sciences are utilised for problem-solving in a multidisciplinary manner
>
> - stakeholders, especially those who were formerly marginalised, are part of a more inclusive and consultative approach to policy decision-making and resource allocation for science and technology (S&T) activities;

ii)   the development of a culture within which the advancement of knowledge is valued as an important component of national development;

iii)   improved support for all kinds of innovation which is fundamental to sustainable economic growth, employment creation, equity through redress and social development.

The strategy outlined in the White Paper is, of necessity, not sector specific, and it is left to other public policy instruments, government line Departments, and industry itself to give effect to sector specific interests.

From an analysis of current public policy instruments impacting on innovation in the construction industry, this paper leads to the following conclusions.

- While the South African government recognises that innovation is fundamental to sustainable economic growth, employment creation, equity through redress and social development, this paper has shown that in comparison with other sectors, the focus of public policy instruments on the construction industry has to date been very limited. (In this regard, the pending establishment of the CIDB, with its specific focus on construction industry development, is seen as a very positive development.)

- However, while there may not be a specific focus on the construction industry in public policy instruments on the construction industry to date, opportunities do in fact exist for the private sector and tertiary education institutions to respond to broad public policy instruments aimed at enhancing innovation. However, the response of industry to the available instruments, which generally require consortia-based approaches and industry co-funding, has been very limited.

- Following on from the above, and as is well know in most countries, private sector investment in the construction industry in formal R&D is very small. The present strategy of encouraging collaborative partnerships between the private sector and tertiary education institutions and/or research organisations is therefore unlikely to be particularly successful in the construction industry.

- The government's strategy of shifting the Parliamentary Grant away from the CSIR to competitive funding and the increasing dependence of the CSIR on contract income has seen a shift in the use of the CSIR's Parliamentary Grant away from investments in public and private good, and towards investments that support the CSIR's own competitive advantage through specialised services and contract research – resulting a lessening of the CSIR's impact on innovation in construction.

- Public sector R&D funding policies, and rapid changes in these policies, can have a significant impact on developing and maintaining the national R&D capacity, or destroying this capacity. Rapid changes in funding policies and a lack of concentration of this capacity has a negatively affect on innovation.

- The public sector plays an important role in influencing innovation in construction through the development of standards and certification, through the statutory organisations of the South African Bureau of Standards and Agrément South Africa. Furthermore, several government departments have played an important role in the development of sector specific guidelines and standards, including the Departments of Housing, Transport, and Water Affairs.

- Notwithstanding that private sector investment in formal R&D is small, the role of industry associations needs to singled out – from the perspective of the investment that these associations make in formal R&D (albeit that it is generally very small), and their role in the development of best practice, standards and specifications which impact on innovation in construction.

- Although still under development, the Construction Industry Development Board (CIDB) has significant potential to impact on innovation in construction.

In conclusion, within the context of the above trends, it is essential that public policy instruments:

- adopt an holistic approach and understanding of R&D, recognising that short-term expedient decisions can destroy an R&D capacity that can take up to a decade or more to re-develop, and
- recognise that the characteristic low investment of the private sector in formal R&D, including collaborative partnerships with public sector organisations, can limit the effectiveness of public sector interventions in the construction industry.

Finally, against the background of a changing policy environment, it is essential that improved monitoring and evaluation instruments be developed and implemented to systematically evaluate the range of policy instruments being implemented, based on socio-economic criteria and with resource allocation and priority-setting as goals.

CHAPTER 17

INNOVATION IN THE BRITISH CONSTRUCTION INDUSTRY: THE ROLE OF PUBLIC POLICY INSTRUMENTS

UK TG 35 Team, c/o Graham M. Winch[1]

EXECUTIVE SUMMARY

This paper reviews the public policy instruments deployed by the UK government to stimulate the rate and direction of innovation in the construction industry. The instruments are analysed in terms of a model of the innovation and diffusion process in construction as a complex systems industry. The model allows the clearer identification of the varying aims and objectives of the different public policy instruments. In line with the model, a tabular presentation of the objectives and means of each of the instruments is also presented.

	Name	Resources in Construction 1999-2000	Objectives	Means	Contribution to Innovation
Programmes to support R&D	Foresight	Not relevant	To determine future UK R&D strategy	Expert reviews and Delphi studies	Providing the long vision
	EPSRC Response Mode Research	£11.5 million	To fund basic scientific research	100% funding of university-based research teams	Generating new ideas, Technology orientated
	EPSRC managed programmes (inc. Foresight Challenge)	£3.6 million	To develop applied research matched to industry needs	Challenged – based collaborative research programmes between universities and industry	Generation of new ideas and diffusion to advanced practice. Process Orientated

1   UK TG 35 Team is a collaborative writing effort. It is coordinated by Graham M. Winch. Roger Courtney, David Gann, Graham Ive, John Mead, and John Stambollouian have all contributed material, while Anne Beesley, John Connaughton, and Charles Egbu have also participated in the discussion of this text.

	Name	Resources in Construc- tion 1999-2000	Objectives	Means	Contribu- tion to Innovation
Programmes to support advanced practices and experimen- tation	Partners in Innovation	£7.5 million	To develop applied research matched to industry needs	Challenge- based collaborative research programmes between research institutes and industry	Developing advanced practice. Process orientated
	EPSRC IGDS courses		To train in advanced practices	Part-time masters level education	Diffusion of new ideas into advanced practice
	Movement for Innovation	£0.6 million	To identify and diffuse advanced practice	Demonstrati on projects proposed and reviewed fand innovations identified. Campaign style.	Diffusion of advanced practice. Process orientated.
Programmes to support performance and quality improve- ment	BRE Framework Agreement	£10.3 million	To support the BRE's research in the immediate post-privat- isation period	Declining funding with negotiated research objectives. Contracts offered on an exclusive basis.	Develop- ment and validation of best practice, and the develop- ment of standards. Technology orientated
Programmes to support taking up of systems and procedures	Construction Best Practice Programme	£2.3 million	To identify and diffuse best practice	Authoritative source of best practice methods and cases	Collation and diffusion of best practice. Process orientated
	Government Procurement Initiatives	Not relevant	To ensure that the state and its agencies are best practice clients	Reforming procurement procedures	Providing the context, which facilitates innovation. Process orientated.
	Regulatory system	Not relevant	To ensure the implement- ation of standard practice	Inspection and education	Diffusion of standard practice. Technology orientated

17.1   INTRODUCTION

The aim of this paper is to identify the public policy instruments, which play a role in innovation in the British construction industry as a contribution to the work of TG35 on systems of innovation in construction. Public Policy Instrument is here defined as *a government initiated measure that influences the rate and direction of innovation by construction firms*. The paper will first present a conceptual framework for exploring the range and types of public policy instrument, before identifying some of the most important ones currently operating in the UK. It focuses on the period since 1994 as the innovation system in British construction has adapted to new approaches to funding and changing strategic objectives. Aspects of the system in the period prior to 1994 are discussed in reports such as Construction Forecasting and Research (1996), and Innovation Policy Research Associates (1992). An overview of recent changes is provided in Cooper (1997).

The institutional structure of the British construction industry generally – what Bowley (1996) called the *contracting system* - is presently undergoing considerable change. The industry is castigated as being poor in terms of levels and predictability of performance, leaving clients deeply dissatisfied with their experience of commissioning built products. There is now a strong appetite for change in the British construction industry, which is prompting a re-examination of traditional ways of doing things, leading to considerable experimentation and innovation. The agenda for change set by the Latham Report of 1994 and reinforced by the report of the Construction Taskforce of 1998 – the Egan Report - which emphasised a stronger sense of industry ownership of the change process, is reviewed in Winch (2000). This paper focuses upon those instruments specifically designed to promote and diffuse innovations in the context of these broader changes in the institutional structure of the industry. (The information in this chapter is accurate up until June 2000.)

17.2   UNDERSTANDING PUBLIC POLICY INSTRUMENTS

We suggest that public policy instruments can be most usefully classified in two dimensions, as proposed in Figure 17.1. The first dimension is whether the instrument is a general part of public policy, but with a significant impact upon construction innovation, or whether its intentions are largely restricted to the construction industry. The second dimension is whether the instrument is explicitly aimed at stimulating innovation, or whether it has other policy objectives, but has the consequences of stimulating or stifling innovation.

The range of measures under the rubric of general indirect instruments is very broad, ranging from fiscal and monetary policy to attempts to meet the Kyoto targets for environmental sustainability. This range is too broad to be addressed here, and so this paper will focus upon the other three categories. The analysis also excludes any programmes funded by the European Union, which will be common to its member nations.

Figure 17.1 Public policy instruments for construction innovation

	direct	indirect
general	instruments developed for a number of sectors which are available to construction firms	Fiscal, monetary, industry and employment policy
construction specific	instruments explicitly aimed at construction firms	public policies which have an incentive/ disincentive affect on innovation

17.3 DIRECT GENERAL PUBLIC POLICY INSTRUMENTS

The current framework for science and technology policy in the UK is based on the 1993 White Paper – *Realising Our Potential* – which is widely seen as a new start in this area. The main general direct instruments operate within the scope of the policies inaugurated since then, and have a major impact upon the rate and direction of innovation in the construction industry. The most important of these are:

17.3.1 Foresight
(http://www.foresight.gov.uk/)

This is a major national undertaking, based principally on Delphi studies, which aims to identify the key technologies in the medium term and hence to guide national science and technology policy. The process split into a number of sector specific panels. The last round with 15 panels started in 1994, and was completed in 1995, and the Construction Panel report (Office of Science and Technology 1995) identified as key "opportunities":

- customised solutions from standardised components
- advanced business processes
- constructing for life
- benefiting society
- a competitive infrastructure.

And as "engines for change":

- learning networks
- the use of IT for innovation dissemination
- fiscal policy for the long term
- a culture of innovation.

As a result of the first round, a competition – Foresight Challenge – for the funding of research centres was launched. Through this mechanism, the Virtual Reality Centre for the Built Environment (http://www.vr.ucl.ac.uk/) was established at University College London with a budget of £1.2 million, which is matched by industry contributions. The second round was launched in April 1999, with a Built Environment and Transport panel to include the remit of the former Construction Panel. While the Construction Panel focused very much upon the process of delivery of the built product, the new Panel is focused much more upon the built product itself, particularly at the urban level. Its vision is "to provide clean, safe, efficient and fair access for people and goods to the front door of society's activities". Construction interests are represented in the Construction Associate Programme within the Panel's work.

17.3.2 University research and graduate programmes

The UK university funding system has a twin flow – the Higher Education Funding Councils fund staff to teach and engage in scholarship, while research is funded principally through the research councils. An overview of the contribution of the university sector to construction research is provided by Lansley and his colleagues (1994). The principal research funding council for construction is now the Engineering and Physical Sciences Research Council (EPSRC - http://www.epsrc.ac.uk/) which has a number of mechanisms of support. Academics in the construction management area may also bid for funds to the Economic and Social Research Council (ESRC).

The *responsive mode,* where academics bid for funding for the projects that they wish to pursue, is the principal method of research support, by the EPSRC – taking around three quarters of all funding. There are, broadly, three programmes under the responsive mode:

- general engineering
- science
- technology (materials/IT).

The contract value of general engineering research grants as of October 1999 was £23.1 million, with a strong bias towards the built product, particularly the engineering of structures, infrastructure, and water installations. As the typical research grant is awarded for a 2 year period, the annual spend can be put at roughly £11.5 million. The EPSRC also runs a number of collaborative *managed programmes* grouped as either Engineering for Manufacturing, or Engineering for Environment, Infrastructure and Health. These are challenge based, in that they must all match public funds with at least a 50% contribution in kind from industry. LINK programmes are part funded by the Department of the Environment, Transport and the Regions (DETR). The principal programmes are:

- Innovative manufacturing initiative (£4.1 million)
    - construction as a manufacturing process (CMP)
- LINK (£1.94 million)
    - construction maintenance and refurbishment
    - meeting client's needs through standardisation (MCNS)
    - integration of design and construction (IDAC)
- Extend quality life (EQUAL)
    - EQUAL in the built environment.

The total value of grants awarded as at October 1999 is given in brackets. These figures are complemented by a minimum of 50% in-kind industry contribution. With the exception of EQUAL in the Built Environment, which is focused upon design for the less-abled, the managed programmes are more construction process orientated. A conference was held in September 1999 with the objectives of reviewing CMP, MCNS, and IDAC to establish the progress made they had made, and identify the work remaining to be done. The report of this conference is currently under preparation.

The research councils also support a number of *doctoral programmes*, awarding studentships to departments. These are, again, largely responsive in that it is up to the student and his or her supervisor to define the topic of study. In addition, the large numbers of self-funded doctoral students – principally from overseas - also make an important contribution to UK construction research activity. More collaborative awards are also available under the Co-operative Awards in Science & Engineering scheme. In an important new development, a collaborative Engineering Doctorate programme has recently been established at Loughborough University's Centre for Innovative Construction Engineering which will provide 10 places per year for the next five years.

MSc programmes also form an important, if diminishing, part of the provision. The Integrated Graduate Development Scheme is aimed at pump-priming the development of modular programmes in specific fields which provide stand-alone modular courses which can build into an MSc qualification. There are currently a number of programmes focused on the construction industry:

University of Salford:	Construction IT
University of Reading:	Intelligent Buildings: Design, Construction and Operation Corporate Real Estate and Facilities Management Non-Handicapped Environments – Design and Management
Glasgow Caledonian and Strathclyde Universities:	Construction and Development Innovation by Collaborative Competition
Nottingham Trent University:	Construction Engineering Design and Management

These specifically funded programmes complement a much longer established body of research-led masters programmes, aimed at diffusing research into practice. These include internationally renowned programmes in the construction management, geotechnics, architectural design, and many more areas.

17.3.3   Sector challenge

There are various programmes run by the Department of Trade and Industry (DTI) – Sector Challenge is the most important of these for innovation purposes. Some 15 projects in the construction industry were awarded in the last (1997) round.

17.4   DIRECT SPECIFIC RESEARCH PROGRAMMES

The sponsoring ministry for the UK construction industry is the Department of the Environment, Transport and the Regions (DETR) - it is therefore principally responsible for the direct specific research programmes in construction. DETR's Construction Directorate is divided into the following divisions (http://www.construction.detr.gov.uk/):

- Construction industry sponsorship (CIS)
- Construction innovation and research management (CIRM)
- Export promotion and construction materials
- Construction market intelligence
- Building regulations.

Following a strategic review in 1997, the Construction Directorate's research programme has become increasingly strategy led, structured into five Business Plans. These are updated in the annual *Construction Research and Innovation Business Plan*. There are five Business Plans:

- Sustainable construction: "to help the industry meet the obligations of the sustainable construction strategy"
- Safety and health in buildings: "to enable policy with respect to safety and health in buildings to be based on a firm scientific footing"
- Technology and performance: "to improve the technological performance of UK construction"
- Construction process: "to develop new and improved approaches to construction processes"
- Best practice: "to promote beneficial changes in management practice and business process in the construction industry".

The Business Plans are based on wide consultation with the industry and reflect strategic advice from the Construction Research and Innovation Strategy Panel (CRISP). As a result of the strategic review, the emphasis in funding is planned to shift between 1996/7 and 2000/1 with a view to obtaining broad parity between each Business Plan, implying reductions in spending areas such as Technology and Performance and Safety and Health, and an increase for Construction Process and Best Practice.

Table 17.1 shows the overall expenditure of the Construction Directorate on public policy instruments to stimulate research and innovation in UK construction for 1998-9 and 1999-2000 year. The principal programmes supported by CIRM are :

- Partners in innovation
- Building Research Establishment Framework Agreement
- Construction Best Practice Programme
- LINK Programme

In addition, a relatively small amount of competitively tendered research is commissioned to meet specific requirements, including the funding of the CRISP secretariat. The Movement for Innovation is the responsibility of the CIS division.

Table 17.1 DETR Construction directorate research expenditure (source DETR)

PROGRAMME	1998-99	1999-00
PII	7.6	7.5
BRE Framework	13.5	11.3
of which Construction Best Practice Programme is	*0.9*	*1.0*
Construction Best Practice Programme: external sub-contracts	0.5	1.3
Movement for Innovation	0.0	0.6
Directly commissioned research (mainly research management)	1.5	1.8
LINK	0.3	0.3
Total	23.4	22.7
PII gearing is 45% DETR 55% industry		
total projects costs are	16.9	16.7
LINK gearing is 25% DETR, 25% EPSRC, 50% industry		
total project costs are	1.2	1.2

note 1: Movement for Innovation costs are outside the £22 million baseline

note 2: £22 million is a baseline, actual expenditure differs with End Year Flexibility

note 3: figures may not appear to add correctly due to rounding errors

note 4: costs exclude VAT

17.4.1   Construction Research and Innovation Strategy Panel

The Construction Research and Innovation Strategy Panel (CRISP), (www.ciboard.org.uk/crisp.htm) for which Davis Langdon Consultancy provides the secretariat, is concerned with industry research strategies and promoting research and innovation within the UK construction industry. It was formed in 1995, and its current strategy, launched in April 1999, identifies three responsibilities – industry improvement, construction futures, and the construction research base. For instance, it is currently reviewing the links between providers of research – principally the universities and the research establishments, and the users of such research in the industry. Within the area of industry improvement, the strategic review identified five broad areas of activity:

- the identification and communication of client needs
- design as the integrator between client needs and the constructed product
- technologies and components – the materials and systems that make up the constructed product
- management of the construction process
- the performance of the constructed product in use;

and four cross-cutting themes, each the responsibility of a theme group:

- the effect of the regulatory framework on innovation
- achieving sustainable construction
- motivation and communication within the industry
- the role of information technology in product and process.

17.4.2   Construction Best Practice Programme

The Construction Best Practice Programme (CBPP) (http//www.cbpp.org.uk) was established in 1998 with the mission to raise awareness of the benefits of best practice and provide guidance and advice to UK construction and client organisations so that they have the knowledge and skills required to implement change. Its budget rose from £1.4 million for 1998-9 to £2.3 million in 1999-2000. Its management team – mainly secondees from industry – is located at the Building Research Establishment, and it is partially funded under the BRE's Framework Agreement.

The main focus of the CBPP's work is the transformation of outmoded management practices and business cultures. The key objectives are to:

- create a desire for improvement by identifying, publicising and supporting the use and benefits of adopting improved business practices
- offer an initial point of contact for organisations wishing to improve
- facilitate links between such organisations and those with the knowledge of how to improve
- provide techniques, advice and knowledge about and tools for best practice.

It is also responsible for the dissemination of the Key Performance Indicators, which were specified in the Egan Report and are presented in Table 17.2.

Table 17.2 UK key performance indicators (percentage annual improvement)
(source: Construction Task Force, 1998)

Key Performance Indicator	Improvement Target
Capital cost of buildings	- 10
Delivery time	- 10
Predictability	+ 20
Defects	- 20
Accidents	- 20
Productivity	+ 10
Turnover and profits of construction firms	+ 10

Two of the CPBB's more specific services are the Construction Productivity Network (CPN), which is managed by the Construction Industry Research and Innovation Association (CIRIA) and the Inside UK Enterprise – Construction programme The CPN promotes the sharing of knowledge amongst all those involved in construction, including architects, designers, consultants, contractors, manufacturers, academics and clients.   CPN workshops are a vehicle for discussion between delegates and speakers, and among delegates themselves.   A report of every workshop is produced and distributed to attendees and network members so that a wider audience can benefit the issues raised.   CPN's programme of workshops covers a broad range of issues including:

- contractual / relationship issues: supply chain management, partnering forms of contract
- the Construction Process: briefing, value management, risk management
- business improvement techniques: benchmarking, TQM
- People and culture: education & training, change management, safety
- communication and IT.

Inside UK Enterprise – Construction offers visits to UK exemplars of best practice in the construction industry. The objective is to transfer best practice concepts and real life experiences from host companies that have implemented them to groups of visitors who wish to do so themselves.

17.4.3   Partners in Innovation (PII)

PII is an annual competition that co-funds collaborative research and innovation projects. It is one of the principal mechanisms for meeting the aims of DETR's five Business Plans for Construction Research & Innovation. PII also complements programmes sponsored by other Government Departments and Agencies.

PII is a *collaborative* scheme which provides up to half the costs of research and innovation projects of critical benefit to the construction sector and the built environment. This means that the DETR spend of £7.5 million in 1999/2000 is levered up to an effective spend of £16.7 million. PII facilitates the pursuit of common goals through active partnerships between government, industry and research organisations. PII is open to all UK construction businesses, industry bodies, institutions, research and technology organisations and universities; participation by SMEs is encouraged. Non-UK organisations can participate, apart from being Lead Partners, provided they add significant value to the project.

PII now has a two-stage application procedure. The first round is open to everyone with innovative ideas applicable to construction: from business enterprises (including Small & Medium-sized Enterprises) pursuing enhanced competitiveness for their sector to common interest groups seeking to improve processes and systems relevant to the built environment. The first stage invites innovative ideas, simply yet powerfully written, that will address the challenges facing construction in the future. Submissions must persuade DETR that a project is feasible and holds real promise of a worthwhile advance. The second stage is open only to those organisations that were successful at the first stage.

The scheme is based on a competitive bidding process within a budgetary ceiling. Proposals offering the best value in meeting the requirements set out in the Business Plans, after taking into account existing work, are recommended for funding. PII also supports proposals that promise breakthroughs in areas not anticipated by the Business Plans, but it is a strategic scheme first and foremost. The majority of awards go to work that addresses DETR priorities, demonstrates the potential for take-up by beneficiaries and offers significant impact in improving the performance of the industry.

17.4.4 The Movement for Innovation

In July 1998, the Construction Task Force published its report entitled Rethinking Construction – popularly known as the Egan Report. This client-led task-force recommended the establishment of a Movement for Innovation (M[4]I), (*http://www.m4i.org.uk/frames.htm*), which was launched in November 1998, and reported back to the industry at conferences held in July 1999 and May 2000. M[4]I aims to lead radical improvement in construction in value for money, profitability, reliability and respect for people, through demonstration and dissemination of best practice and innovation. It reports to a DETR appointed Steering Group, and is run by a Board, supported by a Team of active monitors - its structure is shown in Figure 17.2. Associated with M[4]I, and operating along similar principles is the separate Housing Forum, focusing solely upon the construction of housing, an area which received special attention in the Egan Report.

Figure 17.2 M⁴I movement structure (source: M⁴I team)

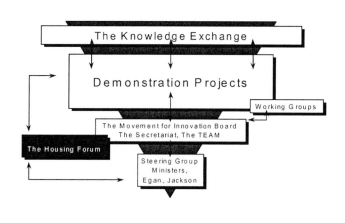

A notable feature of M⁴I is the high-level political support which it receives – including the patronage of the Deputy Prime Minister. Its Steering Group includes two ministers. This gives it a level of political legitimacy which has facilitated the mobilisation of a large number of client and industry representatives.

Figure 17.3 The 5-4-7 diagram (source: M⁴I team)

The Movement for Innovation is an open movement in which companies are invited to put forward current projects which they believe to be innovative forward as Demonstration Projects. These projects are grouped in regional clusters which are facilitated by the M⁴I Team – typically young managers seconded to M⁴I from leading construction companies for this role.

Mead (1999) provides an analysis of this process from a Team perspective. It has summarised the principles of the Egan report in the widely diffused 5-4-7 diagram presented in Figure 17.3. This identifies the five *drivers of change*, the "four P" *business processes* to enable the achievement of an integrated project process, and the 7 *key performance indicators* defined in Table 17.2, thereby providing and overall summary vision for the Movement. More specifically, the Movement aims to:

- bring together clients and all involved in the construction supply chain or in innovation, best practice, or research, who are committed to change and innovation in construction. Along with the Housing Forum it reports regularly to the M[4]I Steering Group
- provide leadership, share experience and work together to create an open, co-operative, no-blame, non-adversarial, team approach to innovation
- drive forward by example and persuasion the changes needed to create an industry in which the norm will be committed leadership, a focus on the customer, a process and team integrated around the product, a quality driven agenda, and a commitment to and respect for people
- achieve, through sustained improvements and innovation in product design and development, in project implementation, in partnering the supply chain and in production of components
- measure, quantify and disseminate, experience and achievements from demonstration projects through the CBPP and the Knowledge Exchange.

The M[4]I processes aims to *inform* by inviting firms to nominate Demonstration Projects which display potential innovations. These projects – none have been rejected to date – are then grouped into clusters where a process of peer review and debate *verifies* the innovations proposed by the Demonstration project nominators. Once verified, the innovation can then be written up and *disseminated* as a case study through the Knowledge Exchange, the CBBP, and other means. The Knowledge Exchange was launched in September 1999, initially as an intranet modelled on the successful RIBANet used by architects. It is intended to open access more widely during 2000.

The M[4]I Board has approved the first wave of the Demonstration Projects that were called for in the Egan Report. Over 80 projects with a total value of £1.7 billion were approved as being suitable to demonstrate the findings contained in *Rethinking Construction* in the first two rounds of nomination. In the third round, launched in October, 1999, Demonstration projects that particularly emphasised commitment to people were encouraged.

17.4.5   The Housing Forum

The Housing Forum is a company limited by guarantee. Its Board members are senior directors of private and public sector housing organisations, design and supply-chain companies. The Forum is funded through sponsorship from the DETR, The Housing Corporation (the central agency responsible for funding social housing) and membership fees. It has a three- year life and a current budget of around £1 million.

The Forum aims to lead innovation in housing by demonstrating how to improve delivery, focusing on providing better value to customers. The main activity is therefore support for demonstration projects via regional cluster groups, and KPIs, Partnering and Customer Satisfaction working groups. The Board is responsible to the Minister for delivering on overall targets as specified in the Egan Report.

Figure 17.4 The demonstration project process (source: M⁴I team)

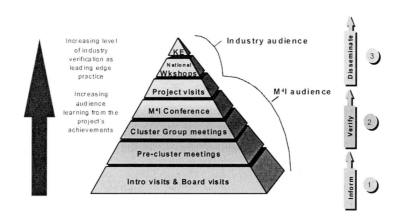

The selection process for forum projects is similar to that for other Demonstration Projects, under the guidance of the Housing Forum Demonstration Project Panel. The results from Rounds 1 and 2 are:

*   39 projects selected: 2,500 units, £300 million value, public and private sector, partnering and consortia, new-build and refurbishment, regional spread
*   35 projects nominated for Round 3.

The next steps are to:

*   establish monitoring procedures and data collection
*   launch regional cluster groups to encourage sharing of experience
*   evaluate completed projects
*   document case-studies
*   publicise through CBPP and Housing Innovation Knowledge Bank.

17.4.6   Building Research Establishment 'Framework' Agreement

The BRE 'Framework' Agreement was a key part of the sale of BRE to a management team in 1997. Under it, DETR undertook to offer to BRE work to a defined value for each of the first five years following privatisation. This is given effect by the Construction and Housing Directorates of DETR issuing 'work specifications' to which BRE responds, indicating how it will undertake the work and providing a price. Provided that DETR judge the responses acceptable in terms of approach and value for money, the work is then placed with BRE. But it is possible for DETR, following discussions, to decide that the BRE response is not acceptable and to offer the work to other contractors through competitive tender.

     The annual income to BRE covered by this agreement remains confidential to DETR and BRE. However, it is widely known that the income diminishes over the five year period, but remains quite substantial even in year 5. The agreement was a fundamental component of the sale since, in conjunction with contracts placed by DETR with BRE which carried over to the new company, it provided reasonable assurance of income during the critical initial years after privatisation. It therefore enabled a business case to be made to financial interests by prospective purchasers of BRE.

     DETR have recently consulted research and industry interests over options for the development of the construction research programme as their obligation to offer work to BRE unwinds and they have more flexibility in the use of their resources.

17.5   INDIRECT SPECIFIC POLICY INSTRUMENTS

These policy instruments are aimed at achieving other public policy objectives, but also have important implications for the rate and direction of innovation in the construction industry. There are two main areas of public policy that can be defined as indirect specific. The first is government procurement. The ways in which the state and its agencies choose to procure its construction services can have a profound, and often unintended impact upon innovation in the industry. The second is the regulatory environment, which covers a broad area from specifying what kinds of buildings can be built where, to ensuring that they are built with hardwoods from sustainable sources.

17.5.1   Government procurement initiatives

Following and extensive review of government procurement practice, a number of different initiatives have been launched to ensure that the government is a best practice client. As some 40% of all construction spending has the government or its agencies as a client, the potential to change the industry is significant. This client role is now largely indirect since:

- most procurement is devolved from Ministries to national Executive Agencies or to local / regional bodies

-     an increasing proportion of that 40% consists of Private Finance Initiative (PFI) projects - a form of Build-Own-Operate-Transfer (BOOT) - which places ownership and financing of built assets with the private sector, and in which public sector procures final services, not assets.

PFI is sufficiently distinct and important that even this brief review will have to distinguish between PFI and other (public-sector owned and financed) procurement in other words, between:

-     procurement of services by the public sector from the private sector, requiring private sector procurement of built assets in order to provide these services
-     provision of those services by means of public sector operation of built facilities owned by the public sector, requiring public sector procurement of built assets.

The central organisations responsible for improving practice and performance in public sector procurement are: HM Treasury Procurement Group [PG], formerly the Central Unit on Purchasing; and Government Construction Client Panel [GCCP] (http://www.hm-treasury.gov.uk/gccp/index.html). These central bodies have recently committed government, on paper, in a sweeping new set of procurement guidelines setting out a new best practice, to a radical and fundamental change in aims and objectives for public sector bodies acting as clients procuring construction.

The most important themes of these changes are:

-     shift from 'lowest cost' to 'weighted average of value and cost' appraisal of bids; i.e. towards a value-for-money (VFM) approach
-     shift from compulsory competitive tendering to use of negotiation, including partnering and framework agreements, where appropriate in order to achieve VFM
-     shift from attempts at partial optimisation (e.g. least construction cost) to total project optimisation (e.g. life-cycle VFM, or at least life-cycle cost)
-     shift of risk allocation: in asset procurement, towards greater risk bearing and risk management by public sector; in service procurement, towards risk transfer to private sector
-     shift to use of 'non-traditional' procurement routes: both Construction Management (greater risk bearing and control by public sector clients) and Integrated (not only Design-and-Build but also Design-Build-Finance-Operate in PFI)
-     shift to performance and functional (i.e. not prescriptive or descriptive) briefs and specifications, based upon clarity of project objective outputs, not of required asset inputs.

Themes which are notably absent from the new agenda include:

- commitment to medium-term investment budgets and programmes
- ensuring predictable 'flow through' of projects once brought to market, i.e. smooth flow from OJEC notification to contract signature; PFI initially made 'flow' much less predictable and slower - recent counter-measures to correct this include pre-vetting of projects prior to OJEC notification to the industry to ensure funds are available and business case appraisal has been successfully completed.

Overall, the aim of the PG and GCCP is to achieve Latham / Egan Report targets for continuous improvement of performance on public sector projects, and to do this by encouraging 'creative' and 'innovative' approaches. They explicitly criticise the past 'obsession' with an ethos solely of accountability (of administration; of rules and regulations to be followed for 'correct', and therefore non-blameable, procedures) and seek to replace this by a new dominant ethos of performance - with rewards for good performance becoming more important than blame for incorrect procedure to the officials in charge of projects. To achieve these new objectives, new structural arrangements within public sector procurement include:

- strengthening of the role of H.M. Treasury, and especially its Procurement Group, in issuing guidance promoting best practice and establishing new codes and standards
- strengthening the powers, authority and expertise of project sponsors; creating a cadre of increasingly well-trained and developed specialists to perform this key role within the civil service
- new forms of employment contract and career development for these specialists, giving performance incentives, and making them stakeholders in final VFM performance of projects for end-users
- new authority to project sponsors to 'manage the end user', to force the latter to clarify their requirements and prioritise these, recognise and understand trade-offs and make timely choices.

Some public sector executive agencies have been encouraged and allowed to act as 'demonstration clients', consciously pioneering best practice in application. These include the Defence Estate Organisation (DEO) with its role in the Construction Supply Networks demonstration project. Other agencies have simply acquired levels of expertise and learnt as organisations from their experience in ways which are recognised to make them 'model clients' for the public sector. These include the Highways Agency and the Scottish Office.

In terms of Figure 17.6, the top-down, centre-driven nature of the new procurement regime has created a widening gap between documented best practice (as embodied, for example, in PG guidelines or in demonstration projects) and 'average' or median practice in a highly diffused and fragmented public sector. At the same time the public sector has recognised that it has to date failed to develop the potential expertise of its specialist staff and organisations to the point of having public sector advanced practice researcher-practitioners. Indeed in the 1970s and 1980s the public sector dismantled what it had of this kind (e.g. researcher-practitioner organisations fusing research into hospital use and impact of design on use with practice in design of NHS hospital projects).

17.5.2   The building regulations

Table 17.3 Principal regulatory instruments

Building Regulations/ Approved Documents
British Standards
European Standards
Codes of Practice
Planning Guidance (PPGs, MPGs)
Pollution Control legislation
Health and Safety legislation
Inland Revenue and Contributions Agency policies
Product sourcing agreements - notably for sustainably harvested timber

Construction in the UK is in many respects a highly regulated industry. These regulations do not have as their policy objectives the direct promotion of innovation in construction, but aim to achieve other policy objectives, such as the health and safety of the public or the workforce, the conservation of natural resources, or the quality of the urban and rural environment. However, there has been some debate regarding the extent to which codes and regulations can stifle or stimulate innovation in construction (Gann et al, 1998), but there is little systematic understanding regarding how the regulatory framework affects innovation processes in construction. Table 17.3 summarises the principal sets of regulations focused upon the UK construction industry.

These can be divided into three main groups:

- *Technical regulations* which affect both product and process
- *Planning regulations* which principally affect the constructed product
- *Labour market regulations* which principally affect the construction process.

17.6   A SUMMARY PERSPECTIVE

Table 17.4 Indirect and direct specific public policy instruments in the UK

	Name	Resources in Construction 1999-2000	Objectives	Means	Contribution to Innovation
Programmes to support R&D	Foresight	Not relevant	To determine future UK R&D strategy	Expert reviews and Delphi studies	Providing the long vision
	EPSRC Response Mode Research	£11.5 million	To fund basic scientific research	100% funding of university-based research teams	Generating new ideas. Technology orientated
	EPSRC managed programmes (inc. Foresight Challenge)	£3.6 million	To develop applied research matched to industry needs	Challenged – based collaborative research programmes between universities and industry	Generation of new ideas and diffusion to advanced practice. Process Orientated
Programmes to support advanced practices and experimentation	Partners in Innovation	£7.5 million	To develop applied research matched to industry needs	Challenge-based collaborative research programmes between research institutes and industry	Developing advanced practice. Process orientated
	EPSRC IGDS courses		To train in advanced practices	Part-time masters level education	Diffusion of new ideas into advanced practice
	Movement for Innovation	£0.6 million	To identify and diffuse advanced practice	Demonstration projects proposed and reviewed fand innovations identified. Campaign style.	Diffusion of advanced practice. Process orientated.

Table 17.4 (Continued) Indirect and direct specific public policy instruments in the UK

	Name	Resources in Construc- tion 1999-2000	Objectives	Means	Contribution to Innovation
Programmes to support performance and quality improvement	BRE Frame- work Agree- ment	£10.3 million	To support the BRE's research in the immediate post- privatisa- tion period	Declining funding with negotiated research objectives. Contracts offered on an exclusive basis.	Development and validation of best practice, and the development of standards. Technology orientated
Programmes to support taking up of systems and procedures	Construc- tion Best Practice Programme	£2.3 million	To identify and diffuse best practice	Authorit- ative source of best practice methods and cases	Collation and diffusion of best practice. Process orientated
	Government Procurement Initiatives	Not relevant	To ensure that the state and its agencies are best practice clients	Reforming procure- ment procedures	Providing the context which facilitates innovation. Process orientated.
	Regulatory system	Not relevant	To ensure the implemen- tation of standard practice	Inspection and education	Diffusion of standard practice. Technology orientated

The current situation in the UK with regard to public policy instruments and construction innovation is fluid and dynamic. Table 17.4 presents an overview of the major indirect and direct specific public policy instruments aimed at stimulating innovation in UK construction. The vertical dimension takes the type of programme to support the different levels of practice identified by TG35, and cross-tabulates them with information on the budget, means and objective of the programme. The sum of the figures in the budget column should not be taken as a total for the public expenditure on construction R&D – only the principal programmes are covered. However, what the figures do show is the relative significance of the indirect specific funding through the universities, particularly in supporting advanced practice. While general direct instruments remain important in influencing the climate of innovation in the UK construction industry, much further research is required before we can begin to understand their implications for construction innovation.

CHAPTER 18

THE U.S. FEDERAL POLICY IN SUPPORT OF INNOVATION IN THE DESIGN AND CONSTRUCTION INDUSTRY

Richard A. Belle, Harvey M. Bernstein, and André Manseau

18.1   GENERAL CONTEXT

The enhancement of innovation systems is critical to maintaining and improving economic competitiveness and sustained capacity for further innovation. This effort will be successful only if all critical sectors of the economy are revitalised. In particular, any national innovation policy must recognise that the quality of the built environment affects all other aspects of the economy. A modern, durable, and flexible physical infrastructure is a necessary precondition for sustained innovation and competitive capacity; a decayed and unresponsive infrastructure will retard development throughout the economy.

The U.S. design and construction industry is the nation's largest manufacturing activity, contributing about 13% of the Gross Domestic Product (GDP), and employs 7.5 million people. It is the largest economic sector after health care. In this role, the industry is the nexus for materials flow in the economy and the key to bringing advanced materials and processes into the marketplace. It impacts virtually all aspects of the U.S. economy from the private and public infrastructure such as roads, shipyards, public works, buildings, and housing to environmentally sound development. The transport of goods and services by all industries is directly affected by the quality of the physical infrastructure and working to improve this infrastructure will have a positive affect on all industries. It is thus no exaggeration to link explicitly the vigour of the design and construction industry to the overall strength of the nation's economy. Simply put, a vigorous and innovative design and construction industry will lead to more jobs and a greater variety of products for all.

Moreover, the nation's quality of life directly reflects the quality of the physical infrastructure. The Clinton Administration has emphasised the criticality of infrastructure development, sustainability, and environmental consciousness in all areas, including construction, intelligent transportation systems, telecommunications, water and power, and facilities management. It has brought attention to important quality of life issues, such as urban sprawl and the loss of time in commuting that affects virtually everyone. As the industry rebuilds and revitalises the physical infrastructure, it is critical that the focus of *sustainability* is supported within the built environment.

18.1.1   The U.S. construction industry

New construction economic activities, with renovation, maintenance and repair were estimated to be $995 billion in 1997 (current US$), accounting for about 13% of the GDP. The largest construction market is residential, accounting for 44% of the total, followed by commercial and institutional (32%), public works (19%), then industrial (5%) (National Science and Technology Council, 1999).

Statistics on the overall construction sector are often difficult to obtain. Official data shared by OECD countries are limited to builders of construction, new and renovation, and exclude design, manufacturing of building products, and property management services. Contribution of builders to GDP can be estimated and compared to other OECD countries. In 1996, the U.S. ranked first in the world on construction added value to GDP, with a value of 265 US$ billion (in constant US$ of 1990), followed by Japan with 239 US$ billion (OECD, 1999). However, Table 18.1 shows that in terms of the share of the construction added value to GDP, U.S. ranked at the bottom of G-7 countries, with France.

Table 18.1 Share of the construction added value to GDP for G-7 countries, 1996
(source: OECD, 1999)

G-7 Country	Share of Construction to GDP (%)
Japan	9.72
Germany	5.38
United Kingdom	5.25
Italy	5.14
Canada	4.75
France	4.16
U.S.A.	4.16

The value of construction activities has increased by an average annual of 7% between 1994 and 1998. It is the private non-residential (business and commercial) that has had the best performance, with an average growth of 13% annually. Other sectors have increased at about the same rate than the overall economy (4 to 5%).

A recent study showed that productivity appears to have increased substantially in construction in the last two decades. However, technological advances were likely not a leading contributor, the majority of the increase appears to have resulted from depressed real wages (Allmon et al, 2000).

18.1.2   R&D and innovation

Yet despite its unquestioned importance and impact on all other sectors of the economy, the design and construction industry as a whole fails to lead other economic sectors in the development and implementation of innovation. Its suffers from one of the lowest overall rates of research and development (R&D) investment compared with other industrial sectors. While the mean R&D investment among all mature U.S. industries is approximately 3.5% of the total output, the level of investment by the design and construction industry is estimated to only 0.5% (CERF, 1993).

As it is shown in Table 18.2, the U.S. R&D effort in construction ranks 4th in the G-7 countries.

Table 18.2   Construction R&D as % of construction added value for G-7 countries, 1996
(source: OECD, 1999 and 2000)

G-7 Country	R&D as % of Construction Added Value
Japan	0.467
France	0.244
Germany	0.097 (in 1997)
U.S.A.	0.093
Canada	0.055
United Kingdom	0.021
Italy	0.026

This very conservative investment strategy is not surprising. Despite its huge size, the U.S. design and construction industry is greatly fragmented, with over 80% of its 1.2 million firms consisting of ten employees or less. Profit margins are very low and the industry is often affected by large seasonal cycles, such that most firms consider R&D investment, and innovation in general, a luxury that they cannot afford. The emphasis by many owners on "low bid" provides even less margin for experimentation, and thus serves as a strong disincentive for innovative policies. In practice, innovation may take place in the process of completing a specific contract, but is not seen as an objective in itself. These factors have helped to make the industry "risk-averse," preferring to utilise the proven and known, and eschewing the alternatives that might offer greater productivity, but could also create greater uncertainty and possible economic dislocation.

The industry practitioner does not operate in a vacuum, however. For many of the industry's products, the government, on both the federal and state level, is a critical actor. The government often is the sole or primary owner of the structure to be built or repaired. The government often participates in the financing of the project or the development of the contract. Perhaps most importantly, the government often assumes part of the risk and liability associated with the physical assets that it owns (bridges, highways, buildings, etc.). In the interest of public safety and open competitiveness, federal and local governments have developed, over time, a variety of policies. While intended for the best of reasons, many of these strategies have had the effect of reinforcing the industry's conservative inclination outlined above, thus making innovative approaches even less likely.

Innovation is often used interchangeably with "new." But innovation does not necessarily mean state-of-the-art or cutting-edge technologies *per se.* Innovation often means the process of employing new combinations of existing, demonstrated, and market-ready technologies, processes, and methods to improve the quality of the built environment. Federal policy towards enhancing design and construction industry innovation should be not to champion newness *per se,* but to help make it easier to commercialise and implement those practices and products whose widespread use would lead to improvement in the built environment.

Innovation is doing something different—breaking tradition—because there is greater perceived value. In the context of the nation's physical infrastructure, innovation pertains to methods and processes, as well as materials and technologies. For example, new and more efficient design methods, performance standards, or procurement practices may be as beneficial to bridges or schools of the future as advanced plastics or structural systems.

18.2   FEDERAL PUBLIC INTERVENTIONS

President Clinton established the National Science and Technology Council (NSTC) in 1993. This Cabinet-level Council is the principal means for the President to co-ordinate science, space, and technology of the diverse parts of the Federal research and development enterprise. The Committee on Technology's Subcommittee on Construction and Building (C&B) was organised in 1994 to co-ordinate and focus activities of 14 federal agencies, and to work in cooperation with industry, labour and academia to improve the life cycle performance, sustainability, efficiency, effectiveness and economy of constructed facilities.

C&B, in conjunction with private industry, developed a plan to identify technologies and practices capable of achieving the following seven goals in the construction industry over the next decade. (National Science and Technology Council, 1999 and Bernstein, 1999). Taking 1994 as the baseline, the goals are:

1. 50% reduction in delivery time. The time from the decision to construct a new facility to its readiness for service is vital to industrial competitiveness and project cost reduction.
2. 50% reduction in the cost of operation, maintenance and energy over the life of the facility.
3. 30% increase in productivity and comfort of the occupants of industrial facilities and in the processes housed by the facility.
4. 50% fewer occupant related illness and injuries caused by improper or poor building design, fire or natural hazards, slips and falls, and illnesses associated with a workplace environment.
5. 50% less waste and pollution at every step of the delivery process, from raw material extraction, the construction process, to final demolition and recycling of the shelter and its contents.
6. 50% more durability (the capability of the constructed facility to continue to function at its initial level of performance over its intended service life) and flexibility (the owner's capability to adapt the constructed facility to changes in use or users' needs).
7. 50% reduction in construction work illness and injuries.

Therefore, reliable baselines and measurement tools are being developed by different industry sectors. Industry representatives identified a number of changes needed to remove barriers to private sector investments in technology required to meet the goals (National Science and Technology Council, 1999 and Bernstein, 1999).

- A speed-up in the regulatory process
- tort reform to avoid unreasonable liability for using innovations
- performance standards and conformance assessment mechanisms to enable users and regulators to asses and accept new materials, products, and systems
- education of builders, managers, regulators in information systems and data, and training of craft workers to increase the pool of skilled labour and to promote safe operating practices
- a closer working relationship between all parties in the facility design and construction process, particularly in the early stages of planning and design, and
- formation of a construction coordination council that would guide private activities and speak for the industry to bring about some of the needed changes in the system.

In response to these issues, a number of initiatives have been developed:

- Partnership for Advancing Technology in Housing (PATH):
  Administered by the National Institute of Standards and Technology Administration (NIST) with participation of Housing and Urban Development, the Department of Commerce, ten additional federal agencies and industrial representatives (home builders, product manufacturers, insurance and financial companies and regulatory bodies) to promote technological innovation in the housing industry. In 1999 it has granted $1.1 million to six industrial projects on energy savings and homebuilding technologies in the framework of a new multiyear cooperative R&D program the PATH Cooperative Research Program (PATH CoRP). Public funding will cover 70% of projects' costs.
- Design Excellence Program:
  Created and administered by the General Services Administration's (GSA) Public Building Service (PBS) it is meant to encourage designers to enhance the use of innovative technologies.
- Development of Private Sector Evaluation Centres:
  - National Evaluation Service Building Innovation Center (NESBIC): for the assessment of safety, functional and environmental qualities for national and international acceptance of innovative building products and services
  - Highway Technology Evaluation Center (HITEC) for the development of information and evaluation of products, materials, services equipment and systems for highways
  - Environmental Technology Evaluation Center (EvTEC) for the promotion of environmental management technologies
  - Civil Engineering Innovative Technology Evaluation Center (CEITEC) for the evaluation of various technologies, particularly those linked to public works and the military.
- A program streamlining the building regulatory system:
  NIST's Subcommittee on Construction and Building (C&B) organised and funds with state and local governments and industry a program for the enhancement of economic development, public safety and environmental quality through improved management and practices of the regulations for siting, design and construction of all types of buildings. The objective is to reduce delays and costs involved in compliance, a major complaint of industry and inhibitors to private sector investment in innovation. The program is administered by National Conference of States on Building Codes and Standards (NCSBCS).
- The High Performance Construction Materials and Systems Program (CONMAT):
  A ten-year, $2 billion, R&D program designed to accelerate the commercialisation of innovative materials and systems for a

revitalised infrastructure capable of withstanding the demands of the twenty-first century. The program promotes material innovation in new construction, repair, rehabilitation, and retrofit technologies. CONMAT is a consortium of over a dozen construction material industry representatives working in close liaison with government and academia. Membership and potential collaboration is open to all members of the design and construction community, including designers, fabricators, equipment manufacturers, material suppliers, architects, contractors, and owners.

- Mechanical and Electrical Systems Council:
  A forum designed to support National Construction Goals by facilitating research, technology transfer and cooperative interaction between organisations and individuals.
- Partnership for Advancing Infrastructure and its Renewal (PAIR):
  Supported by C&B and administered by CERF, its objective is to put an end to the management by crisis approach to infrastructure repair and renewal. PAIR works with leaders from both the private and public sectors to form collaborative partnerships that bring the very best construction technologies and processes to the marketplace. The Partnerships seeks to shorten the long time frame currently needed to take state-of-the-art construction technologies and deploy them on a broad scale. Current focus areas include: transportation infrastructure, school repair and construction, information technology, water surety, among others.

Federal agencies invest approximately a total of $500 million per year in R&D relevant to the industries of construction. Table 18.3 presents the types of support provided by these 14 Federal agencies with regard to the innovation stage.

Table 18.3 Federal agencies participating in the construction and building sub-committee and their types of support to construction innovation

Agency	R&D support	Experimental support	Performance & quality improvement support	Take-up and standards support
Department of Agriculture - Forest Products Laboratory	Wood building products performance	Economic feasibility and minimise environmental impact	Facilitating implementation of new wood building products	Developing codes & standards for wood products
Department of Commerce – Building and Fire Research Laboratory of NIST		Developing evaluation models and measurements for infrastructures and fire safety	Demonstration and case study	

Table 18.3 (Continued) Federal agencies participating in the construction and building
sub-committee and their types of support to construction innovation

Agency	R&D support	Experimental support	Performance & quality improvement support	Take-up and standards support
Department of Defense – Defense Research & Engineering	High performance material and systems, Ground and coastal structures, Sustainable facilities			
Department of Housing and Urban Development – Policy Development & Research	Affordable Housing Urban redevelopment			Developing codes for equitable access to housing
Department of Energy - Office of Building Technology	Energy efficiency Renewable energy		Promote energy codes Upgrading programs	
Department of Health and Human Services - National Institute for Occupational Safety & Health	Develop and demonstrate technologies to reduce occupational hazards		Demonstrations and training programs	
Department of Education - National Library			Information and technical assistance	
Department of Labor - Occupational Safety & Health Administration			Inspections, training and benchmarking	Set standard for construction safety
Department of Transportation - Federal Highway Administration	High performance materials (durability) Non-destructive monitoring Intermodal effectiveness			
Department of Veterans Affairs - Office of Construction Management			Best practices for managing and operating buildings	
US Environmental Protection Agency		Sustainable development Air pollution	Promote energy conservation and sustainable development	
General Services Administration - Public Building Service		Workplace productivity, security and energy efficiency	Demonstration, show cases, and Design Excellence Program	

Table 18.3 (Continued) Federal agencies participating in the construction and building sub-committee and their types of support to construction innovation

Agency	R&D support	Experimental support	Performance & quality improvement support	Take-up and standards support
National Science Foundation	Material/structure Non-destructive evaluation methods Civil infrastructure life cycle			
Federal Emergency Management Agency - Federal Center Plaza			Prevention to minimise the effects of national disasters	

18.3   EFFECTIVENESS AND POSSIBLE REFORMS

Many of the trends that discourage innovation could be reversed, or at least minimised. It is essential that private industry work in concert with the public sector in carefully reforming those federal policies that serve as disincentives to innovation. The NSTC, particularly through its Committee on Technology, is viewed as one of the key government groups that could play a leadership role in this area. Reform is appropriate in the following five areas.

1.   Changing regulatory policies. The reliability and safety of our national physical infrastructure directly affect all citizens. It is only prudent that the federal government assume leadership in adhering to strong standards for safety in construction, maintenance, and repair of built physical assets. But many current regulations are superfluous or redundant, and regularly stifle the entrepreneur's incentives to develop systems and build infrastructure that is cheaper, safer, or quicker to construct and maintain. Regulations should be effective, efficient, and facilitate innovations. Some of the regulatory issues that should receive high priority from the federal government are listed below.

- The industry must redefine professional/product liability for the construction, repair, and/or maintenance of built facilities (litigation expenditures are increasing at 10% per year).
- The legal system should undergo tort reform that results in recognisable limits or caps for potential lawsuits. Currently the litigious nature of the construction industry is a major deterrent to innovation, making the U.S. less competitive with its European and Japanese counterparts.
- Federal agencies should consult outside, independent technical experts through the peer review process prior to promulgation of environmental regulations.

- The federal government should help streamline the application process and increase the availability of waivers for regulations.

2.   *Supporting approval of new materials and technologies.* To truly improve the way new facilities are constructed and repaired and existing ones maintained, the purchasers of such services should adopt policies that forcefully support the use of innovative materials and technologies. Currently, the system is biased *against* the use of innovative approaches, particularly when no standard or specification exists. The government can foster the use of pre-approval, and pre-qualification of materials and technologies in order to reduce risk and liability and ensure public safety. This approach is being increasingly used by Japanese and European competitors to great success. The government should encourage strategies that decrease the approval time for the acceptance, and rapid commercialisation of promising new construction materials and technologies. There are a number of areas where leadership by the public sector can improve the approval process:

- Government agencies acting as owners should fully support and participate in evaluation strategies for which no recognised standard or specification exists. Federal policy should encourage federal, state, and local governments to utilise honest-broker non-profit institutions that can evaluate new materials and technologies for which no standards exists. Such approaches rely on a peer-reviewed process in cooperation with stakeholders to pre-qualify and pre-approve products. A number of such consensus-based evaluation centres have been developed in the past few years, with federal agencies working closely with non-profit institutions, including FHWA support for the Highway Innovative Technology Evaluation Center (HITEC), EPA support for the Environmental Technology Evaluation Center (EvTEC), among others. These efforts have permitted entrepreneurs to expedite their innovative technologies into practice through a single venue, rather than repeat the process for every state or local approval agency.
- Moreover, the federal government should encourage the development and adoption of consensus-based standards and codes.
- The government should consider constituting an authority with a single administrative processing/permitting agency to shorten and improve the approval process.
- The government could take the lead in recognising the enormous potential of advanced, engineered construction materials. Both public and private sectors should embrace the activities of the high-performance CONstruction MATerials initiative, known as CONMAT, that is demonstrating the power and effectiveness of these materials in repairing and revitalising the infrastructure.

3.   *Revising procurement strategies.* All owners, whether in the private or the public sector, should aggressively negotiate for the most cost-effective relationship when purchasing construction processes, materials, or facilities. Unfortunately, over the years this objective has frequently become synonymous with endorsing the lowest initial bid on a contract, regardless of the implications for installation, operations and maintenance, or long-term durability. The federal government, often the prime owner of a facility, should help set an example by recognising that value and cost of a facility must take into account the entire full term cost of the said facility, from initial design to its end use and eventual removal, the "cradle-to-grave cycle".   Policies or strategies that facilitate innovations should encourage include approaches such as the following:

- All owners, including federal, state, and local authorities, should explicitly consider the total life cycle costs when considering the purchasing of services for the construction, repair, or maintenance of facilities. The federal government should take a leadership role in working with state and local authorities in reducing the reflexive reliance on "low-bid" procurement approaches.
- To help encourage the development of a variety of new approaches that place a premium on innovative and cost saving strategies, the federal government should promote the use of incentives for innovative procurement and construction practices.
- Government agencies should take the lead in promoting competitive project delivery systems such as the design-build approach. These delivery systems should be viewed as delivery mechanisms that foster industry R&D investment. The benefits could include the owners getting facilities constructed faster, better, and cheaper; lowered operations and maintenance costs; less resource waste and pollution; and improved bottom-line performance. Innovative procurement practices will respond to market demands for many delivery systems, make industry R&D investment a more rational business decision, and will foster innovation.
- For publicly–funded projects, the selection of consultants for engineering projects should be based on design competitions, with each competing firm compensated for reasonable costs.

4.   *Focusing and co-ordinating R&D investments across federal agencies.* Dedicated support for research and development is the lifeblood of innovation, but has too often been deferred in favour of strategies that appear to favour immediate and short-term payoffs. The NSTC Committee on Technology and other federal authorities should champion the need for a focused construction R&D strategy that fosters innovation as an explicit goal. Federal R&D for construction should be focused and co-ordinated across agencies of cooperation with industry, and supported by OMB and Congress. The government should support some of the approaches outlined below.

- The government serving as construction owner should promote innovative financing including public-private partnerships, capital budgeting, private funding of infrastructure and dedicated user-fees with a research set-aside.
- The government should support a permanent tax credit that rewards R&D investment.
- Future government construction projects should require that a specified element of the project be innovative in materials or process.
- The R&D objectives expressed by "the infrastructure needs of the 21st century cannot be met with 20th century technologies." The government must develop policies that ensure the National Construction Goals remain on target.
- The NSTC concept of co-ordinating and focusing federal R&D across agencies is sound, but needs stronger implementation. OMB and OSTP need to work better together to reflect NSTC plans in agencies' programs and budgets.

5.   *Developing focused national objectives.*   The government has the opportunity to participate in the development of private-public partnership strategies for the advancement of the design and construction industry.   Such an approach cannot be imposed from above; it must involve the enthusiastic support of both private and public sectors:

- Except for funding PATH in the HUD appropriation, Congress has been unresponsive to the National Construction Goals (renamed Construction Technology Goals) effort.   This is particularly troublesome for an industry that is the nation's largest manufacturing sector.   Scarce resources continue to be dissipated in narrowly focused earmarks.   Authorisation and appropriation processes are fragmented among subcommittees and have been incapable of responding to well-coordinated initiatives of industry and federal agencies.   National coordination, through the NSTC or comparable structures, is needed for authorisations and appropriations.
- The government should foster partnerships between private industry, government, and non-government organisations.   Private sector leadership remains essential to avoid stalemate, due to partisan politics, at national and state levels.   The private sector includes business, labour, consumer, environmental, professional and academic interests.   While the private sector must assume political and financial leadership, it can only do so with strong focused support from the public sector.
- Specifically, the government should help ensure that the Partnership for the Advancement of Infrastructure and its Renewal (PAIR), becomes a national initiative.   PAIR's objective is to repair and

revitalise the nation's physical infrastructure through the use of innovative products and practices. PAIR has the potential to become a true national partnership among the federal government, state governments, and the private sector. PAIR can become a vehicle to ensure that future infrastructure revitalisation efforts will be flexible and visionary, with appropriate procurement, delivery, and approval policies conducive to innovation. Without significant federal support, however, this promising initiative will not realise its potential as a major catalyst for infrastructure renewal.

- The government should support the need to continually upgrade the capability and productivity of the work force through tax incentives that encourage and reward continuing education. For example, the government should maintain and expand policies that exclude educational assistance from an employee's gross income.
- The government should support the development of risk management guidelines at all levels of government. It should establish core risk assessment research programs to ensure risk management is based on adequate scientific data.

18.4   CONCLUSION

The design and construction industry has demonstrated in recent years that it has the imagination and intellectual resources to change fundamentally the way we construct and repair our built environment in a sustainable manner. But it cannot do so in a vacuum. The private and public sectors must develop effective partnerships to create a climate more conducive to innovation. Public policies that discourage innovation must be changed. The public sector can play a pivotal role in this endeavour.

CHAPTER 19

PUBLIC POLICY CONTEXT AND TRENDS

André Manseau and George Seaden

The Table 19.1 presents a summary of the most relevant or interesting trends in public instruments of each country, as reported by members of the international task group – TG 35.

Table 19.1: Most interesting trends in public intervention for each participating country

Country	Trends
Argentina	- Increased diffusion of information on current programmes - Support training in SMEs
Australia	- Renewed and innovative efforts to improve the construction industry's performance over a long-term planning horizon
Brazil	- Support quality certification programme for builders based on agreements with key industry actors - Update regulation and building codes to sustain innovation
Canada	- Strategic intelligence on innovation opportunities, processes and resources - Facilitate export through networking and international agreements - Development of objective-based codes
Chile	- Increasing public concern on the importance of an integrated approach for innovation, with the support of an active industry association (Chilean Construction Institute) and a new programme to create strong research centers
Denmark	- Public intervention reaffirmed with a variety of actions involving co-ordination among the various agents - Focus on change from product to process innovation, with a broad approach addressing the overall industry
France	- Development of public-private co-operative R&D projects, particularly for IT and material innovations - Focus on process innovation and co-ordination between procurement, design, construction and maintenance - Focus on reduction of work time and its impact on site organisation and labour management

Table 19.1: (Continued) Most interesting trends in public intervention
for each participating country

Country	Trends
Finland	- Importance of "neutral' evaluation of technology - Increased importance of life cycle and usability (from product to use of)
Germany	- Changing government support from R&D to strategic innovative projects in partnerships between universities, government and industry
Japan	- Development of new procurement approaches to improve effective use and outcomes from innovations to the private sector - Innovation leadership to the industry by changing public institutions status and National Account Law
Portugal	- Increased university-industry collaborative R&D, with greater industry participation - Government maintains an important role as a major buyer/owner (infrastructure, utilities)
South Africa	- Shift from public sector grant R&D funding and public R&D organisations to competitive bidding; and - Increasing emphasis on encouraging collaborative partnerships between private sector, university and research organisations
The Netherlands	- Innovation brokering and knowledge assistance to promote adoption and diffusion of innovations - Procurement policies to actively promote innovativeness and efficiency
United States of America	- Improved quality and clear responsibility with neutral evaluation of new materials and technologies, public sector to be a model owner - Update and follow-up of the seven Construction National Goals that aim to reduce delivery time, reduce life cycle cost, increase productivity and comfort of the occupants, reduce occupant building related illness and injuries, reduce waste, increase durability and reduce construction work illness and injuries.
United Kingdom	- Rapid development of partnering and private finance providing conditions for sharing risks and seeking new solutions

Following the review of various public policy initiatives, of current practices and of their effectiveness, in the fifteen participating countries of the TG 35, we have been able to establish certain number of common innovation trends as well as some more specific national differences.

Contrary to the available general research on potential differences in the national treatment of innovation (see the previous section "Influence of Political and Social Structures"), significant number of identified common trends suggests that the political/constitutional structure of countries and/or their social system of innovation do not create radical differences in their national approaches to their construction industry. However, the government structure, the type of national innovation system and the existence of institution(s) clearly representing industry's needs does seem to influence the relative level of public concern with productivity of construction and the choice of specific policy instruments.

The general situation, which can be observed in most countries and could be considered as the current general context for innovation in the construction industry, is as follows:

- National governments are showing increasing desire to encourage industry-led, longer-term goal setting, and to facilitate cooperation among construction firms for greater effectiveness.

- The most common innovation policy instrument is through public investment in the R&D effort at universities or government institutions. The linear, technology push model of investigator-led research remains well entrenched as the principal mechanism believed to advance innovation.

- Mission-oriented research policies are becoming less popular due to the perception that "government does not always know best". Directly funded construction research institutions are seeing their budget base eroded and have moved out of necessity and/or by policy directive into more collaborative arrangements with the industry.

- Emphasis in the directed, publicly funded research is shifting away from the product and component orientation to process related issues. It may be expected that ideas from advanced manufacturing and information technology will migrate more rapidly to the construction sector and that novel approaches for site assembly will be developed. There is some evidence of construction related policies now being focused on the management of supply chain and on-site problems.

- Public policy instruments in a majority of TG 35 countries have moved towards support of research through cost sharing with the industry or tax relief on R&D investments. This is an indication of increasing acceptance that technology push, science based innovation policy may not lead to increased innovation and that market pull model, with industry initiated research, may be more productive. Response from the construction industry to this collaborative type of programs has not been encouraging. Firms don't seem to take advantage of this type of government aid.

- Information and communication technologies are perceived to have significant impact on the construction processes.

- Environmental issues and compatibility with community interests are increasingly influencing all construction phases, from initial site approval to eventual deconstruction and recycling.

Beyond these commonly observed approaches, we observed other trends in certain countries identified in their reports, which may reflect their different socio-economic, cultural or institutional contexts:

- The construction industries in less industrially developed countries with significant economic challenges face important problems of providing large volumes of new housing and infrastructure at reasonable cost. Maintenance-renovation component is lower, in comparison with more industrially developed countries. Governments of those countries are usually the most important clients for construction, but public policies towards innovation do not appear to address problems related to infrastructure development. Few programs to support innovation and/or R&D are offered and they are often based on public policies for the high-technology sectors in wealthy OECD countries.

- Few countries have now recognized that the fragmented, small enterprise nature of construction companies requires custom designed support through networks of dedicated best practice centres.

- Government sponsored demonstration projects are sometimes used to stimulate innovative ideas, which are then disseminated to the industry. The effectiveness of these projects has not been fully documented to date, but some expert opinions suggest that the results so far do not meet the expectations.

- Government managed/supported agencies for the evaluation of new products and systems are generally considered as having positive effect on innovation. Government can also provide innovation leadership, incorporating new products/systems in its building programs and/or providing unbiased public assurance as to their "fit for use" and thus enabling more rapid introduction and commercialization.

- Recent reviews of building regulations in some countries have shown greater concern with potential innovation impact. Public regulatory policies in the area of safety, energy conservation or environment can have significant positive/negative innovation effects on the construction

- There are some suggestions that lack of innovation in the construction industry is systemic, due to the inability to appropriate tangible benefits from the introduction of new ideas. Policies aimed at changing public sector procurement system through greater emphasis on value rather than cost are being introduced in some countries. This could enhance potential corporate benefits. Large changes, in some countries, in the structure of their "public" works towards greater private ownership and concessions or through promotion of concepts of whole life costs, could also stimulate innovative behavior.

Finally, we identified some interesting national differences among TG 35 participants, as follows:

- Japan's construction expenditure and its investment in R&D are the highest of OECD countries. It has developed, through historical evolution, several large, vertically integrated construction companies that believe in technology as a major competitive advantage. Large, in-house research institutes support corporate activities. Thus, in contrast with other countries, most of construction research is concentrated in the private sector. Government has been concerned with the future of Japanese construction and is currently promoting increased collaboration in R&D efforts, however agreement on shared goals among the key stakeholders remains a challenge.

- Australia's construction expenditure per capita is average by OECD standards and its investment in related research, primarily in the public sector, is considered very low. Government is seeking to create a more internationally competitive industry through its Action Agenda, which focuses on education, greater diffusion of technology and enhanced innovation. Collaboration among all actors is being promoted, with public sector taking modest initiatives and private sector lagging behind.

- United Kingdom has seen during the past decade a sequence of high-level comprehensive reports that examined its construction industry, and found it wanting. Several public/private innovation initiatives are now taking place (with the endorsement of senior politicians and corporate leaders) to achieve ambitious performance goals. Focus is on the supply chain, best practice and sharing knowledge. Public sector is encouraging innovation through proposed changes to its procurement practice, moving from "lowest cost" to "value-for-money" through business-like continuous improvement rather than more traditional accountability ("non-blameable") values.

- Denmark's promotes innovation in a context of an "organization society" which encourages socio-political negotiations on shared objectives. In the post-war period, it was able to implement successfully, lasting public policy of general use of prefabricated elements. Currently, government-led demonstration projects attempt to develop process innovation that will achieve "twice the value for half the price". Results, so far, have been less than anticipated and more intensive coordination effort of multiple stakeholders and of various policy instruments is required.

- Finland is considered as having a very effective innovation system (5th best in the world) and it has experienced recently high rate of economic growth. Public policy on construction innovation is to strongly involve the end-users (real-estate sector) and to encourage rapid commercialization of ideas. There is a comprehensive array of public policy instruments to encourage innovation in construction at all stages of development, nevertheless construction SME's are perceived as lagging others, reluctant to obtain outside financing to expand their operation.

- Netherlands' public policy stresses pre-competitive technological cooperation and special aid to SME's. There has been a relatively positive response from the construction industry, with inter-firm alliances of 11% (24% for all sectors) and use of government innovation subsidies of 11% (25% for all). Local enforcement of land use and of stringent environmental requirements are seen as inhibiting innovative practices.

CHAPTER 20

CONCLUSION

André Manseau and George Seaden

This volume, produced with the help of leading experts in 15 participating countries of the CIB TG 35 has contributed to the development of a framework for better understanding of innovation in construction and to identify major issues and trends on government support in this sector.

Most of TG 35 countries are in a midst of an evolution from a more Keynesian, public intervention attitude towards its construction activity to a greater "laissez faire" market discipline. At the same time, the government's role as the major buyer of construction output and thus its ability to influence directly innovation has generally been eroding.

There is no doubt that construction is a very significant element of economy of every country. Statistical measures, which identify "construction" with on-site value added, underscore the relative importance of the sector and its influence on the overall cost of fixed assets. In fact, in its expanded definition, which would include manufacturing of building products and equipment as well as various services involved in the design, operation and maintenance of buildings and infrastructure, it deeply influences the productive capacity of every country.

Construction almost everywhere is perceived as being "in trouble", with low margins of profit, high costs of production and lack of concern for the end-user. It is seen as relatively slow in adoption of new knowledge. Yet, innovation processes and their impact on the efficiency and the productivity of the industry are poorly understood. The sector is very complex and the role of governments as well as the effectiveness of public interventions has not been addressed until now.

Current governmental reaction has been varied, from indifference or "benign neglect" to undertaking of in-depth reviews of national construction industry to enhance productivity, performance and quality, with greater public policy emphasis on diffusion systems of existing technology and best practice.

While large, vertically integrated global firms exist and more are emerging, their market share is relatively minor. Small, specialized construction firms continue to dominate national markets. It is in this context that successful public policies to encourage innovation in construction must evolve. More than anything, being able to accommodate the fragmentation and the diversity of interest of various SME participants, is required.

It appears that most of currently available public policy instruments in support of innovation have not been of great use to the construction industry, however:

- Programs with greater local presence, focussed on access to technology, promoting collaborative arrangements, seem to be more successful as well as institutions that are able to evaluate new products or processes before market launch.
- Governments remain major buyers of construction services and more open acquisition policies that promote long-term value and performance rather than the initial cost, appear to stimulate innovation.
- Greater emphasis on performance against defined objectives is likely to enhance innovation. Such objectives, that govern safety of occupants or users of buildings and infrastructure as well as compatibility with community values and longer-term sustainability, need to be introduced in regulatory measures.

Work of TG 35 seems to suggest certain necessary pre-conditions for the development of successful innovation policies for this industry.

- Comprehensive country-specific analysis of its issues must be undertaken based on extensive consultation with participants.
- Broad recognition of innovation issues needs to prevail within the industry.
- An organization must exist (or be created) to represent innovation interests of the industry.
- Balance between short and long term, public (safety, sustainability) and private (profit, market dominance) interests must be maintained, through negotiations between senior government and industry leaders. Discussions on purely technological level do not seem to yield satisfactory results.
- Policies must reflect shared objectives of all stakeholders.

Finally, the principal drivers for innovation are often created at the firm level, within a stimulating macro-economic context. It is also known that in most of industrialized countries a few construction companies have achieved superior market position through the use of innovative practices. There is great need for empirical studies to determine the key success factors of these firms to allow others to emulate their example.

KEYWORD INDEX

BIBLIOGRAPHY

This list also includes papers and other documents that are relevant to the topic of innovation systems inconstruction.

Abrahamson, N. et al. (1997). Technology Transfer Systems in the United States and Germany, Washington D.C.

ABS (1996-97) Australia Business Statistics. Cat. 8772.0. Private Sector Construction Industry Australia.

Advisory Council on Science and Technology, Canada, Expert Panel on the Commercialization of University Research (1999). Public Investment in University Research: Reaping the Benefits. Ottawa, published report

AEGIS (Australian Expert Group in Industry Studies) (1999). Mapping the Building and Construction Product System in Australia. Canberra: ISR.

Afuah (1998). Innovation Management: Strategies, Implementation and Profits, Oxford UP: New York.

Agopyan, V., Souza, U.E.L., Paliari, J.C., and Andrade, A.C. (1998). Alternativas para a redução de desperdício de materiais nos canteiros de obras. Relatório Final. PCC-USP/FINEP/ITQC, setembro de 1998. 5 volumes. 1355P.

AI Group (1999). Further Attack on Australian Business R&D. Media Release. 22 December.

Allmon, E., Haas, C.T., Borcherding, J.D., and Goodrum, P. M. (2000). U.S. Construction Labor Productivity Trends, 1970-1998, Journal of Construction Engineering and Management, March-April, pp. 97-104.

Amable, A., Barre, R. and Boyer, R. (1997). Les systèmes d'innovation à l'ère de la globalisation. Ed. Economica, Paris, France

Anderson, F. and Manseau, A. (1999). A Systematic Approach to Generation/Transmission/Use of Innovation in Construction Activities. Paper presented to the Third International Conference on Policy and Innovation, August 1999, Austin, Texas, USA, unpublished document

Apertura (1998). Interview to the President of the Chilean Construction Chamber. Apertura, Vol. 3, N° 13, January/February, p. 14-17.

Archambault, G. (1995). Certification qualibat. Le tournant décisif (The Qualibat Certification. The Decisive Turn) in: Sycodés informations Qualité Construction, n° 29, Mars-April 1995, pp.11-17.

ARTB (1998). ARTB Bouwvisie 2015: The Hague.

Arthur D. Little (1993). Management erfolgreicher Produkte, Wiesbaden

Atkin, B. (1999). Innovation in Construction Sector. ECCREDI Study (European Council for Construction Research, Development and Innovation), June, 58 p.

Australian Expert Group in Industry Studies (AEGIS) (1999). Building & Construction Product System: Public Sector R&D and the Education and Training Infrastructure. Department of Industry, Science and Resources, Australia

AVBB (1998). De Bouw in Cijfers: The Hague.

AWT (1999). Tussenrapportage AWT Verkenningscommissie Bouw: The Hague.

Bang, Henrik (1997). Byggesektoren og teknologisk service. Ministry of Housing and Urban Affairs, Copenhagen.

Barrctt, D. (1998). The Renewal of Trust in Residential Construction, Commission of Inquiry into the Quality of Condominium Construction in British Columbia. Ministry of Municipal Affairs, Victoria, Canada

Barros, Mercia M. S. B. (1996). Metodologia para implantação de tecnologias construtivas racionalizadas na produção de edifícios. São Paulo, 420p. PhD Thesis - Escola Politécnica, Universidade de São Paulo.

Beise, M., Licht, G. und Spielkamp, A. (1995). Technologietransfer an kleine und mittler Unternehmen: Analysen und Perspektiven für Baden-Württemberg. Schriftenreihe des ZEW, Bd. 3, Baden-Baden.

Beise, M., Stahl, H. (1999). Public research and industrial innovations in Germany. Research Policy, Vol. 28, p. 397-422.

Beise, M. et al. (1999b). Innovationsverhalten im Verarbeitenden Gewerbe. Erhebung 1997. Baden-Baden.

Bell, M. and Pavitt, K. (1995). The Development of Technological Capabilities. In: Trade, Technology, and International Competitiveness. World Bank, Washington, USA

Belle, Richard A. (1999). The PAIR Initiative: Repairing and Revitalizing Our Nation's Physical Infrastructure. Public Roads, November/December, p. 12-19.

Bernstein, Harvey M. (1999). Priorities for Federal Innovation Reform. Prepared for the National Science and Technology Council, Civil Engineering Research Foundation.

Bernstein, H. M., Lemer, A. C. (1996) "Solving the Innovation Puzzle – Challenges Facing the US Design & Construction Industry", American Society of Civil Engineers, 127 p.

Bertelsen, Sven (1997). Bellahøj, Ballerup, Brøndby Strand – 25 år der industrialiserede byggeriet. SBI, Statens Byggeforskningsinstitut.

Bertelsen, Sven; Nielsen, Jørgen (1999). The Danish Experience from 10 Years of Productivity Development. The Danish Building Research Institute.

BmbF (1998). Faktenbericht, Bonn.

BmbF (1999). Zur technologischen Leistungsfähigkeit Deutschlands. Zusammenfassender Endbericht, Bonn.

Bon, R. (1994). Whither Global Construction? (Part one). Building Research and Information, Vol. 22, N. 2, London, UK

Bon, R. and Pietroforte, R. (1993). New Construction versus Maintenance and Repair Construction Technology in the US since World War II. Construction Management and Economics, Vol. 11, London, UK

Bonke, Sten and Levring, Peter (1996, Vol. 1). The Contracting System in Danish Construction: Pinning Down Autonomy. University College London & Plan Construction et Architecture, Paris.

Bonke, Sten and Levring, Peter (1996, Vol. 2). Building in a Market Economy – Reviewing the Danish Model. The Danish Building Development Counsil, Copenhagen.

Boutek (2000). Housing is not about houses: The Boutek experience. Division of Building and Construction Technology, CSIR, Pretoria.

Bowlby, R.L. and Schriver, W.L. (1986). Observations on Productivity and Composition of Building Construction Output in the United States, 1972-82. Construction Management and Economics, Vol. 4, N.1, London, UK

Bowley, M. (1996). The British Building Industry London, Cambridge University Press.

Brasil (1991). Ministério da Ação Social. Programa Nacional de Tecnologia da Habitação. Brasília, Secretaria Nacional de Habitação, 31p.

Brown, D. (1998). What Makes Firms Innovative. Arthur. D. Little, Cambridge, UK, private communication

Business Council of Australia (1999). New Times Call for New Ideas. BCA Papers, 1(2): 19.

Campagna, E. (1998). National system of innovation in France: Plan Construction et Architecture. Building Research and Information

Campagnac, E. (2000). The contracting system in the French construction industry: actors and institutions. Building Research and Information, Vol. 28 (2), 131-140

Canadian Foundation for Innovation (1999). Annual Report.

Carassus, J. (1998). Production and management in construction, an economic approach. Cahiers du CSTB No. 395, Paris, France

Carassus, J. (1999). The economic analysis of the construction industry. Cahiers du CSTB No. 405, Paris, France

Cardoso, Francisco F. (1993). Novos enfoques sobre a gestão da produção. Como melhorar o desempenho das empresas de construção. In : ENTAC93 / 17 a 19 novembro, São Paulo. Avanços em tecnologia e gestão da produção de edificações, pp. 557-569.

Cardoso, Fransisco F. (1996). Stratégies d'entreprises et nouvelles formes de rationalisation de la production dans le Bâtiment au Brésil et en France. Thèse de doctorat. Paris, ENPC, 478 p.

Cardoso, Fransisco F. (1997). QUALIHAB: un programme pour améliorer le logement social au Brésil (QUALIHAB: a Program to Improve the Social Housing in Brazil). Sycodés Informations - Qualité Construction, Paris, November-December, pp. 12-16.

Cardoso, Francisco F., Vivancos, Adriano G., Silva, Fred. B. & ALBUQUERQUE NETO, Edson T. (2000). The Qualihab Program and the New Contracts and Contractual Relationships between Firms in Brazil. In : SERPELL, Alfredo (edited by). Information and Communication in Construction Procurement. PUC. Proceedings of the CIB W92 Procurement System Symposium, Santiago, Chile, April 24 – 27, pp. 233-247.

Carlsson, Bo (1995). Technological Systems and Economic Performance: The Case of Factory Automation. Kluwer Academic Publishers, Netherlands.

Castro, Carolina M. P. de (1986). Papel da tecnologia na produção de habitação popular - estudo de caso: C.H. José Bonifácio. São Paulo, 473p. MSc Thesis - Escola de Engenharia de São Carlos, Universidade de São Paulo.

CChC (1998). Boletín Estadístico (Statistical Bulletin), Chilean Construction Chamber, January.

Cebon, P., Newton, P. and Noble, P. (1999) Innovation in Building and Construction: Towards a Model for Indicator Development. Canberra: ISR.

CIB (1994). Transfer of Construction Management Best Practice Between Different Cultures.

CIB (1999). Agenda 21 on Sustainable Construction. Report Publication n° 237.

Civil Engineering Research Foundation (CERF) (1993). A Nation-wide Survey of Civil Engineering-related R&D. Report No. 93-5006. American Society of Civil Engineers, Washington, USA

Clausen, Lennie (1998). Innovation in the Construction Industry – a review of the literatures. Working paper, Department of Planning, Technical University of Denmark & The Danish Building Research Institute, Copenhagen.

CONICYT (1999). Panorama Científico, Vol. 14, May, Santiago, Chile.

Construction Forecasting and Research Limited (1996). The Funding and Provision of the Research and Development in the UK Construction Sector, 1990-1994. UK Department of the Environment, unpublished report

Construction Industry Development Board (1984). Canada Constructs. Ottawa, Canada

Construction Research and Innovation Strategy Panel (CRISP) (1997). Creating Climate of Innovation in Construction. CRISP Motivation Group, London, UK, draft working document

Construction Task Force (1998). Egan Report - Rethinking Construction. London, HMSO.

Cooper, I. (1997). The UK's Changing Research Base for Construction the Impact of Recent Government Policy. Building Research and Information, Vol. 25, pp. 292-300.

Cooper, R.G. (1998). Product Leadership: Creating and Launching Superior New Products. Addison-Wesley, Reading, USA

CPB (1999). Woningbouw: tussen markt en overheid, Sdu Publishers: The Hague.

DAEI (French ministry of construction) (1997). Enquêtes annuelles d'entreprises. Comptes de la Nation, Direction de la Comptabilité publique.

DACST (1996). White Paper on Science and Technology. Department of Arts, Culture, Science and Technology, Pretoria, http://www.dacst.gov.za/default science technology.htm

DACST (1998). National Research and Technology Foresight Project. Department of Arts, Culture, Science and Technology, Pretoria. http://www.dacst.gov.za/ default_science_technology.htm

Danish Energy Agency (1998). Energy in Denmark: Development, policies and results. Danish Energy Agency, Ministry of Environment and Energy, Copenhagen, Denmark.

Danmarks Statistik (1996, 1997, 1998). Statistisk Tiårsoversigt. Copenhagen.

De Oliveira, Roberto, and Cardoso, Francisco (1999). Overview of Brazilian Construction Industry: Issues in Innovation. Synopsis paper CIB, TG-35.

Department of Environment, UK (1995). UK National Technology Foresight Programme, Digest of Statistical Indicators for the Construction Panel. Construction Panel, working document

Department of Environment, UK (1996). Assessing Research Impact, The Construction Research Impact Evaluation System. Report of the Construction Sponsorship Directorate, London, UK

DIW (1998). Vergleichende Branchendaten für das verarbeitende Gewerbe in Ost- und Westdeutschland. Berechnung für 31 Branchen in europäischer Klassifikation (1991 bis 1997), Berlin.

Dixit, A.K. and R.S. Pindyck (1994). Investment under Uncertainty, Princeton UP: Princeton.

DoT (1999). Moving South Africa. Department of Transport, Pretoria. http://www.transport.gov.za/projects/msa/msa.html

DPW (1999). White Paper: Creating an Enabling Environment for Reconstruction, Growth and Development in the Construction Industry. Department of Public Works, Pretoria. http://www.publicworks.gov.za/

Dræbye, Tage (1997). Teknologisk byggeviden. Videnforbrug, videnformidling og videnproduktion – en kortlægning og vurdering, BUR, Copenhagen.

DTI (1998). Department of Trade and Industry Incentive Schemes. Department of Trade and Industry, Pretoria.

Ebling, G. et al. (1999). Innovationsaktivitäten im Dienstleistungssektor. Erhebung 1997. Baden-Baden.

Edquist, Charles (1997). Systems of Innovation – Technologies, Institutions and Organisations. Pinter, London.

EIU (1997) Economist Intelligence Unit. Make or Break: 7 Steps to Make Australia Rich Again. Australian: Metal Trades Industry of Australia.

ENR (1998) Economist Intelligent Unit. World Construction Market Review. Australia, December.

Erhvervsfremme Styrelsen (1993). Bygge/bolig – en erhvervsøkonomisk analyse. Copenhagen.

FAPESP (1999). Inovação tecnológica. Suplemento especial do Notícias Fapesp, no. 46, 76 p.

Farah, Marta Ferreira Santos (1988). Diagnóstico tecnológico da indústria da construção civil: caracterização geral do setor. Tecnologia de edificações, vol. 5, n.119, p.111-116.

FIESP (1999). 3° Seminário da Indústria Brasileira da Construção CONSTRUBUSINESS: Habitação, Infra-Estrutura e Geração de Empregos. Federação das Indústrias do Estado de São Paulo – Fiesp. Comissão da Indústria da Construção – CIC. São Paulo, 7 de junho de 1999. Apresentação PowerPoint, 117 transparências.

FINEP (s/d). Programa de Tecnologia de Habitação. Gestão das Ações de Inocação em C&T. FINEP, Rio de Janeiro, s/d. 12p. (texto datilografado).

Finkel, G. (1997). The Economics of the Construction Industry. M.E. Sharpe, New York, USA

Flanagan, R., Ingram, I. and Marsh, L. (1998). A Bridge to the Future. Reading Construction Forum, University of Reading, UK, interactive CD document.

402

Ford, D. and Ryan, C. (1981). Taking Technology to Market. In: <u>Harvard Business Review</u>, 81(2), P. 117-126.

FRI Foreningen af Rådgivende Ingeniører (1990). Forskning og udvikling i byggesektoren. Situationen ved 80'ernes slutning, Copenhagen.

FRI Foreningen af Rådgivende Ingeniører (1991). Dobbelt-op – Perspektivplan for dansk byggeri. Copenhagen.

Fundação João Pinheiro (1992). Desenvolvimento da indústria da construção em Minas Gerais: impacto na evolução tecnológica e na qualificação da força de trabalho. Belo Horizonte, Centro de Estudos Econômicos, 375p.

Gann, D., Wang, Y. and Hawkins, R. (1998). Do Regulations Encourage Innovation? – The Case of Energy Efficiency in Housing. <u>Building Research and Information</u>, Vol. 26 280-296.

Gann, D.M. (1998a). Learning and Innovation Management in Project Based Firms., 2nd International Conference on Technology Policy and Innovation, August, Lisbon, Portugal

Gazmuri, P., De Solminihac, H., Alarcón, L.F., Robles, J., and Serpell, A. (1993). Plan Estratégico, Cámara Chilena de la Construcción, Santiago, Chile.

Gyles, R. (1992). Royal Commission into Productivity in the Building Industry in New South Wales. Sydney: ABPS.

Hampson, K. (1993). Technology Strategy and Competitive Performance: A Study of Bridge Construction. Unpublished Doctor of Philosophy Dissertation, Stanford University, CA.

Hampson, K. (1997). Construction Innovation in the Australian Context. Paper presented at the International Workshop on Innovation Systems and the Construction Industry, Montreal, January. National Research Council of Canada, Institute for Research in Construction.

Hampson, K. and T. Kwok (1997a). Strategic Alliances in Building Construction: A Tender Evaluation Tool for the Public Sector. <u>Journal of Construction Procurement</u>, 2(1).

Hampson, K. (1998). The Effectiveness of Links Between Industry, Universities and Public Sector Research Organisations. Paper presented at the Federal Government Building for Growth Innovation Forum, May.

Hanna, A.S. and Heale, D.G. (1994). Factors Affecting Construction Productivity: Newfoundland versus Rest of Canada. <u>Canadian Journal of Civil Engineering</u>, Vol. 21, N.4

Harhoff, D. and H. Koenig (1993). Neuere Ansätze der Industrieökonomik-Konsequenzen für eine Industrie- und Technologiepolitik, in: F. Meyer-Krahmer (Hrsg.): Innovationsökonomie und Technologiepolitik, Heidelberg, pp. 47-67.

Henry L. Michel (1998). The Next 25 Years: The Future of the Construction Industry. <u>Journal of Management in Engineering</u>, Sept-Oct, p.27-31.

Hobday, M. (1998). Product complexity, innovation and industrial organisation. <u>Research Policy</u>, 26, pp 689-710

Hoffman & Sønner A/S (1994). Annual Report and Accounts. Copenhagen.

IBGE (1989). Anuário Estatístico do Brasil, IBGE, 1980-87.

Industry Canada (1995). The Construction Industry, Main Messages. The Economic Situation, Third Quarter, unpublished working document

Industry Canada (1996). Sector Competitiveness Frameworks Series – Consulting Engineering, Overview and Prospects.

Industry Canada (1998). Sector Competitiveness Frameworks Series – Construction, Changing Conditions and Industry Response.

Industry Canada, Service Industries and Capital Projects Branch (1998). Sector Competitiveness Frameworks, Construction, Part 1-Overview and Prospects. Ottawa, Canada

Innovation Policy Research Associates (1992). Construction R&D London, Department of the Environment.

Instituto Superior Técnico (2000). Engenharia & Tecnologia 2000. Research Project, Lisbon.

IPT (1987). Diagnóstico tecnológico da indústria da construção civil. Fase I - Relatório Final. Rapport n° 25,464/87. São Paulo, Instituto de Pesquisas Tecnológicas do Estado de São Paulo,106 p.

IPT (1988). Diagnóstico tecnológico da indústria da construção civil. Fase II - Relatório de andamento. Rapport n° 26,457/88. São Paulo, Instituto de Pesquisas Tecnológicas do Estado de São Paulo, 80 p.

IRC (1997a). Discussion Paper: Characteristics of the Innovation Process and System in the Canadian Construction Sector. Prepared by the Nordicity Group Ltd. for the Institute for Research in Construction, National Research Council of Canada.

IRC (1997b*).* Discussion Paper: Nature of Innovation in the Canadian Construction Industry. Prepared by the ARA Consulting Group Inc. Vancouver, Canada for the Institute for Research in Construction, National Research Council of Canada.

Ireland, V. (1994). Process re-engineering in construction. CIB T40 Report. Fletcher Constructions, Sydney, Australia, May.

ISR (Federal Government Department of Industry, Science and Resources) (1999) Building for Growth: An Analysis of the Australian Building and Construction Industries. Canberra.

ISR (Federal Government Department of Industry, Science and Resources) (1999b) Commonwealth and State Government Programs Supporting Innovation in Firms, At October. Canberra.

Jacobs, D. (1997). Overheid, R&D en Internationalisatie. NRLO-report, 97-13.

Jacobs, D., J. Kuijper and B. Roes (1992). De Economische Kracht van de Bouw, SMO: The Hague.

Japan Federation of Construction Contractors (1999). What the management of Construction Contractors for 21st Century Should Be – Formation of the Foundation for the Development of Prominent Contractors in Technology and Management (in Japanese).

Keck, O. (1980). Government policy and technical choice in the German reactor program, Research Policy, Vol. 9, p. 302-356.

404

Keys, B.A. and Caskie, D.M. (1975). The Structure and Operation of the Construction Industry in Canada. Economic Council of Canada, Ottawa, Canada

Kline, S.J. (1985). Innovation is Not a Linear Process. Research Management, July-August 1985

Klodt, H. (1995). Grundlagen der Forschungs- und Technologiepolitik. WiSo-Kurzlehrbücher, Vahlen, München.

Klodt, H., Maurer, R. and Schimmelpfennig, A. (1997). Tertiarisierung der deutschen Wirtschaft, Institut für Weltwirtschaft, Kiel.

Koivu T., Mäntylä K., Appel M., Loikkanen K., Sneck T. and Pulakka S. (2000). Developing Innovation in the Real Estate and Construction Sector – Background and Experiments. Draft. VTT Building Technology, Espoo. (in Finnish).

Kwok, T. (1998) Strategic Alliances in Construction: Contracting Relationships in Public Sector Building, Unpublished Doctor of Philosophy Dissertation, Queensland University of Technology.

Langston, C. and de Valence, G. (1999). International Cost of Construction Study, Stage 2: Evaluation and Analysis. University of Technology, Sydney, Australia, unpublished report

Lansley, P. Luck, R. and Lupton, S. (1994). Construction Research in Universities Swindon, EPSRC.

Larsson, Bengt (1992). Adoption av ny produktionsteknik på byggarbetspladsen. Chalmers Tekniska Högskola, Gothenburg.

Latham, M. (1994). Constructing the Team London, HMSO.

Licht, G. et al. (1995). European Innovation Monitoring System. Ref. No. 226. European Innovation Policy Network. Innovation Policy in Germany.

Lundvall, Bengt-Åke (1992). National Systems of Innovation: Towards a Theory of Innovation and Interactive Learning. Pinter, London.

Lundvall, B-A. and Christensen, J.L. (1999). Extending and deepening the analysis of innovation systems-with empirical illustrations from the DISKO-project. Aalborg University, Denmark, unpublished paper

Lundvall, Bengt-Åke (1999a). Det danske innovationssystem (The Danish Innovation System). DISKO-projektet, report No. 9, summary report. The Agency for Development of Business and Industry, Copenhagen.

Manseau, André (1998). Who cares about the overall industry innovativeness? Building Research & Information, Vol. 26(4), 241-245

Marceau, J. and K. Manley (2000, forthcoming). An Examination of Services Provided by Manufacturers Supplying the Building and Construction Industries. Canberra: ISR.

Marceau, J., K. Manley and K. Hampson (1999). Building and Construction Product System: Public Sector R&D. In AEGIS (1999b) Building and Construction Product System: Public Sector R&D and the Education and Training Infrastructure. Canberra: ISR.

March, J.G. (1991). Exploration and Exploitation in Organizational Learning, Organization Science, Vol. 2(1), pp. 71.87.

McKinsey Global Institute (1995). Sweden's Economic Performance. Private communication

McKinsey (1998). Construção residencial. In : Produtividade : a chave do desenvolvimento acelerado no Brasil. McKinsey Brasil. São Paulo, março. 17 p. mais gráficos.

MDIC (2000a). Documento Básico. In : Fórum da Competitividade. Diálogo para o Desenvolvimento. Ministério do Desenvolvimento, Indústria e Comércio Exterior. Secretaria de Desenvolvimento da Produção. Brasília, maio. 16 p.

MDIC (2000b). Cadeia Produtiva da indústria da Construção Civil. Perfil da Cadeia Produtiva. In : Fórum da Competitividade. Diálogo para o Desenvolvimento. Ministério do Desenvolvimento, Indústria e Comércio Exterior. Secretaria de Desenvolvimento da Produção. Brasília, maio. 18 p.

MDIC (2000c). Síntese do Diagnóstico. In : Fórum da Competitividade. Diálogo para o Desenvolvimento. Ministério do Desenvolvimento, Indústria e Comércio Exterior. Secretaria de Desenvolvimento da Produção. Brasília, maio. 20 p.

Mead, J. (1999). The Movement for Innovation : A View from the Team MSc Construction Economics and Management Report, University College London.

Meijaard, J. (1998). Decision Making in Research and Development: A Comparative Study of Multinational Companies, Thesis Publishers: Amsterdam.

Metcalfe, S. (1995). The Economic Foundations of Technology Policy. In: Stoneman, P. (ed.) (1995), Handbook of the Economics of Innovation and Technological Change, Blackwell Publishers: Oxford.

Meyer-Kramer, F. (1990). Science and Technology in the Federal Republic of Germany, Harlow.

Milford, R V and Loots, A (1998). Innovation in the Construction Industry. Conference on Innovative Concepts in Industrial Development, organised by the Industrial Development Engineering Association (IDEA), Midrand, South Africa.

Miller, R., Hobday, M., Lerous-Demers, T. and Olleros, X. (1995). Innovation in Complex Systems Industries: the Case of Flight Simulation. Industrial and and Corporate Change, 7, pp. 311-346

Miller, R., Lessard, D.R., Michaud, P. Floricel, S. (2000). The Strategic Management Management of large Engineering Projects: Shaping Institutions, Risks and Governance. MIT Press, Cambridge, USA.

Ministère de l'Equipement, du logement et des Transports, France (1999). Cahier thématique Chantier 2000: techniques et chantier, PUCA, sous la direction de J.L. Salagnac

Ministère du Logement, France, Plan Construction et Architecture, (1994). L'innovation en chantier : catalogue des recherches et expérimentations. Diffusion: CSTB-Paris

Ministry of Construction - Japan (1999). Construction White Paper

Ministry of Economic Affairs - Netherlands (1998). Innovatief Aanbesteden, MinEZ: The Hague.

Ministry of Economic Affairs - Netherlands (2000). Entrepreneurship in the Netherlands, Dutch Ministry of Economics Affairs and EIM Small Business Research and Consultancy, Zoetermeer/The Hague: EIM/MinEZ.

Ministry of Housing and Urban Affairs - Denmark (1998). Strategic Action Agenda.

Moret, Ernst and Young (1998). Subsidie Info 1998, Delwel: The Hague.

Mustar Ph. (editor) (1999). Les chiffres clés de la science et de la technologie. Observatoire des Sciences et des Techniques (OST), Economica, édition 1998-1999.

Nam, C.H. and Tatum, C.B. (1992). Strategies for Technology Push: Lessons from Construction Innovations. Journal of Construction Engineering and Management, Vol. 118, No. 3, ASCE, USA

National Research Council of Canada – Industrial Research Assistance Program (1999). Annual Report.

National Research Council of Canada – Institute for Research in Construction (1999). Annual Report.

National Science and Technology Council (1995). Construction Building: Federal Research and Development in Support of the U.S. Construction Industry, U.S. Federal Government.

National Science and Technology Council (1999). Construction Building - Interagency Program for Technical Advancement in Construction and Building, U.S. Federal Government.

NBRI (1987). Low-cost building. National Building Research Institute, CSIR, Pretoria.

Observatory for the Science and Technology and Ministry for Science and Technology (1995). Portugal. Statistic Summary – Inquire to the National Potential Technological and Scientific, Lisbon.

OECD (1992) (Organisation for Economic Cooperation and Development). Technology and the Economy, The Key Relationships. Paris, France

OECD (1995). National Systems for Financing Innovation, Paris, France.

OECD (1996). The Knowledge –based Economy. Paris, France

OECD (1997). Flows and Stocks of Fixed Capital, 1971-1996. Paris, France.

OECD/Eurostat (1997a). Proposed Guidelines for Collecting and Interpreting Technological Innovation Data - Oslo Manual. Paris, France

OECD (1997b). National Innovation Systems. Paris, France

OECD (1998). Science, Technology and Industry Outlook-Construction. Paris, France

OECD (1999). International Sectoral Database. Paris, France

OECD (1999a). Managing National Innovation Systems. Paris, France.

OECD (1999b). Main Industrial Indicators Database. Paris, France

OECD (1999c). Boosting Innovation: the Cluster Approach. Proceedings, Paris, France.

OECD (2000). Basic Science and Technology Statistics. Paris, France

Office of Science and Technology (1995). Progress Through Partnership 2: Construction Panel. London, HMSO.

Ofori, G. (1991). Programmes for Improving the Performance of Contracting Firms in Developing Countries: A Review of Approaches and Appropriate Options. Construction Management and Economics, 9, 19-38.

Okamoto, Shin (2000). Current Conditions and Perspective of Medium-sized Construction Firms in Japan – Practices and Strategies of Joint Research. CIB Report.

Okamoto, Shin (1998). Growth and Future Developments of R&D in the Japanese Construction Industry. CIB Report 218.

Paasio A. (1999). Man – environment – technology. The National Innovation System. Seminar presented in March, Helsinki.

Palmberg C., Niininen P., Toivanen H. and Wahlberg T. (2000). Industrial innovation in Finland. First results of the Sfinno-project. VTT, Group for Technology Studies, Working Papers No. 47/00. Espoo.

Pavez, H. (2000). Present and future of the construction industry in Chile, Keynote presentation at the CIB W92 International Symposium on Procurement Systems, Santiago, Chile, April 24-28.

Picchi, Flávio A. (1993). Sistemas da qualidade: uso em empresas de construção de edifícios. PhD Thesis. 2 v. São Paulo, EPUSP, pp. 52-141.

Pleschak, F., Sabisch, H. (1996). Innovations Management, Stuttgart.

Porter, M (1990). The Competitive Advantage of Nations. The Free Press, New York, USA

Porter, M. (1998). On Competition. Harvard Business Review Book Series, Cambridge, USA

Pries, F. and Janszen, F. (1995). Innovation in the Construction Industry: the Dominant Role of the Environment. Construction Management and Economics, Vol.13, No.1, London, UK

Revay and Associates Ltd. (1993). The Top Fifty, Canadian Construction R&D Performers and Funders. Institute for Research in Construction, National Research Council of Canada, Ottawa, Canada

Revay and Associates Ltd. (1999). Canadian Construction R&D Performers and Funders. Institute for Research in Construction, National Research Council of Canada, Ottawa, Canada

Roelandt, T.J.A., P. Den Hartog, J. van Sinderen and N. van den Hove (1999). Cluster Policy in the Netherlands. In: OECD (1999), Boosting Innovation: the Cluster Approach, Paris.

Roelandt, T.J.A., P.W.L. Gerbrands, H.P. van Dalen and J. van Sinderen (1996). Onderzoek naar Technologie en Economie: Over Witte Vlekken en Zwarte Dozen. Beleidsstudies Technologie Economie 31, MinEZ: The Hague.

Seaden, G (1995). Economics of Technology Development for the Construction Industry. 13th CIB World Building Congress, The Netherlands.

Seaden, G. (1996). Economics of Technology Development for the Construction Industry. CIB Report Publication 202, Rotterdam, Netherlands

Seaden, G. (1997). Productive Capacity Through Innovation. Construction and Engineering Leadership Conference, Calgary, Canada

408

Selle, P. (1996). Scandinavian Voluntary Action in Transformation. In: Rasmussen & Koch Nielsen (eds.): The Third Sector in Transformation, pp. 69-83. Socialforskningsinstituttet, Copenhagen.

Serpcll, A. (1997). Innovation and technology development: a competitive strategy for construction companies, Proceedings of the First International Conference on Construction Industry Development: Building the Future Together, 9-11 December, Singapore.

Sherif, M. and I. Davidson (1994). Re-engineering Approach to Construction: A Case Study. CSIRO, Division of building Construction and Engineering Conference Paper, Hyatt, Melbourne Victoria.

Slaughter, S. E. (1993). Builders as Sources of Innovation. Journal of Construction Engineering and Management, Vol. 119, No. 3, ASCE, USA

South African Government (1994). RDP White Paper. http://www.polity.org.za/govdocs/white_papers/rdpwhite.html

South African Government (1996). Growth, Employment and Redistribution: A macrostrategy; http://www.polity.org.za/govdocs/policy/growth.html

South African Reserve Bank (1998). Quarterly Review, December 1998.

Specht, G., Beckmann, C. (1996). F&E-Management Stuttgart.

Statistics Bureau Management and Construction Agency (1998) Ministry of Construction - Japan. Report on the Survey of Research and Development (in Japanese).

Statistics Canada (1999). Canada Year Book, Ottawa.

Statistics Canada (1999a). Science Statistics, Catalogue No. 88-001-XIB, Vol. 23, No.9.

Statistics Finland (2000). Corporate Enterprices and Personal Businesses in Finland 1998. Statistics Finland, Enterprices 2000: 1. Helsinki.

Statistics Netherlands (1998). Kennis en Economie 1998, CBS: Voorburg.

Stiferband für die Deutsche Wirtschaft (Germany R&D Watch Organisation 1997) Forschung und Entwicklung in der Wirtschaft 1995 bis 1997, Bericht über die FuE-Erhbung 1995 und 1996.

Stoneman, P. (ed.) (1995). Handbook of the Economics of Innovation and Technological Change, Blackwell: Oxford.

Sundbo, J. (1995). Innovationsteori - tre paradigmer. Jurist- og Økonomforbundets Forlag, Copenhagen.

Sundbo, J. (1997). Management of innovation in service. The Service Industries Journal, vol. 17, no. 3, pp. 432-455.

Sycodés Informations (1996/97). Critères du référentiel de certification QUALIBAT (I à VII). Sycodés Informations - Qualité Construction, Paris, 1996-97 (34 à 40) mai-juin 1996 à janvier-février 1997.

Tatum, C.B. (1986). Potential Mechanisms for Construction Innovation. Journal of Construction Engineering and Management, Vol. 112, No. 2, ASCE, USA

Tatum, C.B. (1987). Process of Innovation in Construction Firm. Journal of Construction Engineering and Management, Vol. 113, No. 4, ASCE, USA

Taxation Institute of Australia (1998). Industry Research and Development in Australia. Melbourne: Taxation Institute of Victoria.

Technical Research Centre of Finland (1999). Strategic plan 1999. Espoo.

Technology Agency of Finland (1998). Technology and the future. Helsinki. 178p. (in Finnish).

Technology Partnerships Canada (1999). Annual Report.

Tidd, J. (1995). Development of Novel Products Through Intra-Organizational and Inter-Organizational Networks – The Case of Home Automation. Journal of Product Innovation Management, 12 (4), 307-322

Toole, M.T. (1998). Uncertainty and Home Builder's Adoption of Technological Innovations. Journal of Construction Engineering and Management, Vol. 124, No. 4, ASCE, USA

U.S. Census Bureau (2000). Annual Value of Construction Put in Place, U.S. Federal Government.

U.S. Department of Commerce, International Trade Administration (1989). A Competitive Assessment of the U.S. International Construction Industry. National Technical Information Service, Springfield, VA. USA

Uher, T. E. (1994). What is Partnering? Australian Construction Law Newsletter, (34): 49- 60.

Vargas, Milton (1994). Para uma filosofia da tecnologia. São Paulo, Alfa-Ômega. 1994. p.171-286.

VTT Building Technology (1998). Well-being through construction 1998. Tampere. (in Finnish).

VTT Building Technology (1999). Well-being through construction 1999. Tampere. (in Finnish).

VTT Building Technology (2000). Well-being through construction in Finland 2000. Tampere.

Winch, G. and Campagnac, E. (1995). The Organization of Building Projects: an Anglo/French Comparison. Construction Management and Economics, Vol. 13, No.1, London, UK

Winch, G. (1996). The Contracting System in British Construction: The Rigidities of Flexibility. Groupe Bagnolet Working Paper No. 6, University College London, London, UK

Winch, G. (1998). Zephyrs of Creative Destruction: Understanding the Management of Innovation in Construction. Building Research and Information, Vol. 26, pp. 268-279

Winch, Graham (1999). Innovation in the British Construction Industry: The Role of public Policy Instruments, DRAFT, University College London.

Winch, G. M. (2000). Institutional Reform in British Construction: Partnering and Private Finance. Building Research and Information, Vol. 28, pp. 141-155

ZEW (1998). Mannheim Innovation Panel. Federal Ministry of Education, Science, Research and Technology, Germany